# Information Security
## Text & Cases

Edition 2.0

## Gurpreet S. Dhillon
University of North Carolina at Greensboro, USA

Copyright © 2018 Prospect Press, Inc. All rights reserved.

No part of this publication may be reproduced, stored in a retrieval system or transmitted in any form or by any means, electronic, mechanical, photocopying, recording, scanning or otherwise, except as permitted under Sections 107 or 108 of the 1976 United States Copyright Act, without either the prior written permission of the Publisher, or authorization through payment of the appropriate per-copy fee to the Copyright Clearance Center, Inc. 222 Rosewood Drive, Danvers, MA 01923, website www.copyright.com. Requests to the Publisher for permission should be addressed to the Permissions Department, Prospect Press, 47 Prospect Parkway, Burlington, VT 05401 or email to Beth.Golub@ProspectPressVT.com.

Founded in 2014, Prospect Press serves the academic discipline of Information Systems by publishing innovative textbooks across the curriculum including introductory, emerging, and upper level courses. Prospect Press offers reasonable prices by selling directly to students. Prospect Press provides tight relationships between authors, publisher, and adopters that many larger publishers are unable to offer in today's publishing environment. Based in Burlington, Vermont, Prospect Press distributes titles worldwide. We welcome new authors to send proposals or inquiries to Beth.Golub@ProspectPressVT.com.

Editor: Beth Lang Golub
Production Manager: Rachel Paul
Cover Designer: Annie Clark
Cover Photo: © Gurpreet Dhillon

eTextbook • Edition 2.0 • ISBN: 978-1-943153-24-4 • Available from Redshelf.com and VitalSource.com
Printed Paperback • Edition 2.0 • ISBN: 978-1-943153-25-1 • Available from Redshelf.com and CreateSpace.com

For more information, visit http://prospectpressvt.com/titles/dhillon-information-systems-security/

## About the Cover

Veins of a leaf resemble a net, a complex and convoluted structure to support nourishment. Similarly, Information Systems Security is also complex, and many aspects—ranging from computer networks to organizational structures and people—come together to ensure protection. The image is a representation of such complexity.

# Contents

| | | |
|---|---|---|
| *List of Figures* | | *ix* |
| *Preface* | | *xi* |
| *Acknowledgments* | | *xv* |
| *About the Author* | | *xvii* |
| **Chapter 1** | **Information System Security: Nature and Scope** | **1** |
| | Coordination in Threes | 3 |
| | Security in Threes | 5 |
| | Institutionalizing Security in Organizations | 8 |
| | Questions and Exercises | 10 |
| | Case Study: Virtual Companies and Identity Theft | 11 |
| | References | 13 |
| **Part I** | **Technical Aspects of Information Systems Security** | **15** |
| **Chapter 2** | **Security of Technical Systems in Organizations** | **16** |
| | Vulnerabilities | 18 |
| | Methods of Defense | 24 |
| | Historical Review of Models for Security Specification | 26 |
| | Away from the Military | 31 |
| | Toward Integrity: Biba, Clark-Wilson, and Chinese Walls | 33 |
| | Emergent Issues | 36 |
| | Concluding Remarks | 38 |
| | Questions and Exercises | 39 |
| | Case Study: Breaking into Sprint's Backbone Network | 41 |
| | References | 41 |
| **Chapter 3** | **Cryptography and Technical IS Security** | **42** |
| | Cryptography | 44 |
| | Cryptanalysis | 46 |
| | Conventional Encryption Algorithms | 52 |
| | Asymmetric Encryption | 55 |
| | Future of Encryption | 59 |
| | Blockchains | 60 |
| | Questions and Exercises | 63 |
| | Case Study: The PGP Attack | 65 |

iv • Contents

| | | |
|---|---|---|
| | Case Study: Visa Goes Blockchain | 65 |
| | References | 67 |
| **Chapter 4** | **Network Security** | **68** |
| | TCP/UDP/IP Protocol Architecture | 70 |
| | Middleware Devices | 76 |
| | Types of Network Attacks | 79 |
| | New Trends in Network Security | 88 |
| | Discussion Questions | 94 |
| | Exercise | 94 |
| | Short Questions | 94 |
| | Case Study: The Distributed Denial of Service Attack | 95 |
| | References | 96 |
| **Part II** | **Formal Aspects of Information Systems Security** | **97** |
| **Chapter 5** | **Planning for Information System Security** | **98** |
| | Formal IS Security Dimensions | 101 |
| | Security Strategy Levels | 106 |
| | Classes of Security Decisions in Firms | 108 |
| | Security Planning Process | 115 |
| | IS Security Planning Principles | 119 |
| | Summary | 122 |
| | Questions and Exercises | 123 |
| | Case Study: The Hack at UC Berkley | 125 |
| | References | 126 |
| **Chapter 6** | **Risk Management for Information System Security** | **128** |
| | Risk Assessment | 131 |
| | Risk Mitigation | 139 |
| | Risk Evaluation and Assessment | 143 |
| | COBRA: A Hybrid Model for Software | |
| | Cost Estimation, Benchmarking, and Risk Assessment | 143 |
| | A Risk Management Process Model | 145 |
| | Concluding Remarks | 152 |
| | Questions and Exercises | 153 |
| | Case Study: Insiders Play a Role in Security Breaches | 154 |
| | References | 155 |
| **Chapter 7** | **Information Systems Security Standards and Guidelines** | **156** |
| | The Role of Standards in Information | |
| | Systems Security | 158 |
| | Process Improvement Software | 161 |
| | The SSE-CMM | 162 |
| | Key Constructs and Concepts in SSE-CMM | 165 |

|  |  |  |
|---|---|---|
| | SSE-CMM Architecture Description | 170 |
| | From the Rainbow Series to Common Criteria | 175 |
| | ITSEC | 177 |
| | International Harmonization | 179 |
| | Other Miscellaneous Standards and Guidelines | 192 |
| | Concluding Remarks | 197 |
| | Questions and Exercises | 198 |
| | Case Study: Remote Access Problems at the DHS | 200 |
| | References | 201 |
| **Chapter 8** | **Responding to an Information Security Breach** | **202** |
| | Technicalities of a Breach | 206 |
| | Policy Considerations | 207 |
| | Reputation and Responsiveness | 207 |
| | Risk and Resilience | 208 |
| | Governance | 208 |
| | Steps to Avoid a Potential Attack | 209 |
| | How to Respond When a Breach Occurs | 212 |
| | Best Practices: How to Be Prepared for an Intrusion | 214 |
| | Concluding Remarks | 218 |
| | Questions and Exercises | 219 |
| | Case Study: Equifax Breach | 220 |
| | References | 221 |
| **Part III** | **Informal Aspects of Information Systems Security** | **223** |
| **Chapter 9** | **Behavioral Aspects of Information System Security** | **224** |
| | Employee Threats | 225 |
| | Social Engineering Attacks | 227 |
| | Individual Motivation to Prevent Cyberattacks | 229 |
| | Cybercrime: Criminals and Mediums | 230 |
| | Cyberespionage | 233 |
| | Cyberterrorism | 234 |
| | Cyberstalking | 236 |
| | Questions and Exercises | 239 |
| | Case Study: Cyberterrorism—A New Reality | 240 |
| | References | 245 |
| **Chapter 10** | **Culture and Information System Security** | **246** |
| | Understanding the Concept of Security Culture | 249 |
| | Silent Messages and IS Security | 254 |
| | Leadership and Security Culture | 259 |
| | Security Culture Framework | 262 |
| | OECD Principles for Security Culture | 266 |
| | Concluding Remarks | 269 |

|  |  |  |
|---|---|---|
| | Questions and Exercises | 270 |
| | Case Study: The T-Mobile Hack | 271 |
| | References | 271 |

### Chapter 11 Ethical and Professional Issues in IS Security Management — 273

| | | |
|---|---|---|
| | Coping with Unemployment and Underemployment | 274 |
| | Intellectual Property and Crime | 275 |
| | IT and White Collar Crime | 276 |
| | Managing Ethical Behavior | 277 |
| | Codes of Conduct | 279 |
| | Credentialing | 282 |
| | Questions and Exercises | 286 |
| | Case Study: The DoubleClick Case | 287 |
| | References | 288 |

### Part IV Regulatory Aspects of Information Systems Security — 289

### Chapter 12 Legal Aspects of Information System Security — 290

| | | |
|---|---|---|
| | The Computer Fraud and Abuse Act (CFAA) | 293 |
| | The Computer Security Act (CSA) | 294 |
| | Health Insurance Portability and Accountability Act (HIPPA) | 296 |
| | Sarbanes-Oxley Act (SOX) | 302 |
| | Federal Information Security Management Act (FISMA) | 304 |
| | Concluding Remarks | 306 |
| | Questions and Exercises | 306 |
| | Case Study: FTC versus Wyndham Worldwide Corporation | 307 |
| | References | 309 |

### Chapter 13 Computer Forensics — 310

| | | |
|---|---|---|
| | The Basics | 311 |
| | Types and Scope of Crimes | 312 |
| | Lack of Uniform Law | 313 |
| | What Is "Computer Forensics"? | 314 |
| | Gathering Evidence Forensically | 315 |
| | Formal Procedure for Gathering Data | 317 |
| | Law Dictating Formal Procedure | 321 |
| | Law Governing Seizure of Evidence | 321 |
| | Law Governing Analysis and Presentation of Evidence | 334 |
| | Emergent Issues | 337 |
| | Concluding Remarks | 340 |
| | Questions and Exercises | 342 |
| | Case Study: The Deathserv Case | 342 |
| | Case Study: Tradeoffs in Eavesdropping | 343 |
| | References | 344 |

| Chapter 14 | Summary Principles for IS Security | 345 |
|---|---|---|
| | Principles for Technical Aspects of IS Security | 346 |
| | Principles for Formal Aspects of IS Security | 348 |
| | Principles for Informal Aspects of IS Security | 350 |
| | Concluding Remarks | 351 |
| | References | 352 |

| Part V | Case Studies | 353 |
|---|---|---|

## Case Study 1: The Anthem Data Breach — 354

- Anthem's History and Background of the Industry — 355
- Who Is at Stake? — 356
- Lawsuits and Policy Issues — 357
- Significance of the Problem and What Anthem Is Doing About It — 357
- Strategic Alignment—A Possible Cause of the Breach — 359
- Strategic Choices — 361
- Questions — 364
- References — 364

## Case Study 2:
## Process and Data Integrity Concerns in a Scheduling System — 365

- Case Description — 365
- Reform — 368
- Implications — 369
- Summary — 371
- Questions — 371
- Sources Used — 372

## Case Study 3: Case of a Computer Hack — 373

- Computer System — 374
- Changes — 375
- History of the System — 376
- Other Issues — 377
- Hack Discovered — 377
- Immediate Response — 378
- Further Research and Additional Symptoms — 378
- Summary — 382

## Case Study 4:
## Critical Infrastructure Protection: The Big Ten Power Company — 383

- The Big Ten Power Company — 383
- Federal and State Efforts to Secure
- Critical Infrastructure — 387
- Federal Legislation — 389
- Questions — 391

**Case Study 5: The Case of Sony's PlayStation Network Breach** — 392
    Sony Gets Attacked — 393
    Technical Details of the Attack — 394
    The Emergent Challenges — 396
    Questions — 399

*Index* — *401*
*Appendix A: Additional End-of-Chapter Short Questions* — *414*

# Figures

**Figure 1.1.** A "fried egg" analogy of coordination in threes — 5
**Figure 1.2.** Information handling in organizations — 9
**Figure 3.1.** Conventional encryption process — 45
**Figure 3.2.** A simplified Data Encryption Algorithm showing a single cycle — 53
**Figure 3.3.** Details of a given cycle — 54
**Figure 3.4.** Asymmetric encryption — 56
**Figure 3.5.** The digital signature process — 57
**Figure 3.6.** Hierarchical X.509 certificate structure — 58
**Figure 3.7.** Blockchain creation and validation process — 61
**Figure 4.1.** Network signaling and the layers — 72
**Figure 4.2.** Original IPv4 address scheme — 73
**Figure 4.3.** IPv6 address scheme — 73
**Figure 4.4.** The DNS scheme — 74
**Figure 4.5.** System server ports — 75
**Figure 4.6.** The UDP protocol — 76
**Figure 4.7.** The socket arrangement — 76
**Figure 4.8.** Network structures — 77
**Figure 4.9.** Example of a suspect email — 87
**Figure 4.10.** Traditional data center infrastructure — 89
**Figure 4.11.** Virtualized server — 89
**Figure 4.12.** Four directions of network connections — 90
**Figure 4.13.** East-West traffic attack — 91
**Figure 4.14.** A possible secure configuration — 92
**Figure 5.1.** Layers in designing formal IS security — 105
**Figure 5.2.** Strategy levels — 107
**Figure 5.3.** Double loop learning — 111
**Figure 5.4.** Double loop security design process — 111
**Figure 5.5.** A network of IS Security means and fundamental objectives — 114
**Figure 5.6.** A high-level view of the Orion strategy — 117
**Figure 5.7.** IS security planning process framework — 120
**Figure 6.1.** Vulnerability classes — 134
**Figure 6.2.** Classes of controls — 135
**Figure 6.3.** Risk mitigation flow of activities — 141
**Figure 6.4.** Overview of productivity estimation model — 144
**Figure 6.5.** I2S2 Model at level one — 146
**Figure 7.1.** CMM levels — 163
**Figure 7.2.** SSE-CMM levels — 170

**Figure 7.3.** Level 2 with project-wide definition of practices  173
**Figure 7.4.** Level 3 with project-wide processes are integrated  173
**Figure 7.5.** Level 4 with management insight into the processes  174
**Figure 7.6.** Level 5 with continuous improvement and change  175
**Figure 7.7.** Evolution of security evaluation criteria  179
**Figure 7.8.** The evaluation process  181
**Figure 7.9.** Requirements specification, as espoused by CC  184
**Figure 7.10.** Asset management template  186
**Figure 8.1.** Continuous diagnosis and mitigation framework  211
**Figure 8.2.** How to respond to a breach  214
**Figure 9.1.** A typical cyberespionage attack  234
**Figure 9.2.** Percentage of hits from W32.Stuxnet by country  243
**Figure 9.3.** Overview of Stuxnet  244
**Figure 10.1.** Elements of culture and security interpretations  251
**Figure 10.2.** Web of culture  254
**Figure 10.3.** Playing for crisis management  258
**Figure 10.4.** Role of leadership in cybersecurity  261
**Figure 10.5.** Types of culture  263
**Figure 10.6.** Radar map for confidentiality culture in a medical clinic  267
**Figure 10.7.** A call for an informal security culture developing campfire  269
**Figure 14.1.** Ideal and overengineered solutions  349
**Figure C.1.** Different cybersecurity postures  362
**Figure C.2.** Screenshot of the VistA scheduling software user interface  366

# Preface

Information is the lifeblood of an organization. Every decision made by an organization is based on some form of information. Decisions result in more information, which thus informs subsequent decisions. Therefore a lack of information can result in serious problems and may also hinder the success of an organization. Given the overreliance of organizations on interconnected devices, computer systems, and the Internet, the security of information is a nontrivial task. During the nascent stages of the information revolution, locking a computer system in a centralized location could take care of the majority of potential problems. But as computing became decentralized and distributed, simple old-fashioned locks and keys were not enough. Today, security is not just about maintaining the *confidentiality*, *integrity*, and *availability* of data. Security is also not just about *authenticating* individuals or *nonrepudiation*. When a cyberbreach, such as the 2017 Equifax fiasco, takes place, identity is lost. Protecting individual *identity* has become as much a security issue as maintaining the authenticity of individuals. Maintaining security is a multifaceted task.

Security problems can arise from intentional or unintentional abuses. There are numerous examples where systems have failed and caused significant disruptions. Security problems have also occurred because of the sheer complexity and limits of technology (or the human ability to think through various scenarios). In 1988, when the first Internet worm—the Morris worm—infected nearly 6,000 computers in a single day, the cause was linked to the buffer overflow in Unix finger daemon. Today, virus and worm attacks still exist, sometimes exploiting humans to prevent detection or using them for extortion, and at other times relying on technological complexities to remain undetected.

The detection of threats has become increasingly difficult—hence the need for this book. The first version of this book was published in 2007. At that time, we were trying to define a need for security (i.e., information security, information system security, cybersecurity). Today we have moved beyond this, and have a strong need to establish a course in security. It seems natural. It seems to be a given. Still, a challenge exists as to what to cover in an introductory information systems security course. Such a course sits well as an upper division undergraduate course or as an introductory postgraduate course. In this book, we have incorporated the need for a multifaceted, multidisciplinary, eclectic program. The book takes a managerial orientation, concentrating on presenting key security challenges that a manager of information technology (IT) is going to face when ensuring the smooth functioning of the organization. This will help harmonize the organization's objectives, needs, and opportunities, while increasing managerial awareness of weaknesses and threats. The book also provides a balanced view that spans the full run of technological security concerns, managerial implications, and ethical and legal challenges.

## The Nature of This Book

This book has the following features:

- *A managerial orientation.* As noted previously, this book presents key security challenges that a manager of IT will face with respect to the smooth functioning of the organization. This will help managers focus on the objectives, needs, and opportunities presented by their organizations, while being aware of any weaknesses and threats.
- *An analytical approach.* This book provides a high level framework to conceptualize about IS security problems.
- *A multidisciplinary perspective.* Since IS is multidisciplinary in nature, this book draws on a range of informing disciplines, such as computer science, sociology, law, anthropology, behavioral science. *Computer science* provides the fundamental tools for protecting the technological infrastructure. *Sociology* provides an understanding of the social aspects of technology use and adoption. *Law* provides a legal framework, within which a range of compliance issues are discussed. *Anthropology* provides an understanding of the culture that should exist in order to ensure good security. *Behavioral science* provides insight into the intentions and attitudes of stakeholders.
- *Comprehensive and balanced coverage.* This book provides a balanced view of technological IS security concerns, managerial implications, and ethical and legal challenges.

## The Scope of This Book

The book is organized into four parts. *Part I* deals with technical aspects of IS security. Issues related to formal models, encryption, cryptography, and other system development controls are presented. *Part II* presents the formal aspects of IS security. Such aspects relate to the development of IS security management and policy, risk management and monitoring, and audit control. *Part III* is concerned with the informal aspects of IS security. Issues related to security culture development and corporate governance for IS security are discussed. *Part IV* examines the regulatory aspects of IS security. Various IS security standards and evaluation criteria are discussed. Issues related to computer forensics are presented.

## Changes and Enhancements in Edition 2.0

The book was originally published in 2007, with a minor update in 2016. Edition 2.0 is a complete rework of the original text. While some sections, chapters, and structure of the original text have been maintained, there are many new sections and chapters. New chapters and topic areas for edition 2.0 include

- New material on blockchain technologies
- A complete rewrite of the network security chapter
- A reconfiguration of the material on IS security standards, with a focus on international harmonization efforts

- A new chapter on responding to an IS security breach
- A new chapter on the behavioral aspects of IS security
- A new chapter on the ethical and professional issues in IS security management
- Updated and reconfigured material on the legal aspects of IS security
- Updated end-of-chapter case studies
- Four new case studies about maintaining process integrity, critical infrastructure protection, and the Anthem and Sony cybersecurity breaches

# Pedagogical Aids

This book is organized to present materials over a given semester. The 14 chapters can be covered in anywhere between 10 and 12 weeks, leaving time for guest speakers and case studies. The instructor manual (available from the publisher) presents a sample syllabus, which can aid instructors in deciding how the material should be presented.

- *In brief.* Each chapter ends with an "in brief" section. This section summarizes key points covered in the chapter. While students can use this section to revise concepts, the section can also be the basis for in-class discussion.
- *Questions and exercises.* At the end of each chapter there are three types of questions and exercises:
  - **Discussion questions** are critical-thinking assignments. These can be used for take-home assignments or for in class, small group discussions.
  - **Exercises** at the end of the chapter are small activities that can be used either at the start of the class or as a wrap-up. These can also be used as an icebreaker.
  - **Short questions** are largely for testing the understating of key concepts. These can be used either by students to revise key concepts or even as part of a pop quiz by the instructor.
- *Case studies.* At the end of each chapter there is a short case study. The cases typically pertain to materials presented in the chapter. These can be used in multiple ways—as a homework assignment, in-class discussion, and critical thinking.

# Instructor Resources

Instructor resources can be accessed through: http://prospectpressvt.com/titles/dhillon-information-systems-security. The following materials are included in the instructor resources:

- *Hands-on projects.* There is a companion exercise handbook that can be requested from the publisher. The hands-on projects use contemporary cybersecurity software to understand basic security principles.
- *PowerPoint presentations.* The instructor manual comes with PowerPoint presentations that instructors can use as is or customize to their liking.
- *Test questions.* The book comes with a test bank, which includes multiple choice, true-false, and short answer questions. Solutions to the questions are also provided.

- *Videos.* Each chapter is accompanied by one or more videos. The videos provide supplemental material to be used face-to-face, in flipped-classroom sessions, or during online lectures.

I hope you enjoy reading and using the book as much as I enjoyed writing it. I welcome feedback and am willing to schedule guest appearances in your classes.

*Gurpreet S. Dhillon*
*The University of North Carolina*
*Greensboro, NC, USA*

# Acknowledgments

Each time I have written a book, I have been blessed with the support of my family. It goes without saying that this book, and the others that I have authored, would not have been possible without the relentless support of my wife and my children. Simran, Akum, and Anjun have suffered the most in terms of dealing with the upheavals, trials, and tribulations. Writing a book is not an easy task. It has consumed my mornings and nights, days and years. And I most sincerely acknowledge my family for their support.

My students deserve a special mention. My doctoral classes have always sparked new thoughts, new considerations, and ideas for new research. More specifically, my past and current doctoral students have played a special role in the shaping of this book. What I present in this book is a result of years of discourse with my doctoral students, which is synthesized into the chapters. I thank you for being part of the discourse.

I also want to thank the individuals who have helped me write specific parts of this book. I acknowledge their support and help most sincerely. Michael Sosnkowski for Chapter 4. Kane Smith for his contributions to Chapter 9. Currie Carter for his contributions to Chapter 13. Several cases studies were developed by my students. Thanks to Sam Yerkes Simpson for developing and writing "Process and Data Integrity Concerns in a Scheduling System." Sharon Perez for developing and writing "Case of a Computer Hack." Isaac Janak for developing and writing "Critical Infrastructure Protection: The Big Ten Power Company." Kevin Plummer for developing and writing "The Case of Sony's PlayStation Network Breach." Two other individuals deserve special mention: Md Nabid Alam for his help in preparing the instructor support materials for the book and Yuzhang Han for his assistance in preparing hands-on exercises accompanying this book.

There are many other individuals who have contributed by planting the seed of an idea or in reviewing drafts of the work. Stephen Schleck very diligently reviewed an earlier draft of the book, and later expressed that the collective body of work helped him prepare and succeed in the CISSP exam. Many reviewers provided feedback on the draft chapters. Thank you for taking time out to help me improve the book and its presentation. The following reviewers specifically helped in improving edition 2.0:

Ella Kolkowska, Örebro University, Sweden
Dawn Medlin, Appalachian State University, United States
Oleksandr Rudniy, Fairleigh Dickinson University, United States
Roberto Sandoval, University of Texas at San Antonio, United States

Finally, I would like to thank Beth Lang Golub for having faith in me. I first met Beth at a Decision Science Institute conference in Orlando, Florida, to discuss the first edition of this book. We are now in edition 2.0 and hopefully many more to come! Beth's persistence and meticulousness ensured that the project remained on track. It has been a pleasure working with Beth.

# About the Author

**Gurpreet S. Dhillon** is a professor and head of Information Systems and Supply Chain Management at the University of North Carolina, Greensboro. He has a PhD in Information Systems from the London School of Economics, UK. He also has an MSc from the Department of Statistical and Mathematical Sciences at the London School of Economics and an MBA specializing in Operations Management. His research has been published in many of the leading journals in the information systems field. He is also the editor-in-chief of the *Journal of Information System Security* and has published more than a dozen books. His research has been featured in various academic and commercial publications, and his expert comments have appeared in the *New York Times*, *USA Today*, *Business Week*, CNN, NBC News, NPR, among others. In 2017, Dhillon was listed as the "most frequently appearing author in the 11 high-impact IS journals" between 1997 and 2014. In a study published in the *ACM Transactions on Management Information Systems*, he was listed among the top 90 researchers in the world. The International Federation for Information Processing awarded Gurpreet a Silver Core Service Award for his outstanding contributions to the field. Visit Gurpreet at http://www.dhillon.us.com, or follow him on Twitter @ProfessDhillon.

# CHAPTER 1

# Information System Security: Nature and Scope

*Our belief in any particular natural law cannot have a safer basis than our unsuccessful critical attempts to refute it.*

—Sir Karl Raimund Popper, *Conjectures and Refutations*

Joe Dawson sat in his office and pondered as to how best he could organize his workforce to effectively handle the company operations. His company, SureSteel Inc., had grown from practically nothing to one with a $25 million annual turnover. This was remarkable given all the regulatory constraints and the tough economic times he had to go through. Headquartered in a posh Chicago suburb, Joe had progressively moved manufacturing to Indonesia, essentially to capitalize on lower production costs. SureSteel had an attractive list of clients around the world. Some of these included major car manufacturers. Business was generally good, but Joe had to travel an awful lot to simply coordinate various activities. One of the reasons for his extensive travel was that Joe could not afford to let go of proprietary information. Being an industrial engineering graduate, Joe had developed some very interesting applications that helped in forecasting demand and assessing client buying trends. Over the years, Joe had also collected a vast amount of data that he could mine very effectively for informed decision-making. Clearly there was a wealth of strategic information on his stand-alone computer system that any competitor would have loved to get their hands on.

Since most of Joe's sensitive data resided on a single computer, it was rather easy for him to ensure that no unauthorized person could get access to the data. Joe had given access rights to only one other person in his company. He could also, with relative ease, make sure that the data held in his computer system did not change and was reliable. Since there were only two people who had access to the data, it was a rather simple exercise to ensure that the data was made available to the right people at the right time. However, complex challenges lay ahead. It was clear that in order for SureSteel to grow, some decision-making had to be devolved to the Indonesian operations. This meant that Joe would not only have to trust some more people, perhaps in Chicago and Indonesia, but also establish some form of an information access structure. Furthermore, there was really no need to give full access to the complete data set. This meant that some sort of a responsibility structure had to be established. Initially the Chicago office had only 10 employees. Besides himself and his executive assistant, there was a sales director, contract negotiations manager, finance manager, and other office support staff. Although the Indonesian operations were larger, no strategic planning was undertaken there.

Joe had hired an information technology (IT) specialist to help the company set up a global network. In the first instance Joe allowed the managing director in Indonesia to have exclusive access to some parts of his huge database. Joe trusted the managing director and was pretty sure that the information would be used appropriately. SureSteel continued to grow. New factories were set up in Uzbekistan and Hungary. Markets expanded from being primarily US based to Canada, the UK, France, and Germany.

All along the growth path, Joe was aware of the sensitive nature of the data that resided on the systems. Clearly there were a lot of people accessing it in different parts of the world and it was simply impossible for him to be hands-on as far as maintaining security was concerned. The IT specialist helped the company implement a firewall and other tools to keep a tab on intruders. The network administrator religiously monitored the traffic for viruses and generally did a good job keeping malicious code at a distance. Joe Dawson could at least for now sit back and relax, confident that his company was doing well and that his proprietary information was indeed being handled with care.

---

Every human activity, from making a needle to launching a space shuttle, is realized based on two fundamental requirements: coordination and division of labor [5]. Coordination of various tasks towards a purposeful outcome defines the nature of organizations. At the crux of any coordinating activity is communication. Various actors must communicate in order to coordinate. In some cases, computers can be used to coordinate. In other situations, it is perhaps best not to use a computer. At times coordination can be achieved by establishing formal rules, policies, and procedures. In yet other situations it may be best to informally communicate with other parties to achieve coordination. However, the end result of all purposeful communication is coordination of organizational activities so as to achieve a common purpose.

Central to any coordination and communication is information. In fact, it is information that holds an organization together [7]. An organization can therefore be defined as a series of information-handling activities. In smaller organizations it is relatively easy to handle information. However, as organizations grow in size and complexity, information handling becomes cumbersome and yet increasingly important. No longer can one rely on informal roles to get work done. Formal systems have to be designed. These preserve not only the uniformity of action, but also the integrity of information handling. The increase in size of organizations also demands that a vast amount of information be systematically stored, released, and collected. There need to be mechanisms in place to retrieve the right kind of information, besides having an ability to distribute the right information to the right people. Often, networked computer systems are used to effect information handling. It can therefore be concluded that information handling can be undertaken at three levels—technical, formal, and informal; and the system for handling information at these three levels is an organization's information system.

For years there has been a problem with defining information systems. While some have equated information systems to computers, others have broadened the scope to include organizational structures, business processes, and people into the definition. Whatever may be the orientation of particular definitions, it goes without saying that the wealth of our society is a product of our ability to organize and this ability is realized through information handling. In many ways the systems we create to handle information are the very fabric of organizations. This is conventional wisdom. Management thinkers ranging from Mary Parker Follett to Herbert Simon and Peter Drucker have all brought to the fore this character of the organization.

This book therefore treats organizations in terms of technical, formal, and informal parts. Consequently, it also considers information handling in terms of these three levels. This brings us to the question as to what might information system security be? Would it be the management of access control to a given computer system? Or, would it be the installation of a firewall around an organization's network? Would it be the delineation of responsibilities and authorities in an organization? Or, would it be the matching of such responsibilities and authorities to the computer system access privileges? Would we consider inculcating a security culture to be part of managing information system security? Should a proper management development program, focusing on security awareness, be considered part of information system security management? I am inclined to consider all these questions (and more) to be part of information system security. Clearly, as an example, restricting access privileges to a computer system will not work if a corresponding organizational responsibility structure has not been defined. However, when we ask information technologist in organizations to help us in ensuring proper access, the best they can come up with is that it is someone else's responsibility! They may be right in pointing this out, but doesn't this go against the grain of ensuring organizational integrity? Surely it does. However, we usually have difficulty relating loss of organizational integrity with lost confidentiality or integrity of data.

This book considers information system security at three levels. The core argument is that information systems need to be secured at a technical, formal, and an informal level. This classification also defines the structure of the book. In the first chapter the nature and scope of the information system security problem is discussed. Issues related to the technical, formal, and informal are explored. It is shown how a coordination of the three helps in maintaining information system security. The subsequent sections in the book then explore the intricate details at each of the levels.

## Coordination in Threes

Most people, when asked about an organization's information systems and their security, would talk about only one kind of a system—**the formal system**. This is because formal systems are characterized by great tenacity [3]. It is assumed that without tenacious consistency organizations cannot function. In many ways it is important to religiously follow the formal systems. Clearly over a period of time such ways of working get institutionalized in organizations. However, any misinterpretation of the formal system can be detrimental to an organization. In a classic book, *The Governing of Men*, Alexander Leighton [4] describes how a misinterpretation of the formal system stalled a Japanese intern program during World War II. A formal system gets formed when messages arrive from external parties, suppliers, customers, regulatory agencies, financial institutions. These messages are usually very explicit and are transcribed by an organization to get their own work done. Not only is it important to protect the integrity of these messages; it is also essential to ensure that these are interpreted in a correct manner and the resulting outcomes are adequately measured.

Messages from external parties are often integrated into internal formal systems, which in turn trigger a range of activities: proper inventory levels may be established; further instructions may be offered to concerned parties; marketing plans may be formulated; formal policies and strategies may be developed. Much of the information generated by the formal system is stored in ledgers, budget statements, inventory reports, marketing plans, product development plans, work in progress plans, compensation plans, etc.

The information flow loop, from external to internal and then to external, is completed when messages are transmitted by the organization to external parties from which it originally received the messages, or other additional parties. Such messages usually take the form of raising invoices, processing payments, acknowledging receipts, etc. Traditional systems development activities have attempted to map such information flows as one big "plumbing" system. Business process reengineering advocates adopt a process orientation and map all activities in terms of information flows. Information technologists then computerize all or a majority of information flows, such that the operations are efficient and effective. Although all this may be possible, there are a lot of informal activities and systems in operation within and beyond the formal systems.

**The informal system** is the natural means to augment the formal system. In ensuring that the formal system works, people generally engage in informal communications. As the size of the organization grows, a number of groups with overlapping memberships come into being. Some individuals may move from one group to another, but are then, generally, aware of differences in attitudes and objectives. The differences in opinions, goals, and objectives are usually the cause of organizational politicking.

Tensions do arise when an individual or group has to conform to rules established by the formal system, but may be primarily governed by the norms established in a certain informal setting. Clearly formal systems are rule based and tend to bring about uniformity. Formal systems are generally insensitive to local problems and as a consequence there may often be discordance between rules advocated by the formal system and realities created by cohesive informal groupings. The relevant behavior of the informal groupings is the tacit knowledge of the community. Generally, the informal system represents a subculture where meanings are established, intentions are understood, beliefs are formed, commitments and responsibilities are made, altered, and discharged.

The demarcation between formal and informal systems is interesting. It allows us to examine the real-world situations to see which factors can best be handled by the formal system. There would certainly be aspects that should be left informal. The boundary between the formal and informal is best determined by decision-makers, who base their assessment on identifying those factors that can be handled routinely and those which it would be best to leave informal. It is very possible to computerize some of the routine activities. This marks the beginning of **the technical systems**.

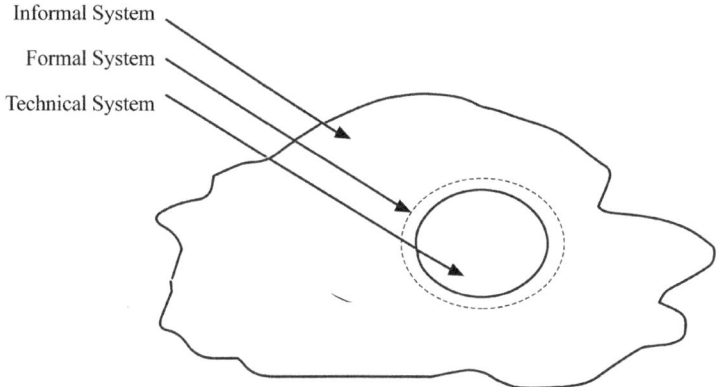

**Figure 1.1.** A "fried egg" analogy of coordination in threes

The technical system essentially automates a part of the formal system. At all times a technical system presupposes that a formal system exists. However, if it does not, and a technical system is designed arbitrarily, it results in a number of problematic outcomes. Case studies in this book provide evidence of this situation. Just as a formal system plays a supportive role to the largely informal setting, similarly the technical system plays a supportive role to the formal, perhaps bureaucratic, rule-based environment.

Clearly there has to be good coordination between the formal, informal, and the technical systems. Any lack of coordination either results in substandard management practices or it opens up the organization to a range of vulnerabilities. An analogy in the form of a *fried egg* may be useful to describe the coordination among the three systems. As represented in Figure 1.1, the yolk of the egg represents the technical system, which is firmly held in place by the formal system of rules and regulations (the vitelline, yolk membrane, in our analogy). The informal system is represented by the white of the egg.

The *fried egg* is a good way to conceptualize the three coordinating systems. It suggests the appropriate subservient role of the technical system within an organization. It also cautions about the consequences of over-bureaucratization of the formal systems and their relationship to the informal systems. The threefold classification forms the basis on which this book is developed.

## Security in Threes

As has been argued elsewhere [2], managing information system security to a large extent equates to maintaining integrity of the three systems—formal, technical, and informal. Any discordance between the three results in potential security problems. Managing security in organizations is the implementation of a range of controls. Such controls could be for managing the confidentiality of data, or for maintaining the integrity and availability of data. It is essentially the controls that are instituted.

Control is "the use of interventions by a controller to promote a preferred behavior of a system being controlled" [1]. Thus, organizations that seek to contain opportunities for security breaches, would strive to implement a broad range of interventions. In keeping with the three systems, controls themselves could either be technical, formal, or informal. Typically, an organization can implement controls to limit access to buildings, rooms, or computer systems (technical controls). Commensurate with this, the organizational hierarchy could be expanded or shortened (formal controls) and an education, training, and awareness program put in place (informal controls). In practice, however, controls have dysfunctional effects. The most important reason is that isolated solutions (i.e. controls) may be provided for specific problems. These "solutions" tend to ignore other existing controls and their contexts. Thus, individual controls in each of the three categories, though important, must complement each other. This necessitates an overarching policy that determines the nature of controls being implemented and therefore provides comprehensive security to the organization.

Essentially, the focus of any security policy is to create a shared vision and an understanding of how various controls will be used such that the data and information is protected in an organization. The vision is shared among all levels in the organization and uses people and resources to impose an environment that is conducive to the success of an enterprise. Typically an organization would develop a security policy based on sound business judgement, the value of data being protected, and the risks associated with the protected data. It would then be applied in conjunction with other enterprise policies: *viz.* corporate policy on disclosure of information and personnel policy on education and training. In choosing the various requirements of a security policy, it is extremely difficult to draw up generalizations. Since the security policy of an enterprise largely depends upon the prevalent organizational culture, the choice of individual elements is case specific. However, as a general rule of thumb all security policies will strive to implement controls in the three areas identified above. Let us now examine each in more detail.

## Technical Controls

Today businesses are eager to grasp the idea of implementing complex technological controls to protect the information held in their computer systems. Most of these controls have been in the area of access control and authentication. A particularly exciting development has been smart card technology, which is being used extensively in the financial sector. However, authentication methods have made much progress. It has now been recognized that simple password protection is not enough, and so there is the need to identify the individual (i.e., is the user the person he/she claims to be?). This has to some extent been accomplished by using the sophisticated "challenge-response box" technology. There have been other developments such as block ciphers, which have been used to protect sensitive data. There has been particular interest in message authentication, with practical applicability in the financial services and banking industry. Furthermore, the use of techniques such as voice analysis and digital signatures has further strengthened technology-oriented security controls. Ultimately, implementation of technological solutions is dependent upon cost justifying the controls.

Although technological controls are essential in developing safeguards around sensitive information, the effectiveness of such technological solutions is questionable. The perpetrators "generally stick to the easiest, safest, simplest means to accomplish their objectives, and those means seldom include exotic, sophisticated methods" [6]. For instance, it is far easier for a criminal to obtain information by overhearing what people say or finding what has been written on paper, than through electronic eavesdropping. In fact, in the last four decades there has hardly been any proven case of eavesdropping on radio frequency emanations. Therefore, before implementing technological controls business enterprises should consider instituting well-thought-out baseline organizational controls (e.g., vetting, allocating responsibilities, awareness).

## Formal Controls

Technological controls need adequate organizational support. Consequently, rule-based formal structures need to be put in place. These determine the consequences of misinterpretation of data and misapplication of rules in an organization and help in allocating specific responsibilities. At an organizational level, development of a "task force" helps in carrying out security management and gives a strategic direction to various initiatives. Ideally the task force should have representatives from a wide range of departments such as audit, personnel, legal, and insurance. Ongoing support should be provided by computer security professionals. Besides these, significant importance should be given to personnel issues. Failing to consider these adequately could result in disastrous consequences. Thus, formal controls should address not only hiring procedures but also the structures of responsibility during employment. A clearer understanding of the structures of responsibility helps in the attribution of blame, responsibility, accountability, and authority. It goes without saying that the honest behavior of the employees is influenced by their motivation. Therefore, it is important to foster a subculture that promotes fair practices and moral leadership. The greatest care, however, should be used for the termination practices of the employees. It is a well-documented fact that most breaches of information system security occur shortly before the employee leaves the organization.

Finally, the key principle in assessing the amount of resources to allocate to security (technical or formal controls) is that the amount spent should be in proportion to the criticality of the system, the cost of remedy, and the likelihood of the breach of security occurring. It is necessary for the management of organizations to adopt appropriate controls to protect themselves from claims of negligent duty and also to comply with the requirements of data protection legislation.

## Informal Controls

Increasing awareness on security issues is the most cost-effective control that an organization can conceive of. It is often the case that information system security is presented to the users in a form that is beyond their comprehension, and this ends up being a demotivating factor in implementing adequate controls. Increased awareness should be supplemented with an ongoing education and training program. Such training and awareness programs are extremely important in developing a "trusted" core of members of the organization. The emphasis should be on building an organizational subculture where it is possible to understand the intentions of the management. An environment should also be created that is conducive to developing a common belief system. This would ensure that members of the organization are committed to their activities. All this is possible by adopting good management practices. Such practices have special relevance in organizations these days since they are moving towards outsourcing key services and thus having an increased reliance upon third parties for infrastructural support. This has consequences of increased dependency and vulnerability of the organization, thereby increasing the probability of risks.

The first step in developing good management practices and reducing the risk of a security breach is to adopt some baseline standards, the importance of which was highlighted earlier. Today the international community has taken concrete steps in this direction and has developed information security standards. Although issues of compliance and monitoring have not been adequately addressed, these are certainly first steps in our endeavor to realize high-integrity reliable organizations.

## Institutionalizing Security in Organizations

We have, so far, developed an understanding of organizations as constituted of technical, formal, and informal systems. We also have an appreciation of security controls and their systematic position at the three levels. We have also argued that overall information system security comes into being by maintaining the overall integrity of the three systems and the corresponding controls. We shall now use the three levels and the corresponding controls to show how security in organizations would be compromised, which will establish an agenda for the rest of the book.

Figure 1.2 (a) represents the organization in terms of formal information handling. Many organizational theorists call this the organizational structure. Although such charts are often considered to be a useful means to communicate what happens in organizations (and are generally handed to anyone who inquires about the structure), there are differences in opinion as to their utility. As was mentioned previously, this difference in opinion is the consequence of our difficulty in representing the informal aspects of information handling. In an era when organizations did not rely on computers to undertake work, to a very large extent we relied on formal bureaucratic information-handling procedures. As a matter of fact, the British ruled over a large part of the world without computers and in the 18th century the cotton trade thrived between England and the US without the use of any technical systems. In order to enable the functioning of organizations such elaborate structures were necessary. Security of information handling at this level largely related to linking access rights to the hierarchical level. This worked very well since structures remained stable for extended periods of time. For example, it was always the chief accountant who had the key to the company vault. So it was the chief accountant who was responsible and accountable if any money was lost. Information system security therefore had more to do with locks and keys.

Figure 1.2 (b) shows how certain formal information-handling activities can be made technical, essentially for efficiency and effectiveness purposes. Obviously it is important to bring in efficiencies in the working of an organization. Initially some of the formal activities were made technical by merely using a telephone or a telex. More recently computers have been used. With time these computers were networked, first within the confines of the company and later across companies. Initial security of the technical structures was also lock and key based. Essentially, all computing power was concentrated into an information center, which was locked and only authorized persons could gain access. However, with time

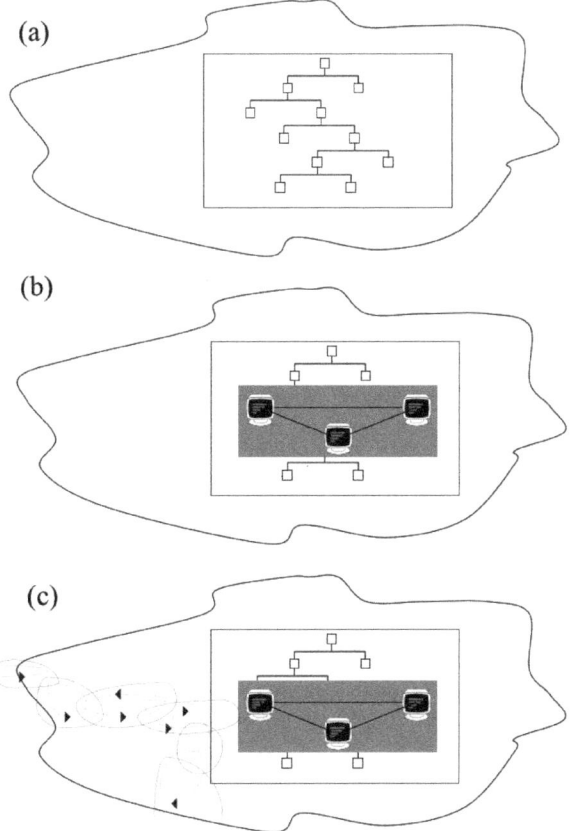

**Figure 1.2.** Information handling in organizations: (a) Exclusive formal information handling; (b) Partly computerized formal controls; (c) The reality, with technical, formal, and informal information handling.

this changed as distributed systems came into being. In terms of security, however, we still relied on simple access control mechanisms such as passwords.

Organizational life would be much simpler if we had to deal with only the formal or the technical aspects of the system. However, as shown in Figure 1.2 (c), there are numerous social groups, which communicate with each other and have overlapping memberships. It is not only important to delineate the formal and informal aspects, but also to adequately position the technical within the broader scheme of things. Security in this case is largely a function of maintaining consistency in communication and ensuring proper interpretation of information. Besides, issues related to ethics and trust become important.

A further layer of complexity is added when organizations establish relationships with each other. However, at the core of all information handling, coordination in threes still dominates and remains a unique aspect of maintaining security. With this foundation of coordination in threes, and our argument that information system security to a large extent is the management of integrity between the three levels, we can begin our journey to explore issues and concerns in information system security management. We start by considering the information security issues in the technical systems and systematically move to the formal and informal levels. We also present case studies at each level to ponder on the relevant issues.

> **In Brief**
> - An organization is defined as a series of information-handling activities.
> - Information handling can be undertaken at three levels—technical, formal, and informal. The system for handling information at these three levels is an organization's information system.
> - Security can only be achieved by coordinating and maintaining the integrity of operations within and between the three levels.
> - It is important to remember the subservient role of the technical system within an organization. Over-engineering a solution or over-bureaucratization of the formal systems have consequences for security and integrity of operations.
> - Management of information system security is connected with the appropriate use of technology in the organization and right design of the business processes.
> - Information system security has to be understood appropriately at all three levels of the organization: **technical, formal, and informal.**

# Questions and Exercises

**Discussion questions.** These are based on a few topics from the chapter and are intentionally designed for a difference of opinion. These questions can best be used in a classroom or a seminar setting.

1. Even though information system security goes way beyond the security of the technical edifice, applications and organization resources can only be protected by using the latest security gadgets. Isn't this a contradiction in itself? Discuss.

2. The advent of inter-networked organizations and the increased reliance of companies on the Internet to conduct their business has increased the chances of abuse. Is this the case? Discuss this issue using examples from the popular press.

3. Do we really need to understand and place great importance on the informal controls prior to establishing security rules? If so, why? If not, why not?

4. Over-engineering a solution or over-bureaucratization of the formal systems have consequences for security and integrity of operations. Comment.

**Exercise.** Look at various cybersecurity surveys over the past 10 years—CSI/FBI, Verizon, etc.—what are some of the noticeable changes? Besides the increase in number of incidents, what do the surveys report in terms of nature of threats?

## Short Questions

1. Coordination in threes refers to what three aspects of information security?
2. When an organization implements controls to limit access to buildings, rooms, or computer systems, these are referred to as _____ controls.
3. The organizational hierarchy can be considered a part of _____ controls.
4. Training and an employee awareness program could be considered a part of what type of control?
5. The first step in developing good management practices and reducing the risk of a security breach is by adopting some _____ _____ standards.
6. Most breaches of information system security occur shortly _____ the terminated employee leaves the organization.
7. Formal controls should not only address the hiring procedures but also the structures of _____ during employment.
8. Training and awareness programs are extremely important in developing a _____ core of members of the organization.
9. Coordination in threes still applies, but a further layer of _____ is added when organizations establish relationships with each other.
10. A firewall is an example of a _____ control.

# Case Study: Virtual Companies and Identity Theft

JetBlue was compromised by hackers for three years from 2008 to 2011 and lost over 160 million credit card numbers, which in itself is not unusual. However, due to the unique structure of the company and seemingly casual corporate response, the attack highlights a major breakdown of accountability between the client and JetBlue. The attack started with malware being placed on legacy computer systems which were to be replaced in future upgrades. The thieves appeared to be working out of Russia or Ukraine as server interconnections led back into this region, and while two of the defendants were caught three more are still at large. The goal of the attack was clear from testimony: to obtain credit cards, passwords, names, and personal information. Once the malware was found the efforts by the company were subdued and amounted to notifying their staff after a two-month investigation as well as reporting to authorities in the form of notifications to state attorneys general in Maryland and New Hampshire that the breach had occurred. Passengers and the public on the other hand were not notified of any issues and only a limited communication was eventually provided to the public in which it was reiterated that JetBlue staff had been provided with credit monitoring for one year at no cost.

Why is this attack and fallout any different from others around the world? One hint is the way JetBlue is structured—they are a virtual company, which has become a growing trend in the hyper-competitive global economy. The airline business has many overheads which are not limited only to fuel prices, unionized labor, expensive fixed assets, and the need for highly trained staff. Likewise, while many nations subsidize or own their country's airline, in the United States airlines are private for-profit enterprises competing for the same domestic and international customers who may also be serviced by companies located outside of the States. To cut costs while at the same time providing a high-quality customer experience with fast bookings, ease of access, and low price, JetBlue aggressively removed fixed costs, such as centralized call centers, which are not directly related to servicing the airplanes. Therefore, in place of call centers the agents work from home, and administration tasks are automated, or provided remotely from computer centers. Since the goal is 100% up time this means not only spreading out the work force geographically but interconnecting them as well with robust telecommunications and services. This may explain why older systems were in production and therefore vulnerable to be breached as they were in constant use.

Corporate culture now comes into play with a virtual company's design. The logic used here is as follows: if a company has little or no physical contact between employee and employer, as is the case with the agents in JetBlue, then the staff and customer base become an abstraction to management, removing the employee's and client's welfare as a tangible reality to management. The analogy is that of a fighter pilot dropping bombs on a city in contrast to a foot soldier coming face to face with an enemy. The pilot has limited feedback from the horror he commits on the ground, while the soldier has immediate and sensory information of fear, anger, and despair.

The foot soldier lives the outcome of the effect he has caused while the pilot is insulated and does not. When such insulation is extended to the customer and worker relation with management then poor choices and behaviors by management can occur.

The question then is whether a virtual company can provide real care and concern for its staff and clients. From the actions of JetBlue in this case one would have to say "no" since JetBlue's reaction was not to immediately inform the customer base and only later to provide rudimentary credit checks for the affected staff. How then to counteract such a situation since virtual companies, due to their flexibility and cost savings, will continue to grow and shape the future business world? The solution proposed is not to return to 19th- or 20th-century corporate structures but rather to look at the loss of customer or staff information through error, neglect, or poor design as a natural aspect of business. The example used will be for the loss of personal information but it could also be expanded into other areas such as credit and medical information.

The proposed solution is that of a self-healing system which is hidden in the actions taken to rectify the breach but visible in the communication to the public, authorities, and personnel affected by the breach. Returning to the case of JetBlue, and using the same compromise as an example, the malware is placed on older information systems and then is discovered three years later. At the point of discovery and confirmation that a breach has occurred automated procedures and systems would go into action. Credit checks would be done automatically for each affected employee on the backend; new passwords and digital certificates would be issued for all users and clients sites; public notification and government involvement would be instigated via predefined memorandums of understanding to bring full awareness of the breach. At the same time backend systems would be automatically scanned and reviewed with a short list of end-of-life equipment to be replaced on an accelerated schedule. Budgetary funds which had been automatically accumulating for just such an event would be freed to provide for the above corrections with predefined authority for expenditures on the various projects needed to rectify the vulnerabilities. Finally, personal information gathered on the clients would be reviewed for tampering, or potential identity theft, and anomalies would trigger automatic notification to those customers who may have been attacked. At this point one could envision a further stage of automation in which bank records, credit history, and medical records are readjusted to ensure that any criminal identity misuse is removed, thus allowing for a better credit score or refund of fees garnered during the period of theft.

1. What is the goal of such self-healing systems?
2. What steps can be taken to ensure security and integrity of virtual companies?

Case prepared by Michael Sosnkowski and materials drawn from:

Schaal, D. 2013. Hack of JetBlue network for crew members went undetected for 3 years. *Skift.com*. July 26. Retrieved from http://goo.gl/u3SQtu on Apr. 3, 2017.

Young, M., & Jude, M. 2004. *The case for virtual business processes: Reduce costs, improve efficiencies, and focus on your core business*. Indianapolis, IN: Cisco Press.

# References

1. Aken, J.E. 1978. *On the control of complex industrial organisations*. Leiden: Nijhoff.
2. Dhillon, G. 1997. *Managing information system security*. London: Macmillan.
3. Hall, E.T. 1959. *The silent language*. 2nd ed. New York: Anchor Books.
4. Leighton, A.H. 1945. *The governing of men*. Princeton: Princeton University Press.
5. Mintzberg, H. 1983. *Structures in fives. designing effective organizations*. Englewood Cliffs, NJ: Prentice-Hall.
6. Parker, D. 1991. Seventeen information security myths debunked. In *Computer security and information integrity*, ed. K. Dittrich, S. Rautakivi, and J. Saari. Amsterdam: Elsevier Science Publishers, 363–370.
7. Stamper, R.K. 1973. *Information in business and administrative systems*. New York: John Wiley & Sons.

# PART I

# TECHNICAL ASPECTS OF INFORMATION SYSTEMS SECURITY

# CHAPTER 2

# Security of Technical Systems in Organizations

> *Physics does not change the nature of the world it studies, and no science of behaviour can change the essential nature of man, even though both sciences yield technologies with a vast power to manipulate their subjective matters.*
>
> —Burrhus Frederic Skinner, *Cumulative Record*, 1972

As the size of Joe Dawson's company, SureSteel, Inc., grew, he became increasingly concerned about security of data on his newly networked systems. Part of the anxiety was because Joe had recently finished reading Clifford Stoll's book *Cuckoo's Egg*. He knew that there was no hardware or software that was foolproof. In the case described by Stoll, a German hacker had used the Lawrence Berkeley Laboratory computer systems to systematically gain access to the US Department of Defense computer systems. Clifford Stoll, a young astronomer at the time, converted the attack into a research experiment and over the next 10 months watched this individual attack about 450 computers and successfully enter more than 30. What really concerned Joe was that the intruder in *Cuckoo's Egg* did not conjure up new methods for breaking operating systems. He repeatedly applied techniques that had been documented elsewhere. Increasingly Joe began to realize that vendors, users, and system managers routinely committed blunders that could be exploited by hackers or others who may have an interest in the data. Although the nature of sensitive data held in SureSteel computers was perhaps not worthy of being hacked into, nevertheless the ability of system intruders to do so concerned Joe Dawson.

Joe Dawson called a meeting with SureSteel's IT staff, which now numbered 10 at the Chicago corporate office. He wanted to know how vulnerable they were and what protective measures they could take. Steve Miller, the technologist who headed the group, took this opportunity to make a case for more elaborate virus and Trojan horse protection. Joe quickly snubbed the request and directed all to focus attention on how best to protect the data residing on SureSteel systems. The obvious responses that came ranged from instituting a policy to change passwords periodically to firewall protection and implementing some sort of an intrusion detection system.

Coming out of the meeting Joe was even more confused than he was prior to calling the meeting. Joe was unsure if by simply instituting a password policy or by implementing an intrusion detection system it would be possible to ensure that his systems would be secure. He had to do some research. Over the years Joe had been reading the magazine *CIO*. He felt that perhaps there was an article in the magazine that would help him understand the complexities of information system security. Fortunately, he was able to locate an article written by Scott Berinato and Sarah Scalet, "The ABCs of Security." As he read this article, he became increasingly unsure as to how he should go about dealing with the problem. He made copies of an excerpt from the article and distributed them to all company IT folks:

> An entire generation of business executives has come of age trained on the notion that firewalls are the core of good security. The unwritten rule is: The more firewalls, the safer. But that's just not true. Here are two ways firewalls can be exploited. One: Use brute force to flood a firewall with too much incoming data to inspect. The firewall will crash. Two: Use encryption, a basic security tool, to encode an e-mail with, say, a virus inside it. A firewall will let encrypted traffic pass in and out of the network.
>
> Firewalls are necessary tools. But they are not the core of information security. Instead, companies should be concentrating on a holistic security architecture. What's that? Holistic security means making security part of everything and not making it its own thing. It means security isn't added to the enterprise; it's woven into the fabric of the application. Here's an example. The nonholistic thinker sees a virus threat and immediately starts spending money on virus-blocking software. The holistic security guru will set a policy around e-mail usage; subscribe to news services that warn of new threats; reevaluate the network architecture; host best practices seminars for users; oh, and use virus blocking software, and, probably, firewalls.

What became clear to Joe was that managing information system security was far more complex than he had initially thought it would be. There was also no doubt that he had to begin thinking about possible avenues as soon as possible. Since his IT people seemed to feel that they could handle the technical aspects of security more easily than others, Joe set out to concentrate efforts in this area first.

---

When dealing with information system security, the *weakest point* is considered to be the most serious vulnerability. When we implement home security systems (ADT, Brink, etc.), the first thing we try and do is to identify the vulnerabilities. The most obvious ones are the doors and windows. The robber is generally not going to access a house via a 6-inch-thick wall. In the information system security field, this is generally termed as the *principle of easiest penetration*. Donn Parker, an information system security guru summarizes the principle as "perpetrators don't have the values assumed by the technologists. They generally stick to the easiest, safest, simplest means to accomplishing their objectives" [4]. The principle of easiest penetration suggests that organizations need to systematically consider all possible means of penetration since strengthening one might make another means more attractive to a perpetrator. It is therefore useful to consider a range of possible security breaches that any organization may be exposed to.

# Vulnerabilities

At a technical level, our intent is to secure the hardware, software, and the data that resides in computer systems. It therefore follows that an organization needs to ensure that the hardware, software, and the data are not **modified, destroyed, disclosed, intercepted, interrupted,** or **fabricated**. The six threats exploit vulnerabilities in computer systems and are therefore the precursors to technical security problems in organizations.

- **Modification** is said to occur when the data held in computer systems is accessed in an unauthorized manner and is changed without requisite permissions. Typically, this happens when someone may change the values in a database or alter the routines to perform additional computations. Modification may also occur when data might be changed during transmission. Modification of data at times can occur because of certain changes to the hardware as well.
- **Destruction** occurs simply when the hardware, software, or the data are destroyed because of malicious intent. Although the remit of our definition here is destruction of data and software held in computer systems, there are many instances where destruction of data at the input stage could have serious implications for proper information processing, be it for business operations or for compliance purposes. Destruction as a threat was evidenced when the US Justice Department found Andersen to have systematically destroyed tons of documents and computer files that were sought in probes of the fallen energy trading giant Enron.
- **Disclosure** of data takes place when data is made available or access to software is made available without due consent of the individual responsible for the data or software. An individual's responsibility for data or software is generally a consequence of one's position in an organization. Unauthorized disclosure has a serious impact on maintaining security and privacy of the systems. Although disclosures can occur because of malicious intent, many times it is the lack of proper procedures that results in data being disclosed. At a technical level disclosure of data can be managed by instituting proper program and software controls.
- **Interception** occurs when an unauthorized person or software gains access to data or computer resources. Such access may result in copying of programs, data, or other confidential information. At times an interceptor may use computing resources at one location to access assets elsewhere.
- **Interruption** occurs when a computer system becomes unavailable for use. This may be a consequence of malicious damage of computing hardware, erasure of software, or malfunctioning of an operating system. In the realm of e-commerce applications, interruption generally equates to denial of service.
- **Fabrication** occurs when spurious transactions are inserted into a network or records are added to an existing database. These are generally counterfeit objects placed by unauthorized parties. In many instances it may be possible to detect these as forgeries, however at times it may be difficult to distinguish between genuine and forged objects/transactions.

Table 2.1 presents a summary of vulnerabilities as these apply to the hardware, software, and data. This summary also maps the domain of technical system security in organizations.

**Table 2.1.** Vulnerability of computing resources

| Computing resource | Type of Vulnerability |
|---|---|
| Hardware | Destruction; Interception; Interruption |
| Software | Modification; Interception; Interruption |
| Data | Destruction; Interception; Interruption; Fabrication; Modification; Disclosure |

Threats to the hardware generally fall into the following three categories: destruction, interception, interruption. Although use of simple lock-and-key precautions and common-sense help in preventing loss or destruction of hardware, there could be a number of situations were locks and keys alone may not help. In many situations—fire, floods, terrorist attacks—situations can arise where hardware can get destroyed or services get interrupted. This was clearly evidenced in the terrorist attacks on the World Trade Center in New York and the Irish bombings of 1992 in London.

In situations where the hardware may be of extreme importance, theft and replication can also lead to serious security vulnerability concerns. For instance, in 1918 Albert Scherbius designed a cipher machine, which in later years was used by the Germans to send encrypted messages. Enigma, as the machine came to be known, was intercepted by the Poles and was eventually decrypted. In 1939 the Poles gave the French and the British replicas of Polish-made Enigmas together with the decryption information.

Threats to software may be a consequence of its modification, interception, or interruption. The most serious of the software threats relates to situations where it is modified because a new routine has been inserted in the software. This routine may kick in at a given time and result in an effect that may be harmful to the data or otherwise entirely cease regular operations. Such routines are termed "logic bombs." In other cases a Trojan horse, virus, or a trapdoor may cause harm to the usual operations of the system.

Clearly the responsibility to protect hardware lies with a limited number of employees of the company. Software protection extends to programmers, analysts, and others dealing directly with it. Loss of data, on the other hand, has a broader impact since a large number of individuals get affected by it. But, data does not have any intrinsic value. The value of data resides in the manner in which it is interpreted and used. However, loss of data does have a cost. It could be the cost to recover or reconstruct it. It could also be cost of lost competitiveness. Whatever may be the cost of data, it is rather difficult to measure it. The value of data however is time sensitive. What may be of value today, may lose value or its charm tomorrow. This is the reason why data protection demands measures commensurate with its value.

One of the key priorities in managing vulnerability of technical systems is therefore to consider the requirements for security, particularly data. This entails ensuring the **confidentiality**, **integrity**, and **availability** of data. These three requirements are discussed in subsequent paragraphs.

## Data Security Requirements

In a classic sense **confidentiality**, **integrity**, **availability** have been considered as the critical security requirements for protecting data. The requirements for confidentiality, integrity, and availability are context dependent. This means that given the nature of use of a system, there is going to be a different expectation for each of the requirements. For instance, there is a greater requirement for integrity in the case of electronic funds transfer relative to maintaining confidentiality of data in a typical defense system, where the need for maintaining confidentiality is higher. However, in the case of computer systems geared towards producing the daily newspaper, the availability requirement becomes important. **Authentication** and **non-repudiation** are two other security requirements that have become important, especially in a networked environment.

> *Confidentiality*: this requirement ensures privacy of data.
> *Integrity*: this requirement ensures that data and programs are changed in an authorized manner.
> *Availability*: this requirement ensures proper functioning of all systems such that there is no denial of service to authorized users.
> *Authentication*: this requirement assures that the message is from a source it claims to be from.
> *Non-repudiation*: this requirement prevents an individual or entity from denying having performed a particular action related to data.

**Confidentiality.** Confidentiality has been defined as the protection of private data, either as it resides in the computer systems or during transmission. Any kind of an access control mechanism therefore acts as a means to protect confidentiality of data. Since breaches of confidentiality could also take place while data is being transmitted, it also means that mechanisms such as encryption that attempt to protect the message from being read or deciphered while in transit, also help in ensuring confidentiality of data. Access control could take different forms. Besides locks and keys and related password mechanisms, cryptography (scrambling data to make it incomprehensible) is also a means to protect confidentiality of data. Table 2.2 presents various aspects of the confidentiality attribute.

Clearly when any of the access control mechanisms fail, it becomes possible to view the confidential data. This is usually termed disclosure. At times a simple change in the information can result in it losing confidentiality. This is because the change may have signaled an application or program to drop protection. Modification, or its lack, may also be a cause of loss of confidentiality, even though the information was not disclosed. This happens when someone secretly modifies the data. It is clear therefore that confidentiality loss does not necessarily occur because of direct disclosure. In cases where inference can be drawn without disclosure, confidentiality of data has been compromised.

The use of the *need-to-know principle* is the most acceptable form of ensuring confidentiality. Clearly both users and systems should have access to and receive data only on a need-to-know basis. Although the application of the need-to-know principle is fairly straightforward in a military setting, its use in commercial organizations, which to a large extent rely on the value of trust and friendliness of relationships, can be stifling to the conduct of business.

For this reason, in a business setting the need-to-withhold principle (the inverse of need to know) is more appropriate. The default situation in this case is that information is freely available to all.

**Table 2.2.** Confidentiality attribute and protection of data and software

|  | Data | Software |
|---|---|---|
| **Confidentiality** | A set of rules to determine if a subject has access to an object | Limited access to code |
| **Kinds of controls** | Labels, encryption, discretionary and mandatory access control, reuse prevention | Copyright, patents, labels, physical access control locks |
| **Possible losses** | Disclosure, inference, espionage | Piracy, trade secret loss, espionage |

**Integrity.** Integrity is a complex concept. In the area of information system security, it is related to both intrinsic and extrinsic factors, i.e. data and programs are changed only in a specified and authorized manner. However, this is a rather limited use of the term. Integrity refers to an unimpaired condition, a state of completeness and wholeness and adherence to a code of values. In terms of data, the requirement of integrity suggests that all data is present and accounted for, irrespective of it being accurate or correct. Since the notion of integrity does not necessarily deal with correctness, it tends to play a greater role at system and user policy levels of abstraction than at the data level. It is for this reason that the Clark-Wilson model combines integrity and authenticity into the same model.

Explained simply, the notion of integrity suggests that when a user deals with data and perhaps sends it across the network, the data starts and ends in the same state, maintaining its wholeness and completeness and arriving in an unimpaired condition. As stated before, authenticity of data is not considered important. If the user fails to make corrections to data at the beginning and end states, the data could still be considered as having high integrity. Typically, integrity checks relate to identification of missing data in fields and files, checks for variable length and number, hash total, transaction sequence checks, etc. At a higher level, integrity is checked in terms of completeness, compatibility, consistency of performance, failure reports. Generally speaking, mechanisms to ensure integrity fall into two broad classes: prevention mechanisms and detection mechanisms.

Prevention mechanisms seek to maintain integrity by blocking unauthorized attempts to change the data or change the data in an unauthorized manner. As an example, if someone breaks into the sales system and tries to change the data, it is an instance of an unauthorized user trying to violate the integrity of data. However, if the sales and marketing people of the company attempt to post transactions so as to earn bonuses, then it is an example of changes made in an unauthorized manner. Detection mechanisms simply report violations of integrity. They do not stop violations from taking place. Detection mechanisms usually analyze data to see if the required constraints still hold. Confidentiality and integrity are two very different attributes. In the case of confidentiality, we simply ask the question, has the data been compromised or not? In integrity we assess the trustworthiness and correctness of data. Table 2.3 presents various aspects of the integrity attribute.

**Table 2.3.** Integrity attribute and protection of data and software

|  | Data | Software |
|---|---|---|
| **Integrity** | Unimpaired, complete, whole, correct | Unimpaired, everything present and in an ordered manner |
| **Kinds of controls** | Hash totals, check bits, sequence number checks, missing data checks | Hash totals, pedigree checks, escrow, vendor assurance sequencing |
| **Possible losses** | Larceny, fraud, concatenation | Theft, fraud, concatenation |

**Availability.** The concept of availability has often been equated to disaster recovery and contingency planning. Although this is valid, the notion of availability of data really relates to aspects of reliability. In the realm of information system security, availability may relate to deliberately denying access to data or service. The very popular denial of service attacks are to a large extent a consequence of this security requirement not having been adequately addressed. System designs are often based on a statistical model demonstrating a pattern of use. The prevalent mechanisms ensure that the model maintains its integrity. However, if the control parameters (e.g., network traffic) are changed, the assumptions of the model get changed, thereby questioning its validity and integrity. Consequently, the data and perhaps other resources do not become available as initially forecasted.

Availability attacks are usually the most difficult to detect. This is because the task at hand is to identify malicious and deliberate intent. Although statistical models do describe the nature of events with fair accuracy, there is always scope for a range of atypical events. Then it is a question of identifying a certain atypical event that triggers the denial of service, which is a rather difficult task. Table 2.4 presents availability attributes for protection of data and software.

**Table 2.4.** Availability attribute and protection of data and software

|  | Data | Software |
|---|---|---|
| **Availability** | Present and accessible when and where needed | Usable and accessible when and where needed |
| **Kinds of controls** | Redundancy, back up, recovery plan, statistical pattern recognition | Escrow, redundancy, back up, recovery plan |
| **Possible losses** | Denial of service, failure to provide, sabotage, larceny | Larceny, failure to act, interference |

**Authentication.** The security requirement for authentication becomes important in the context of networked organizations. Authentication assures that the message is from a source it claims to be from. In the case of an ongoing interaction between a terminal and a host, authentication takes place at two levels. First there is assurance that the two entities in question are authentic, i.e. assurance that they are what they claim to be. Second, the connection between the two entities is assured such that a third party cannot masquerade as one of the two parties.

Besides authenticity of communication, it refers to the extrinsic correct and valid representation of that which it means to represent. Therefore, timeliness is an important attribute of authenticity, since obsolete data is not necessarily true and correct. Authenticity also demands having an ability to trace it to the original source. Computer audit and forensic people largely rely on the authentication principle when tracing negative events. Table 2.5 presents authenticity attributes for protecting data and security.

**Table 2.5.** Authentication attribute and protection of data and software

|  | **Data** | **Software** |
|---|---|---|
| **Authentication** | Genuine. Accepted as conforming to a fact | Genuine. Unquestioned origin |
| **Kinds of controls** | Audit log, verification validation | Vendor assurances, pedigree documentation. Hash totals, maintenance log. Serial checks |
| **Possible losses** | Replacement, false data entry, failure to act, repudiation, deception, misrepresentation | Piracy, misrepresentation, replacement, fraud |

**Non-repudiation.** The importance of non-repudiation as an IS security requirement came about because of increased reliance on electronic communications and maintaining the legality of certain types of electronic documents. This led to the use of digital signatures, which allow a message to be authenticated for its content and origin.

Within the IS security domain, non-repudiation has been defined as a property achieved through cryptographic methods, which prevents an individual or entity from denying having performed a particular action related to data [1]. Such actions could be mechanisms for non-rejection or authority (origin); proof of obligation, intent, or commitment; or for proof of ownership. For instance, non-repudiation in a digital signature scheme prevents person A from signing a message and sending it to person B, but later claiming it wasn't him/her (person A) who signed it after all. The core requirement for non-repudiation is that persons A and B (from our example above) have a prior agreement that B can rely on digitally signed messages by A (via A's private key), until A notifies B otherwise. This places the onus on A to maintain security and privacy and the use of A's private key.

There are a range of issues related to non-repudiation. These shall be discussed in more detail in subsequent chapters. Table 2.6 summarizes some of the non-repudiation attributes for protecting data and software.

**Table 2.6.** Non-repudiation attribute and protection of data and software

|  | Data | Software |
|---|---|---|
| **Non-repudiation** | Genuine, true, and authentic communication | Genuine. True |
| **Kinds of controls** | Authentication, validation checks | Integrity controls, non-modification controls |
| **Possible losses** | Monetary, loss of identity, disclosure of private information | Vulnerability of software code, fraud, misconstrued software |

# Methods of Defense

Information system security problems are certain to continue. In protecting the technical systems, it is our intent to institute controls that preserve confidentiality, integrity, availability, authenticity, and non-repudiation. The sections below present a summary of a range of controls. Subsequent chapters in this section give details as to how these controls can be instituted.

## Encryption

Encryption involves the task of transforming data such that it's unintelligible to an outside observer. If used successfully, encryption can significantly reduce chances of outside interception and any possibility of data modification. It's important to note that the usefulness of encryption should not be overrated. If not used properly, it may end up having a limited effect on security, and the performance of the whole system may be compromised. The basic idea is to take plain text and scramble it such that the original data gets hidden beneath the level of encryption. In principle only the machine or person doing the scrambling and the recipient of the scrambled text (often referred to as ciphertext) know how to decrypt it. This is because the original encryption was done based on an agreed set of keys (specific cipher and passphrase).

It is useful to think of encryption in terms of managing access keys to your house. Obviously the owner of the house has a key. Once locked, the house key is usually carried on the person. The only way someone can gain access to the house is by force, i.e. by breaking a door or a window. The responsibility of protecting the house keys resides with the owner of the house. However, if the owner's friend wants to visit the house while the owner is not at home, the owner may pass along the extra set of keys for the friend to enter the house. Now both the owner and the friend can enter the house. In such a situation, the security of the key itself has been compromised. If the owner's friend makes a copy of the key to pass along to his friends, then the security is further diluted and compromised. Eventually the security of the lock-and-key system would be completely lost. The only way in which security can be recovered is by replacing the lock and the key.

In securing and encrypting communications over electronic networks, there are similar challenges to managing and protecting keys. Keys can be lost, stolen, or even discovered by crackers. Although it is possible for crackers to use a serious amount of CPU cycles to crack a cipher, they may more easily to get access by inquiring about the password from an unsuspecting technician. Crackers may also guess passwords based on common word usage or personal identities. It is therefore good practice to use alphanumeric and nonsensical words as passwords, such as "3to9*shh$dy" or similar.

There clearly are techniques that do not require relying on a key. In such cases a decrypting program is built into the machine. In either case the security of data through encryption is as good as the protection of the keys and the machines. Increasingly security of data is being undertaken through the use of public key encryption. In this case a user has two pairs of keys—public and private. The private key is private to the user while the public key is distributed to other users. The private and public keys of a user are related to each other through complex mathematical structures. The relationship between the private and public key is central in ensuring that public key encryption works. The public key is used to encrypt the message, while the private key is used to decrypt the encrypted message by the recipient.

## Software Controls

Besides communication, software is another weak link in the information systems security chain. It is important to protect software such that the systems are dependable and businesses can undertake transactions with confidence. Software-related controls generally fall into three categories:

1. *Software development controls.* These controls are essentially a consequence of good systems development. Conformance to standards and methodologies helps in establishing controls that go a long way towards right specification of systems and development of software. Good testing, coding, and maintenance are the cornerstones of such controls.
2. *Operating system controls.* Limitations need to be built into operating systems such that each user is protected from others. Very often these controls are developed by establishing extensive checklists.
3. *Program controls.* These controls are internal to the software, where specific access limitations are built into the system. Such controls include access limitations to data.

Each of the three categories of controls could be instituted at an input, processing, and output levels. The details of each of the control types are discussed in a later chapter. Generally, software controls are the most visible since users typically come in direct contact with these. It is also important to design these controls carefully since there is a fine balance between ease of use of systems and the level of instituted security controls.

## Physical and Hardware Controls

As stated earlier, perpetrators generally stick to the easiest and quickest methods to subvert controls. The simplest controls, such as ensuring locks on doors, guards at entry doors, and a general physical site planning that ensures security, cannot be ignored. Numerous hardware devices are also available nowadays that ensure the technical security of computer systems. A range of smart card applications and circuit boards controlling access to disk drives in computers are now available.

## Historical Review of Models for Security Specification

Indeed, designing security of information systems within organizations is a nebulous task. Organizations attempt to make their information assets secure in varying ways—be it by incorporating the latest technology or by applying generally accepted models, criteria, and policies, with the hope that the systems will be secure. In this section we explore the nature and scope of the formal models for the technical specification of information system security.

A formal model for specification of information system security is a process for building security into computer-based systems while exploiting the power of mathematical notation and proofs. A formal method relies on formal models to understand reality and subsequently implement the various components. A model can be construed as an abstraction of reality and a mental construction that is embodied into a piece of software or a computer-based information system.

Any function of a computer-based system can be viewed at two levels [5]:

1. *The user view.* The user view, which is elicited during requirement analysis for a system, records what a system should do. The user view is generally an aggregate of views of various stakeholders and is to a large extent independent of the details on the manner in which it will be implemented. The model that embodies the user view is the specification of the system.
2. *The implementation view.* This view is built during system design and records how the system is to be constructed. The model that embodies the implementation view is commonly referred to as a design. In an ideal state the design and specification should adequately reflect each other.

An example of such formal methods and models is evidenced in the Trusted Computer System Evaluation Criteria (TCSEC) originally developed by the US Department of Defense (DoD). The TCSEC were a means to formalize specifications such that the vendors could develop applications according to generally accepted principles. The criteria were attempting to deal with issues of trust and maintaining confidentiality of data from the perspective of the vendor. Hence the TCSEC represented the user view of the model. Soon many other such criteria were developed, such as the European Information Technology Security Evaluation Criteria, the Canadian Trusted Computer Product Evaluation Criteria, the US Federal Criteria for Information Technology Security, and most recently the Common Criteria. A detailed discussion of all the criteria and related standards is presented in a later chapter.

In the realm of secure systems development, the user and implementation views have largely been overlooked and the formal security model tends to embody in itself the security policy. There are various kinds of policies such as the organizational security policy and the technical security policy. The language, goals, and intents of various security policies are different, though the ultimate aim is to secure the systems. An example of a technical security policy is access control. The intent behind restricting access is generally to maintain confidentiality, integrity, and availability of data. Access control could either be mandatory or discretionary. Clearly the policy is motivated by the lack of trust in application programs communicating with each other, not necessarily the people. This would mean that employees of a particular organization could make an unclassified telephone call despite having classified documents on their desks. This necessitates an organizational security policy. However, it might be difficult to describe the desired behavior of the people in formal language. Models tend to be simple, abstract, easy to comprehend and prove mathematically [2] and hence, have limited utility in specifying technical security measures alone.

The various formal methods for designing security tend to model three basic principles: confidentiality, integrity, and availability. These have also often been touted as the core principles in information system security management. In fact, maintaining confidentiality was the prime motivation behind the original TCSEC. The US DoD wanted to develop systems which would allow for only authorized access and usage. For this reason the computer science community created ever so sophisticated models and mechanisms that considered confidentiality as the panacea of security. Clearly security of systems, especially for commercial organizations, went beyond confidentiality to include issues of integrity and availability of data. The notion of integrity deals with individual accountability, auditability, and separation of duties. It can be evaluated by considering the flow of information within a system and identifying areas where the integrity is at risk.

## Evaluation Criteria and Their Context

The Trusted Computer Systems Evaluation Criteria were first introduced in 1983 as a standard for the development of systems to be used within the US government, particularly within the DoD. This document established the DoD procurement standard, which is in use even today, albeit with some modifications. Although the original standards were set within the particular context of the military, their subsequent use has underplayed the importance of contextual issues. The DoD was largely concerned with safeguarding the classified data while procuring systems from vendors. The criteria list different levels of trusted systems, from level D with no security measures to A1 where the security measures are highly regarded. The intervening levels include C1, C2, B1, B2, and B3. As one moves from level D to A the systems become more secure through the use of dedicated policies operationalized by formal methods and provable measures. Formal models such as the Bell–La Padula, the Denning Information Flow model for access control, and Rushby's model provide the basis.

## Bell–La Padula

The Bell–La Padula model, published in 1973, sets the criteria for class A and class B systems in the TCSEC. It deals with controlling access to objects. This is achieved by controlling the abilities to read and write information. The Bell–La Padula model deals with mandatory and discretionary access controls. The two basic axioms of the model are:

1. A subject cannot read information for which it is not cleared (no read up rule)
2. A subject cannot move information from an object with a higher security classification to an object with a lower classification (no write down rule)

A combination of the two rules forms the basis of a trusted system, i.e. a system that disallows an unauthorized transfer of information. The classification and the level in the model are not one-dimensional, hence the entire model ends up being more complex than it appears to be. The system is based on a tuple of *current access set (b)*, *hierarchy (H)*, *access permission matrix (M)*, and *level function (f)*.

The *current access set* addresses the abilities to extract or insert information into a specified object, based on four modes: execute, read, append, and write, addressed for each subject and object. The definitions of each of the four modes are listed below:

1. Execute: neither observe nor alter
2. Read: observe, but do not alter
3. Append: alter but do not observe
4. Write: observe and alter

The level of access by a subject is represented as a triple—*subject, object, access-attribute*. In a real example this may translate to *Peter, Personnel file, read*, which means that Peter currently has read access to the personnel file. The total set of all triples takes the form of the *current access set*.

The *hierarchy* is based on a tree structure, where all objects are organized in a structure of either trees or isolated points, with the condition that all nodes of the structure can only have one parent node. This means that the hierarchy of objects is either that of single isolated objects or one with several children, although a child can have only one parent. This is typically termed a tree structure.

The *access permission matrix* is the portion of the model that allows for discretionary access control. It places objects vs. subjects in a matrix, and represents access attributes for each subject relative to a corresponding object. This is based on the access set modes. The columns of the matrix represent system objects and the rows represent the subjects. The entry in the matrix is the access attribute. A typical matrix may appear as the one in Table 2.7.

**Table 2.7.** A typical access attribute matrix

| Subject | Object | | |
|---|---|---|---|
| | $O_x$ | $O_y$ | $O_z$ |
| $S_1$ | e | r,w | e |
| $S_2$ | r,w | e | a |
| $S_3$ | a,w | r,a | e |

The *level function* classifies the privileges of objects and subjects in a strict hierarchical form with the following labels: top secret, secret, confidential, and unclassified. These information categories are created based on the nature of the information within the organization and are designated a level of access, so that a subject could receive the relevant security designation. With respect to the level function, considering the two classes C1 and C2, the basic theorem is that (C1,A) dominates (C2,B) if and only if C1 is greater than or equal to C2, and A includes B as a subset.

The development of the Bell–La Padula model was based on a number of assumptions. First, there exists a strict hierarchical and bureaucratic structure, with well-defined responsibilities. Second, people in the organization will be granted clearance according to their need-to-know basis in order to conduct work. Third, there is a high level of trust in the organization and people will adhere to all ethical rules and principles, since the model deals with trust within applications, as opposed to people. For example, it is possible to use covert means to take information from one level to the other. The "no read up" and "no write down" rules however attempt to control user actions.

## Denning Information Flow Model

While the Bell–La Padula model focuses attention on the mandatory access control, the Denning Information Flow model is concerned with the security of information flows. The Information Flow model is based on the assumption that information constantly flows, is compared, and merged. Hence, establishing levels of authority and compiling information from different classes is a challenge. The Denning model is defined, first, as a set of objects, such as files and users, that contain information; second, as active agents who may be responsible for information flows; third, as security classes where each object and process are associated with a security class; fourth, as involving a "determination operator," which decides the security of an object that draws information from a pair of objects; fifth, as involving a "flow operator" that indicates if information will be allowed to flow from one security class to another.

Clearly the "flow operator" is the critical part of the model since it determines if information will be allowed to flow from, say, a top-secret file to an existing secret file. The "flow operator" is also the major delimiting factor that prohibits the flow of information within the system. The definition of a secure information flow follows directly from these definitions. The flow model is secure if and only if a sequence of operations cannot give rise to an information flow that violates the flow operation. Together these properties are drawn into a universally bounded lattice. The first set of requirements of this lattice is that the flow operation is reflexive, transitive, and antisymmetric. The reflexive requirement is that information in a specified security class, be it Confidential{cases}, must be able to flow into other information containers within that same security class. The transitive rule requires that if information is allowed to travel from a file with security class Confidential{cases} to another information container file with security class Confidential{case_detail}, and that information is permitted to flow from Confidential{case_detail} to Confidential{case_summary}, then it must be permitted that information from file with clearance Confidential{cases} can flow directly to Confidential{case_summary}. Finally, the antisymmetric requirement is that if information can flow between two objects with different security classes, both from the first to the second, and from the second to the first, then we can set the security classes as equivalent.

The second set of requirements for this lattice is that there exist lower and upper bounds operations. That is, for all security classes, there should exist a security class such that information from an object with that security class would be permitted to flow to any other object. This requires that when information between two classes is merged, that it is possible to select the lowest of the security levels in order to allow the intersection of the information. For example, to use the lower bound operation on Confidential{cases, names} and TopSecret{names} to derive what the intersection of that information would result in, the result would be that the access to the output information would have the classification of Confidential{names}.

The upper bound requirement is already denoted as the flow operation. It is in line with the lower bound, with the exception that the highest of the security levels and the union of the information is chosen. So, if information is merged together, the security level of the merged information assumes the highest previous form. Thus, for Confidential{cases} and TopSecret{results}, when information is merged between these two, the level of the object would be TopSecret{cases, results}.

As a result, the lattice allows for the Bell–La Padula *no read up* and *no write down*, since an object with the highest class within a system can receive information from all other classes within the lattice, but cannot send information to any other object, while the lowest security class can send information to any other security class in the lattice, but cannot receive information from any of them. Together, the upper bound and lower bound provide the secure flow. If two objects with different classes, such as Confidential{case1, names} and Confidential{case2, names} are merged to create a new object, the resulting security level would have to be a more restrictive Confidential{case1, case2, names}. As for the lower bound, it restricts the flow of information downwards, so that objects with security class Confidential{cases, results} and Confidential{cases, names} can only receive an item classified no higher than Confidential{cases}.

Another result is that information cannot flow between objects with incompatible security classes, which returns us to the restrictive nature of strict access control models. So, information within objects of class TopSecret{names} and TopSecret{results} cannot be drawn together unless it is accessed by an object with a higher security level. This maintains the need-to-know nature of strict access controls, so that users and files are given the ability to collect information only for domains to which they are designated.

## The Reference Monitor and Rushby's Solution

To enforce the access control policy of the previously mentioned models, the TCSEC discusses the use of the reference monitor concept. The rationale for this monitor is the efficient and able enforcement of the policy, because there is a need for making sure that all interactions within the computer system occur with some type of mediation that implements the policy at all times. This monitor must be accessed whenever the system is accessed, while it must be small and well identified in order for the system to be able to call on it whenever it is needed. To meet this need, three design requirements were specified by the TCSEC: the mechanism must be tamperproof, the reference validation mechanism must always be invoked, and the reference validation mechanism must be small enough for it to be subjected to analysis, tests, and have completeness that can be assured.

The emergence of the idea of a security kernel is rooted in this need. The security kernel is a small module in which all security features are located, and thus allows for intensive evaluation, testing, and formal verification. Rushby, however, argued against the use of a security kernel because in practice it ended up being inapplicable without the use of trusted processes, which must be permitted to break some of the rules normally imposed by the kernel. The reason for this lies in the restrictive and rigid natures of the security requirements demanded by the aforementioned models, because in the end, there are a number of classifications to deal with, and Rushby outlined a particular situation where they would fail: the print spool.

The print spool reads files and forwards them to the printer. If the printer spool was given the highest possible classification, it could read the user files, and then write them as spool files with the highest level of security classification. However, this requires that users be disallowed to access their own spool files, even to check the status of the printer queue. An option would be to allow spool files to retain the classification of the original user files, so that the spool could still *read down* the files. Then, the inability of the printer spool to *write down* would prevent the spool from even deleting the lower classification files after having been processed. The security kernel solution would be to declare the spooler a *trusted process*, which would allow it to contravene the *no write down* rule.

Rushby's model uses a method called the separation of users. That is, no user would be able to read or modify data or information belonging to another user. Meanwhile, users could still communicate using common files, provided by a file server, where the file server performs the single function of sharing files, unlike the complex operating system. This forms the basis of the Rushby Separation Model.

The reference monitor assumes that users access a common mass of information under the jurisdiction of a single unit. The separation approach assumes that it would be easier to offer users their own domains, thus simplifying all models and policies. However, users do need to cooperate and share data within an information system, and the file server's task is to provide this communication. The purpose of the separation kernel is to create an environment that suggests an appearance of separation amongst machines, and allows only for communication from one machine to the other through external communication lines, even though this physical distribution is not, in fact, in existence. In essence, this method simplifies the need for a reference monitor, and focuses on the need for logical separation while sharing resources, such as processor and communication lines. The end result is that the users' access to their own data needs no security control, thus effectively maintaining security without worrying about the restrictions of the above models, and hence, offering a little more flexibility.

## Away from the Military

The aforementioned models concentrated solely on access controls for a very particular reason: the TCSEC and its variants, and even most of the range of security models prior to the late 1980s were concerned mostly with confidentiality. Even today the Common Criteria are essentially based on the confidentiality principle. Most evaluation criteria (TCSEC, Common Criteria, etc.) are quite overt about the importance of confidentiality; after all, TCSEC was setting the standard for systems that were to be implemented within the Department of Defense. Within a military setting, it is assumed that trust exists amongst various members. Technology, however, is considered "untrustworthy," and it requires effort to make the hardware and the software trustworthy. Thus, the covert channels that the Bell–La Padula model left unguarded were not of grave concern, because it was the trust of the application that was in doubt, not necessarily the users.

The rigidity of the models seems to have complemented the rigidity of the organization, as structures within the military organization remain relatively constant, responsibilities and duties are often clear and compartmentalized much like the security classes, and where the philosophy of *need to know* reigned supreme as the status quo.

It is interesting to see the effectiveness of the TCSEC in what it set out to achieve—a set of standards for a military-type organization. In that sense, it achieved its mission quite simply, yet the effect it had on the field of security is somewhat akin to the chicken and egg conundrum. It was the culmination of years of research into the confidentiality of systems because security was more importantly deemed to be about keeping secrets secret. Meanwhile it also spawned a market acceptance of this type of solution to security, where the TCSEC (and now the Common Criteria) are still considered the ultimate in evaluation criteria. If we are to speak in the language of formal methods and mathematics, this is where the problem arises: the models are wonderful models for the abstraction of the system they were abstracting, yet we are expected to apply the models to different systems with different abstractions. So, we see that the TCSEC and for that matter any other evaluation criteria are not meeting the requirements for what the average organization of today requires, while originally, the models and their abstractions never presumed that they could.

## Military and Non-military: Toward Integrity

Clearly confidentiality is an important trait for organizations, particularly military ones. However, it is evident that non-military organizations secure their systems with another idea in mind: it costs them money if a system has incomplete or inaccurate data [1]. The military organization is built on protecting information from an enemy, hence the *need-to-know* philosophy based on efforts to classify information and maintain strict segregation of people from information they are not allowed to see.

The evaluation criteria-type models have been more interested in preventing unauthorized read access; businesses are far less concerned with who reads information than with who changes it. This demand for maintaining the integrity of information, and in general catering for the real needs of the non-military sector prompted research into other models and criteria. Most evaluation criteria and the related models made clear that they were not well matched with the private sector, because of their lack of concern with the integrity of its information. Although issues of integrity were added in the Trusted Network Initiative, with the allusion to the Biba Model for Integrity, the fact remained the same: the criteria were not all that concerned with integrity. A system that qualifies for TCSEC and Common Criteria scrutiny and has added functionalities for integrity checks would not receive any recognition for this because it is outside the scope of the TCSEC. Even worse, a system that is designed to support integrity in other forms than the restrictive Biba model, such as the Clark-Wilson model, may not even qualify for evaluation.

# Toward Integrity: Biba, Clark-Wilson, and Chinese Walls

## Biba

The Biba model is the Bell–La Padula equivalent for integrity. Objects and subjects have hierarchical security classification related to their individual integrity, or trustworthiness. The integrity of each object and subject can be compared, so long as they follow two security properties:

1. If a subject can modify an object, then the integrity level of the subject must be higher than the integrity level of the object.
2. If a subject has read access to a particular object, then the subject can have write access to a second object only if the integrity level of the first object is greater than or equal to the integrity of the second object.

The parallel is quite clear with the Bell–La Padula model, particularly with its own two axioms. However, confidentiality and integrity seem to be the inverse of each other. In the Bell–La Padula model, the restriction was that a subject cannot read information for which it is not cleared and a subject cannot move information from an object with a higher security classification to an object with a lower classification. In comparison to the Biba model it argues that if a subject can read information then the subject must have a higher security level than the object, and if the subject can move information from one object to another, then the latter object must have a lower security level than the first. Biba's model inverts the latter axiom, and demands that a high-integrity file must not be corrupted with data from a low-integrity file.

The parallels between the latter two axioms of the Bell–La Padula and the Biba models are very much at the heart of their systems. For Biba, this axiom is to prevent the flow of non-trusted information into a file with a high-integrity classification, while for Bell–La Padula, this second axiom tries to prevent the leaking of highly confidential information to a lower classified file. This property for Biba prevents a subject accessing a file from contaminating it with information of lower integrity than the file itself, thus preventing the corruption of a high-integrity file with data created or derived from a less trustworthy one. The first axiom aims to prohibit the modification of a file with a high-integrity classification, unless the subject has a higher integrity classification.

As is the case with the Bell–La Padula model, the Biba model is difficult to implement. Its rigidity on the creation of information based on integrity classes, or levels of trust, although seemingly novel and necessary, in practice is too restrictive; thus, very few systems have actually implemented the model. The model demands the classification of integrity sources and that strict rules apply to this—the integrity policy will have to be at the very heart of the organization; however, integrity policies have not been studied as carefully as confidentiality policies, even though some sort of integrity policy governs the operation of every commercial data-processing system. To demand an integrity policy of this level within an organization is to demand that the organization has a clear vision on the trust mechanisms involved within its own organization—i.e. the organization operates merely on the formal and technical level, without the ambiguity of the informal side of the organization.

## The Clark-Wilson Model

The Clark-Wilson model is based on the assumption that bookkeeping in financial institutions is the most important integrity check. The model recognizes that the recording of data has an internal structure such that it accurately models the real-world financial state of the organization. However, it is noted that the integrity of the integrity check is also a problem, since someone who is attempting a fraud could also create a false sense of financial integrity by altering the financial checks, by such methods as creation of false records of payments and receipts, for example. The solution would be to separate responsibilities as much as possible, disallowing the opportunity for a person to have that much authority over the integrity checks; if a person responsible for recording the receipts of goods is not authorized to make an entry regarding a payment, then the false entry on receipt of goods could not be balanced by the corresponding payment entry [3], thus leaving the books unbalanced, and the integrity check still valid. The Clark-Wilson model attempts to implement this separation and integrity check into an information system, while drawing on the criteria provided by the US Department of Defense in their TCSEC.

There are two key concepts to the model: the Constrained Data Item (CDI) and the Transformation Procedure (TP). The CDI is related to balancing entries in account books, as in the above example. The TP is the set of legitimate processes that may be performed on the specified sets of CDIs, akin to the notion of double bookkeeping, or integrity checks.

To begin with, however, the model does impose a form of mandatory access control, but not as restrictive as the *no read up* and *no write down* criteria of the previously analyzed models: in non-military organizations, there is rarely a set of security classifications of users and data. In this case, the mandatory access control is concerned with the access of users to Transformation Procedures, and of Transformation Procedures to Constrained Data Items. The CDIs may not be accessed arbitrarily for writing to other CDIs—this would result in a decreased integrity of the system. There are, instead, a set of requirements which the CDI can be processed in accordance to. As well, in order to enforce the sense of separation of duties, users may only invoke some Transformation Procedures, and a pre-specified set of data objects or CDIs, as demanded by their duties.

The four requirements of this particular model are as follows:

1. The system must separately identify and authenticate every user.
2. The system must ensure that specified data items can be manipulated only by a restricted set of programs, and the data center controls must ensure that these programs meet the *well-formed transaction rules*, which have already been identified as Transformation Procedures.
3. The system must associate with each user a valid set of programs to be run, and the data center must ensure that these sets meet the separation of duties rule.
4. The system must maintain an auditing log that records every program executed, and the name of the authorizing user.

The four requirements, noticeably, relate heavily to the principles of security. The first alludes to maintaining authorized access, which falls under confidentiality. The second ensures the integrity of the data. The third discusses the need to clearly set out responsibilities. The fourth alludes to the accountability of users and programs. In order to maintain and enforce system security, the system, in addition to the above requirements, must contain mechanisms to ensure that the system enforces its requirements at all times, and the mechanisms must be protected against unauthorized change; both requirements relate to the reference monitor concept from the TCSEC.

The security of the system, however, hinges on the state of the CDIs. Integrity rules are applied to data items in the system, and the outcome are the CDIs; the CDIs must meet the Integrity Validation Procedures, and upon doing so, the system will be deemed secure.

The system has a point of origin where it must be in a secure state. This initial state is ensured by the Integrity Validation Procedures, which will validate the CDIs. From here, all changes to CDIs must be restricted to these well-formed transactions, or Transformation Procedures, which evaluate and preserve the integrity of the information. When data is entered into the system, it is either identified as valid data, and thus granted a state of being secure and is validated as a CDI, or if the data does not meet the requirements of being a CDI, it is rejected and labeled an Unconstrained Data Item. A Transformation Procedure is then called upon to check the data item and transform it into a valid CDI, or reject the input. From this initial secure state, the Transformation Procedure that accepts/rejects the data is part of a set of Transformation Procedures, that when invoked fully, will transfer the system from one secure state to another.

Yet, what is key to this model, as opposed to the previous models, with their rigid controls, is that its certification is application-specific. The process by which Integrity Validation Procedures and Transformation Procedures ensure the integrity of the entire system will actually be defined differently for each person, based on the role they play within the organization, which is based on their duties, and this necessarily enforces the notion of the separation of duties within the computer system. This level of adaptability, user orientation, and application subjectivity are what set this model apart from the rest, along with its accent on data integrity, rather than solely on confidentiality.

This level of subjectivity has a side effect, however. The integrity requirements are specified in terms of the actions of Transformation Procedures on CDIs, which are only used in particular circumstances based on the user and the application, which in turn depends on the organization's procedures for developing the integrity of the data. Because of the bottom-up nature of this approach, in addition to its subjectivity, it is not possible to create a single statement of security policy in the Clark-Wilson model, which the Bell–La Padula model was centered on. The consequence of this dependence on applications and users for the level of controls that are to be used. The Clark-Wilson model, on the other hand, cannot actually be evaluated to guarantee a given level of security, as the various criteria schemes demand.

At this point we face a little standoff between security effectiveness and evaluation criteria: can objective criteria truly assess the security of a system based on differing roles, duties, and applications, which are then based on the organization? Clearly organizations are considered as creators and enforcers of policy based on the models. Users, however, are merely trusted without question. This is much to be desired, but also raises the issue that perhaps objective criteria are ineffective in gauging to what extent the system guarantees security. The Clark-Wilson model provides a strong example of this. Under TCSEC it would not be recognized beyond its confidentiality functionalities. Meanwhile its integrity cannot be gauged due to the fact that there is no guaranteed level of security since different mechanisms are called upon for different tasks. Finally, security in general is not well matched under the criteria since this model cannot even provide a single statement of security, as the criteria dictate.

The organization is more than what sets the security policy—it is the environment that should dictate the entire technical information system. In organizing security in all types of organizations, military and non-military alike, the function of the organization must first be assessed and understood, and then the information system should be drawn from this. This process of deduction should work for security as well, and that is what has been argued continuously in an implicit manner throughout this chapter. Restrictive models often dictate the structure of the organization, and in this we see failures; for this reason the Biba model, despite how efficiently it would maintain integrity, is one that organizations cannot adapt to, and thus, it is not used. It would be possible for organizations to change their structure to cater for the rigid security classifications of the access controls models mentioned under the military organization, but this is not at all recommended because it would result in a loss of functionality and flexibility, traded off for control and security. The loss of functionality and flexibility would prove to be devastating to non-military organizations, particularly commercial organizations.

An example of a model created for a particular organization is the Bell–La Padula model, and that is why it works well for the military organization, because it was developed with that structure and culture in mind, with its emphasis on trust, classifications, and responsibilities. Another example is the Brewer-Nash Chinese Wall Security Policy.

## Emergent Issues

We have been presented with a variety of models, placed within the context of evaluation criteria, principles, and policies. It would be tempting to argue that one model is stronger than another because it better deals with integrity, while another is more valid because it solidly confronts confidentiality. This would be a mistake, and this chapter would not have met its objective if the reader considers that it is really possible to argue for one model against the other.

If anything, this chapter has outlined the beauty of the model: it is an abstraction of an abstraction. It is the abstraction of security measures and a policy. However, the second abstraction is easily forgotten: the measures and policy are in themselves abstractions of the requirements and specifications of the organization. Thus, the context of the abstraction is key, and this context is the environment—the organization, its culture, and its operations.

So, the Trusted Computer System Evaluation Criteria are valid and complete. The Bell–La Padula and Denning models for confidentiality of access controls are valid and complete. Rushby's Separation Model showed that the completeness of the reference monitor could be maintained without the inconsistency of the trusted processes. The Biba model for integrity is valid and complete. The reasons for their validity, however, are not only because they are complete within their inner workings, their unambiguous language and derivations through axioms. Their completeness and validity are due to the fact that the abstraction that they represent, the world that they are modeling, and the organization for which they are ensuring the security policy, are all well defined: they all refer to the military organization. This military organization comes with a culture of trust in its members, a system of clear roles and responsibilities, while the classification of the information security levels within the models are not constructs of the models, but instead reflect the very organization they are modeling. Meanwhile, integrity was never really much of a concern for the US Department of Defense.

This is where the line is drawn between the military and the non-military organization. In the non-military organization, integrity of the information is key to the well-being of the organization. Particularly in the commercial world, what is key to the well-being of the organization is key to its very survival, so integrity cannot be taken lightly. However, the TCSEC, Common Criteria, and the aforementioned models do not reflect this need within this new context: where trust should not be assumed, where information flows freely with a notion of needing-to-withhold rather than needing-to-know, where roles and responsibilities are not static, and where information carries no classification without its meaning. The Clark-Wilson model reflected how integrity was key to the non-military organization, and the consequence of this was that it showed how the TCSEC could not cater for the same priorities as the Clark-Wilson model. Subsequent criteria took on the role of developing non-military criteria, which was where the TCSEC stopped—after all, the TCSEC was only ever a standard for developing systems for the US military complex. Yet even the Clark-Wilson model showed that to attempt to scrutinize systems objectively in general is a problematic task, particularly since the model did not even have a single security policy. This is attributed to the fact that it is heavily decentralized in nature, and criteria cannot be expected to analyze this formally on a wide scale. After all, the model should reflect the organization, and the organization is not generic, while the model may be. This demands further analysis and models, such as the Brewer-Nash Chinese Wall Security Policy, which derives a model for consultancy-based organizations. While this model is not as interesting in its inner workings, it is a step in the right direction, towards a model that is based on its organization, instead of requiring that the organization base itself on a model.

It seems we have come full circle, with a little bit of confusion occurring in the mid-1980s through to the 1990s. When the TCSEC were released, the Bell–La Padula and Denning type access controls were accepted as standard within these criteria, because the criteria and models were based on a specific type of organization. Yet, at some point while this was occurring, the message was lost. The field of computer security began believing that the TCSEC and the new Common Criteria were the ingredients to a truly secure computer system for all organizations, and thus, systems should be modeled on these criteria. Debates have gone on about the appropriateness of the TCSEC for commercial organizations, while this debate should never have happened, because the TCSEC were never meant for non-military organizations. So the Information Technology Security Evaluation Criteria (ITSEC) arrived, along with the CTCPEC and FC-ITS, and started considering more than what was essential for the military organization. Integrity became key, organizational policies gained further importance. The need for considering the organization had finally returned to the limelight. The organization should drive the model, which is enabled by the technology. This is the basic criteria for a security policy. The model should never drive the organization, because this is a failure of the abstraction. However, we have now gone back to the Common Criteria, which are nothing more than an amalgamation of all the disparate criteria.

As the non-military organizations learn this large yet simple lesson, much work is still required. Models are powerful and necessary, but a solid analysis of the environment is also necessary: the culture and the operations need to be understood before the policy is made, greater awareness is needed, and hopefully trust will arise out of the process. In the meantime, progress is required in research into the integrity of the information within the system, as this is, after all, the lifeblood of the organization.

## Concluding Remarks

In this chapter we have sketched out the domain of technical information system security. At a technical level we have considered information system security to be realized through maintaining confidentiality, integrity, availability, authenticity, and non-repudiation of data and its transmission. Ensuring technical security, as described in this chapter, is a function of three confounding principles:

- The principle of easiest penetration
- The principle of timeliness
- The principle of effectiveness

The three principles ensure the appropriateness of controls that need to be instituted in any given setting. The easiest penetration principle lays the foundation for ensuring security by identifying and managing the weakest links in the chain. The timeliness principle triggers the delay in cracking a system, such that the data that a perpetrator might access is no longer useful. The effectiveness principle ensures the right balance between controls, such that the controls are not a hindrance to the normal workings of the business.

Various sections in the chapter have essentially focused on instituting security based on one or more of these principles. The rest of the chapters in this section go into further detail on each of the control types and methods.

> **In Brief**
> - The core information system security requirements of an organization are: confidentiality, integrity, availability, authenticity, and non-repudiation.
> - Data is usually protected from vulnerabilities such as being modified, destroyed, disclosed, intercepted, interrupted, or fabricated.
> - Perpetrators generally stick to the easiest and cheapest means of penetration.
> - Principles of easiest penetration, timeliness, and effectiveness are the basis for establishing information system security.
> - **A note on further reading.** This chapter has introduced a number of formal models, which have been presented in a non-mathematical form. However, an understanding of the detail is important, but is beyond the scope of this book. Readers interested in a fuller description of BLP, Biba Integrity, Cark-Wilson, and other such models are directed to the following texts:
>   1. Bishop, M. 2005. *Computer security: Art and science.* Addison Wesley.
>   2. Pfleeger, C. 2015. *Security in computing.* Prentice Hall.
>   3. Russell, D, and G. Gangemi. 1992. *Computer security basics.* O'Reilly & Associates.
> - Principles of confidentiality, integrity, and availability have their roots in the formal models for security specification.
> - Formal models for security specification originally targeted the security needs of the US Department of Defense.
> - All formal models presume existence of a strict hierarchy and well-defined roles.
> - The basic tenets of the formal models are reflected in the major security evaluation criteria, including the TCSEC, Common Criteria, and their individual country-specific variants.

## Questions and Exercises

**Discussion questions.** These are based on a few topics from the chapter and are intentionally designed for a difference of opinion. These questions can best be used in a classroom or a seminar setting.

1. Although traditionally information system security has been considered in terms of maintaining confidentiality, integrity, and availability of data, comment on the inadequacy of these principles for businesses today.

2. The need-to-know principle has its weaknesses when it comes to security of information in certain contexts. Comment.

3. Discuss the relationship between core security requirements and the principles of easiest penetration, timeliness, and effectiveness.

4. Think of a typical e-commerce application (e.g., online bookstore, online banking) and discuss the significance of formal models in systems design.

5. Information flow and integrity are perhaps important aspects of banking systems. Discuss how such models could be used in a typical transaction processing system.

**Exercise.** Commercial banks usually make their privacy policies publicly available. Locate a privacy policy from your financial institution and evaluate it in terms of core information system concepts introduced in this chapter.

## Short Questions

1. At a technical level, the six threats to hardware, software, and the data that resides in computer systems are?
2. The three critical security requirements for protecting data are?
3. Name two other security requirements that have become important, especially in a networked environment.
4. The use of the *need-to-know principle* is the most acceptable form of ensuring _____.
5. What requirement assures that the message is from a source it claims to be from?
6. Denial of service attacks are to a large extent a consequence of this security requirement not having been adequately addressed.
7. What requirement ensures that data and programs are changed in an authorized manner?
8. Privacy of data is ensured by what requirement?
9. What requirement prevents an individual or entity from denying having performed a particular action related to data?
10. A digital signature scheme is one means to ensure _____.
11. The philosophy of *need to know* is based on efforts to classify information and maintain strict segregation of people, and was developed by the military as a means of restricting _____ access to data.
12. The Biba model is similar to the Bell–La Padula model except that it deals mainly with the _____ of data.
13. An example of a model created for a particular organization is the Bell–La Padula model, and that is why it works well for the _____ organization, because it was developed with that structure and culture in mind.
14. In the non-military organization, _____ of the information is key to the well-being of the organization.

# Case Study: Breaking into Sprint's Backbone Network

Many of the technical controls put into place can be circumvented with a simple phone call. Recently, famed hacker Kevin Mitnick demonstrated this by breaking into Sprint's backbone network. Rather than mounting a buffer overrun or DoS attack, Mitnick simply placed a call posing as a Nortel service engineer and persuaded the staff at Sprint to provide login names and passwords to the company's switches under the guise that he needed them to perform remote maintenance on the system. Once the password information had been obtained, Mitnick was able to dial in and manipulate Sprint's networks at will. Many people believe this was an isolated incident, and they would not fall for a similar act of social engineering, but Mitnick gained notoriety during the 1980s and 1990s by applying similar techniques to computer networks around the world. Mitnick's more notorious crimes included accessing computer systems at the Pentagon and the North American Aerospace Defense Command (NORAD), and stealing software and source-code from major computer manufacturers. Kevin Mitnick was arrested six times, and has been working as a consultant specializing in social engineering techniques, having "gone straight" after serving a five-year sentence for his most recent crime. He even has authored several books regarding social engineering including *The Art of Intrusion* and *The Art of Deception*.

1. What procedures could help prevent a similar breach of security at your organization?
2. Phishing is usually associated with identity theft, but could this tactic also be used to gain information needed to circumvent security controls?
3. Many social engineering breaches involve using what is believed to be insider information to gain the trust of individuals in an effort to obtain confidential information. Test your ability to obtain what some might consider "insider information" using a search engine to find contact or other useful information referencing your organization.

Source: *Information Age* (London, UK), September 10, 2002.

# References

1. Chalmers, L.S. 1986. An analysis of the differences between the computer security practices in the military and private sectors. In *Proceedings of the 1986 IEEE symposium on security and privacy*. Oakland, CA: Institute of Electrical and Electronic Engineers.
2. Gasser, M. 1988. *Building a secure computer system*. New York: Van Nostrand Reinhold.
3. Longley, D. 1991. Security of stored data and programs. In *Information security handbook*, ed. W. Caelli et al. Basingstoke, UK: Macmillan. 545–648.
4. Parker, D. 1991. Seventeen information security myths debunked. In *Computer security and information integrity*, ed. K. Dittrich et al. Amsterdam: Elsevier Science Publishers. 363–370.
5. Wordsworth, J.B. 1999. Getting the best from formal methods. *Information and Software Technology* 41: 1027–1032.

# CHAPTER 3

# Cryptography and Technical IS Security

*Speech was made to open man to man, and not to hide him; to promote commerce, and not betray it.*

—David Lloyd,
*The Statesmen and Favourites of England since the Reformation*

After having heeded Randy's advice, Joe Dawson ordered Matt Bishop's book. As Randy had indicated, it was a rather difficult read. Although Joe did have a mathematics background, he found it a little challenging to follow the various algorithms and proofs. Joe's main problem was that he had to really work hard to understand some basic security principles as these related to formal models. Surely it could not be that tough, he thought. As a manager, he wanted to develop a general understanding of the subject area, rather than an understanding of the details. Joe certainly appreciated the value of the mathematical proof, but how could this really help him ensure security for SureSteel, Inc.? His was a small company and he basically wanted to know the right things to do. He did not want to be taken for a ride when he communicated with his IT staff. So, some basic knowledge would be useful. Moreover, one of the challenging things for Joe was to ensure security for his communications with offices in Indonesia and Chicago.

Years earlier Joe remembered reading an article in the *Wall Street Journal* on Phil Zimmerman, who had developed some software to ensure security and privacy. What had stuck with Joe all these years was perhaps Phil Zimmerman being described as some cyberpunk programmer who had combined public-key encryption with conventional encryption to produce the software PGP—pretty good privacy. Zimmerman had gone a step too far in distributing PGP free of charge on the Internet.

PGP was an instant success—among dissidents and privacy enthusiasts alike. Following the release of PGP, police in Sacramento, California, reported that they were unable to read the computer diary of a suspected pedophile, thus preventing them from finding critical links to a child pornography ring. However, human-rights activists were all *gaga* about PGP. During that time there were reports of activists in El Salvador, Guatemala, and Burma (Myanmar) being trained on PGP to ensure security and privacy of their communications.

Whatever the positive and negative aspects may have been, clearly it seemed (at least in 1994), that PGP was there to stay. And indeed it did become very prominent over the years, becoming a standard for encrypted email on the Internet. Joe was aware of this, essentially because of the extensive press coverage. In 1994 the US government began investigating two software companies in Texas and Arizona that were involved in publishing PGP. At the crux of the investigations were a range of export controls that applied to PGP, as it fell under the same category as munitions.

As Joe thought more about secure communications and the possible role of PGP, he became more interested in the concept of encryption and cryptography. His basic knowledge of the concept did not go beyond what he had heard or read in the popular press. Clearly he wanted to know more. Although Joe planned to visit the local library, he could not resist walking over to his computer, loading the Yahoo page and searching for "encryption." As Joe scrolled down the list of search results, he came across the US Bureau of Industry and Security website (wwwbxa.doc.gov) dealing with commercial encryption export controls. Joe learned very quickly that the US had relaxed some of the export control regulations. He remembered that these were indeed a big deal when Phil Zimmerman first came out with his PGP software. As Joe read, he noticed some specific changes that the website listed:

- Updates License Exception BAG (§740.14) to allow all persons (except nationals of Country Group E:1 countries) to take "personal use" encryption commodities and software to any country not listed in Country Group E:1. Such "personal use" encryption products may now be shipped as unaccompanied baggage to countries not listed in Country Groups D or E. (See Supplement No. 1 to part 740 of the EAR for Country Group listings.)
- Clarifies that medical equipment and software (e.g. products for the care of patients or the practice of medicine) that incorporate encryption or other "information security" functions are not classified in Category 5, Part II of the Commerce Control List.
- Clarifies that "publicly available" ECCN 5D002 encryption source code (and the corresponding object code) is eligible for de minimis treatment, once the notification requirements of §740.13(e) are fulfilled.
- Publishes a "checklist" (new Supplement No. 5 to part 742) to help exporters better identify encryption and other "information security" functions that are subject to U.S. export controls.
- Clarifies existing instructions related to short-range wireless and other encryption commodities and software pre-loaded on laptops, handheld devices, computers and other equipment.

Although Joe thought he understood what these changes were about—perhaps something to do with 64-bit encryption—he really had to know the nuts and bolts of encryption prior to even attempting to understand the regulations and see where he fit in. Joe switched off his computer and headed to the local county library.

## Cryptography

Security of the communication process, especially in the context of networked organizations, demands that the messages transmitted are kept confidential, maintain their integrity, and are available to the right people at the right time. The science of cryptology helps us in achieving these objectives. Cryptology provides logical barriers such that the transmissions are not accessed by unauthorized parties. Cryptology incorporates within itself two allied fields: cryptography and cryptanalysis. Cryptography includes methods and techniques to ensure secrecy and authenticity of message transmissions. Cryptanalysis is the range of methods used for breaking the encrypted messages. Although traditionally cryptography was essentially a means to protect the confidentiality of the messages, in modern organizations it plays a critical role in ensuring authenticity and non-repudiation.

The goal in the encryption process is to protect the content of a message and ensure its confidentiality. The encryption process starts with a plaintext document. A plaintext document is any document in its native format. Examples would be a .doc (Microsoft Word), .xls (Microsoft Excel), .txt (an ASCII text file), and so on. Once a document has been encrypted it is referred to as ciphertext. This is the form that allows the document to be transmitted over insecure communications links or stored on an insecure device without compromising the security requirements (confidentiality, integrity, and availability). Once the plaintext document has been selected it is sent through an encryption algorithm. The encryption algorithm is designed to produce a ciphertext document that cannot be returned to its plaintext form without the use of the algorithm and the associated key(s). The key is a string of bits that is used to initialize the encryption algorithm. There are two types of encryption, symmetric and asymmetric. In symmetric encryption, a single key is used to encrypt and decrypt a document. Symmetric encryption is also referred to as conventional. At a most basic level, there are five elements of conventional encryption. These are described below and illustrated in Figure 3.1.

- **Plaintext:** This is the original message or data, which could be in any native form.
- **Encryption algorithm:** This algorithm performs a range of substitutions and transformations on the original data.
- **Secret key:** The secret key holds the exact substitutions and transformations performed by the encryption algorithm.
- **Ciphertext:** This is the scrambled text produced by the encryption algorithm through the use of the secret key. A different secret key would produce a different ciphertext.
- **Decryption algorithm:** This is the algorithm that converts the ciphertext back into plaintext through the use of the secret key.

In order to ensure the safe transmission of the plaintext document or information, there is a need for two conditions to prevail. First, the secret key needs to be safely and securely delivered to the recipient. Second, the encryption algorithm should be strong enough to make it next to impossible for someone in possession of one or more ciphertexts to work backwards and establish the encryption algorithm logic. There was a time when it was

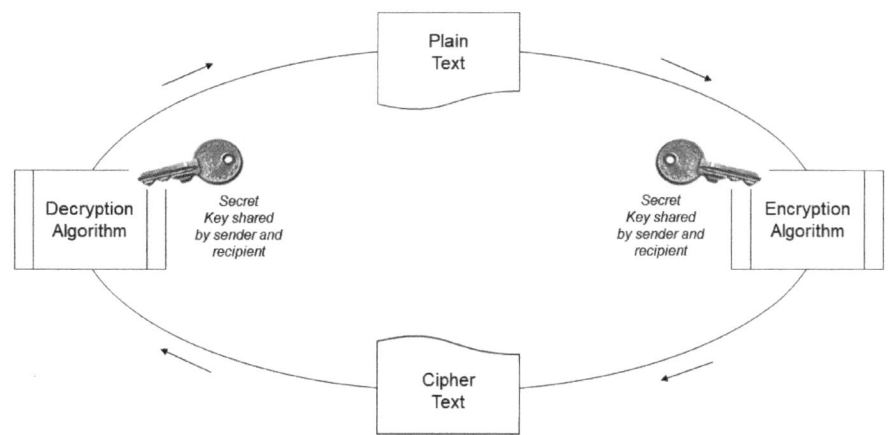

**Figure 3.1.** Conventional encryption process

possible to break the codes by merely looking at a number of ciphertexts. However, with the advent of computers encryption algorithms have become ever so complex and it usually does not make sense to keep the algorithm secret. Rather, it is important to keep the key safe. It is the key that holds the means to decrypt. Therefore it becomes important to establish a secure channel for sending and receiving the key. This is generally considered to be a relatively minor issue since it's easier to protect short cryptographic keys, generally 56 or 64 bits, than a large mass of data.

In terms of handling the problem of secret keys and decrypting algorithms, perhaps the easiest way to deal with this is by not having an inverse algorithm to decrypt the messages. Such ciphers are termed as *one-way functions* and are commonly used in e-commerce. An example of this would be a situation where a user inputs a login password, typically for accessing any account, and the password is encrypted and gets transmitted. The resultant ciphertext is then compared with the one stored in the server. One-way function ciphers, in many cases, do not employ a secret key. It is also possible to provide the user with a *no key* algorithm. In this case both sender and receiver would have an encrypt/decrypt switch. The only way such communications can be broken is through physical analysis of the switches and reverse analysis. This shifts the onus on protecting the security of physical devices at both ends.

In terms of encryption and decryption algorithms (Figure 3.1), there are two possibilities. First, the same key is used for encryption and decryption. Ciphers that use the same key for both encrypting and decrypting plaintext are referred to as *symmetric ciphers*. Second, two different keys are used for encryption and decryption. Ciphers using a different key to encrypt and decrypt the plaintext are termed as *asymmetric ciphers*.

Any cryptographic system is organized along three dimensions. These include the type of operation that may be used to produce ciphertext; the number of keys that may be used; the manner in which the plaintext may be processed. These are described below:

- **Process used to create ciphertext.** All kinds of ciphertext are produced through the process of substitution and transposition. Substitution results in each element of the plaintext being mapped onto another element. Transposition is the rearrangement of all plaintext elements.
- **Number of keys.** As stated previously, if the sender and receiver use the same key, the system is referred to as symmetric encryption. If a different key is used by the sender and receiver, the system is referred to as asymmetric or public-key encryption.
- **Manner in which plaintext is processed.** If a block of inputted text is processed at a time, it is referred to as *block cipher*. If input is in the form of a continuous stream of elements, then it is termed as a *stream cipher*.

## Cryptanalysis

Cryptanalysis is the process of breaking in to decipher the plaintext or the key. It is important to understand cryptanalysis techniques so as to better ensure security in the encryption and transmission process. There are two broad categories of attacks on encrypted messages. These are:

- **Ciphertext attacks.** These are perhaps the most difficult type of attacks and it is rather impossible for the opponent to break the code. The only method that can be used to break the code is brute force. Typically the opponent will undertake a range of statistical analysis on the text in order to understand the inherent patterns. However, it is only possible to perform such tests if the plaintext language (English, French) or the kind of plaintext file (accounting, source listing, etc.) are known. For this reason it is usually very easy to defend against cipher-text-only attacks since the opponent has minimal information.
- **Plaintext attacks.** Usually such attacks may be based on what is contained in the message header. For instance, fund transfer messages and postscript files have a standardized header. Limited as this information may be, it is possible for a smart intruder to deduce the secret key. A variant of the plaintext attack is when an opponent is able to deduce the key based on how the known plaintext gets transformed. This is usually possible when an opponent is looking for some very specific information. For instance, when an accounting file is being transmitted, the opponent would generally have knowledge of where certain words would be in the header. Or in other situations there may be specific kinds of disclaimers and copyright statements that might be located in certain places.

In terms of protecting the transmission and ensuring that encrypted text is not broken, analysts work to ensure that it takes a long time to break the code and that the cost of breaking the code is high. The amount of time it takes and the cost of breaking the code then become the fundamental means of identifying the right level of encryption. Some estimates of time required to break the code are presented in Table 3.1.

**Table 3.1.** Times required for undertaking a key search

| Key size in bits | No. of alternative keys | Time required to decrypt | |
|---|---|---|---|
| | | 1 key per microsecond | Decryption rate: 1 million keys per microsecond |
| 32 | $2^{32}$ | 35.8 minutes | 2.15 milliseconds |
| 56 | $2^{56}$ | 1,142 years | 10 hours |
| 128 | $2^{128}$ | $5.4 \times 10^{24}$ years | $5.4 \times 10^{18}$ years |
| 168 | $2^{168}$ | $5.9 \times 10^{36}$ years | $5.9 \times 10^{30}$ years |

## The Basics of Cryptanalysis

Our aim in an encryption process is to transform any computer material and communicate it safely and securely, whether it is in ASCII characters or binary data. However, in order to simplify the explanation of the encryption process let us consider messages written in standard English. The representation of characters could take the following form:

```
A  0      J   9      S  18
B  1      K  10      T  19
C  2      L  11      U  20
D  3      M  12      V  21
E  4      N  13      W  22
F  5      O  14      X  23
G  6      P  15      Y  24
H  7      Q  16      Z  25
I  8      R  17
```

Representation of a letter by a number code allows for performing arithmetic to the operation. This form of arithmetic is called *modular*, where instances such as P + 2 equals R or Z + 1 equals A would occur. Since the addition wraps around from one end to the other, every result would be between 0 and 25. This form of modular arithmetic is written as mod $n$. And $n$ is a number in the range $0 \leq \text{result} < n$. In net effect the result is the remainder by $n$. As an example, 53 mod 26 and alternative ways of arriving at the result are:

1. 53 mod 26 would be 53 divided by 26, with the remainder being the result, which in this case would be 26 times 2 equals 52 and remainder 1

or

2. Count 53 ahead of A or 0 in the above representation and each time after crossing Z or 25 return to position A or 0. This will result in arriving at B or 1, which is the result.

### *Substitution*

Encryption can be carried out in two forms: **substitution** and **transposition**. In substitution, literally each letter is substituted for the other, while in transposition the order of letters gets rearranged. The earliest known forms of simple encryption using substitution date back to the era of Julius Caesar. Known as the **Caesar Cipher**, each letter is translated to a letter that appears after a fixed number of letters in the text. It is said that Caesar used to shift three letters. Thus a plaintext A would be d in ciphertext (Note: capital letters are generally used to depict plaintext, while ciphertext is in lower case). Based on the Caesar Cipher, a plaintext word such as LONDON would become orqgrq.

An example of another simple encryption would be the use of a key. Here any unique sequence of letters may be used as a key, say richmond. The key is then written beneath the first few letters of the alphabet, as shown below

```
A B C D E F G H I J K L M N O P Q R S T U V W X Y Z
r i c h m o n d a b e f g j k l p q s t u v w x y z
```

Cryptanalysis of simple encryption can typically be carried out by using frequency distributions. There are published frequency distributions of count and percentage time a given letter is used. This forms the basis for interpreting the usage of certain letters in a text and hence the analysis and deciphering of the coded message. For instance in English, the letter E is the most common letter used. Typically in a sample of 1,000 letters, E will be the most frequent. In Russian, the letter O is most common. Similarly, certain pairs of letters have the most frequency. For example in English EN is the most common pair of letters.

A more advanced form of the simple encryption discussed above involves polyalphabetic ciphers. The main problem with simple encryption is that frequency distributions give away a lot of information for breaking the code. However, if the frequencies could be managed so as to be relatively flat, a cryptanalyst would have limited information. If E (a commonly occurring letter) is enciphered sometimes as a and sometimes as b, and Z (a less commonly occurring letter) is also enciphered as a or b, then the mix-up produces a relatively moderate distribution. It is possible to combine two distributions by using two separate encryption alphabets—first for all odd positions and second for all even positions.

**The Vigenère Tableau**

The *Vigenère cipher* was originally proposed by Blaise de Vigenère in the 16th century. This cipher is a polyalphabetic substitution based on the Vigenère Tableau (Table 3.2). It is a collection of 26 permutations. The Vigenère cipher uses the table along with a keyword to encipher messages. Each row of the table corresponds to the Caesar Cipher previously discussed. Each row progressively shifts from 0 to 25.

**Table 3.2.** Vigenère Tableau

|   | A | B | C | D | E | F | G | H | I | J | K | L | M | N | O | P | Q | R | S | T | U | V | W | X | Y | Z |
|---|---|---|---|---|---|---|---|---|---|---|---|---|---|---|---|---|---|---|---|---|---|---|---|---|---|---|
| A | A | B | C | D | E | F | G | H | I | J | K | L | M | N | O | P | Q | R | S | T | U | V | W | X | Y | Z |
| B | B | C | D | E | F | G | H | I | J | K | L | M | N | O | P | Q | R | S | T | U | V | W | X | Y | Z | A |
| C | C | D | E | F | G | H | I | J | K | L | M | N | O | P | Q | R | S | T | U | V | W | X | Y | Z | A | B |
| D | D | E | F | G | H | I | J | K | L | M | N | O | P | Q | R | S | T | U | V | W | X | Y | Z | A | B | C |
| E | E | F | G | H | I | J | K | L | M | N | O | P | Q | R | S | T | U | V | W | X | Y | Z | A | B | C | D |
| F | F | G | H | I | J | K | L | M | N | O | P | Q | R | S | T | U | V | W | X | Y | Z | A | B | C | D | E |
| G | G | H | I | J | K | L | M | N | O | P | Q | R | S | T | U | V | W | X | Y | Z | A | B | C | D | E | F |
| H | H | I | J | K | L | M | N | O | P | Q | R | S | T | U | V | W | X | Y | Z | A | B | C | D | E | F | G |
| I | I | J | K | L | M | N | O | P | Q | R | S | T | U | V | W | X | Y | Z | A | B | C | D | E | F | G | H |
| J | J | K | L | M | N | O | P | Q | R | S | T | U | V | W | X | Y | Z | A | B | C | D | E | F | G | H | I |
| K | K | L | M | N | O | P | Q | R | S | T | U | V | W | X | Y | Z | A | B | C | D | E | F | G | H | I | J |
| L | L | M | N | O | P | Q | R | S | T | U | V | W | X | Y | Z | A | B | C | D | E | F | G | H | I | J | K |
| M | M | N | O | P | Q | R | S | T | U | V | W | X | Y | Z | A | B | C | D | E | F | G | H | I | J | K | L |
| N | N | O | P | Q | R | S | T | U | V | W | X | Y | Z | A | B | C | D | E | F | G | H | I | J | K | L | M |
| O | O | P | Q | R | S | T | U | V | W | X | Y | Z | A | B | C | D | E | F | G | H | I | J | K | L | M | N |
| P | P | Q | R | S | T | U | V | W | X | Y | Z | A | B | C | D | E | F | G | H | I | J | K | L | M | N | O |
| Q | Q | R | S | T | U | V | W | X | Y | Z | A | B | C | D | E | F | G | H | I | J | K | L | M | N | O | P |
| R | R | S | T | U | V | W | X | Y | Z | A | B | C | D | E | F | G | H | I | J | K | L | M | N | O | P | Q |
| S | S | T | U | V | W | X | Y | Z | A | B | C | D | E | F | G | H | I | J | K | L | M | N | O | P | Q | R |
| T | T | U | V | W | X | Y | Z | A | B | C | D | E | F | G | H | I | J | K | L | M | N | O | P | Q | R | S |
| U | U | V | W | X | Y | Z | A | B | C | D | E | F | G | H | I | J | K | L | M | N | O | P | Q | R | S | T |
| V | V | W | X | Y | Z | A | B | C | D | E | F | G | H | I | J | K | L | M | N | O | P | Q | R | S | T | U |
| W | W | X | Y | Z | A | B | C | D | E | F | G | H | I | J | K | L | M | N | O | P | Q | R | S | T | U | V |
| X | X | Y | Z | A | B | C | D | E | F | G | H | I | J | K | L | M | N | O | P | Q | R | S | T | U | V | W |
| Y | Y | Z | A | B | C | D | E | F | G | H | I | J | K | L | M | N | O | P | Q | R | S | T | U | V | W | X |
| Z | Z | A | B | C | D | E | F | G | H | I | J | K | L | M | N | O | P | Q | R | S | T | U | V | W | X | Y |

In describing the use of the Vigenère Tableau to undertake encryption, it is best to use an example. Suppose the plaintext message reads:

```
IT WAS THE BEST OF TIMES IT WAS THE WORST OF TIMES
```

And the keyword used is KEYWORD. We begin the process by writing the keyword above the plaintext as shown below. Ciphertext is derived by referring to Table 3.2 and finding the intersection of the keyword and plaintext letters.

```
Keyword:     KEYWO RDKEY WORDK EYWOR DKEYW ORDKE YWORD KEYW
Plaintext:   ITWAS THEBE STOFT IMESI TWAST HEWOR STOFT IMES
Ciphertext:  sxuwg kkofc ohfid mkagz wgeqp vvzyv qpcww sqco
```

Decryption is a straightforward process where each letter of the keyword is identified in the column and ciphertext letter traced down in the column. The index letter for the row is the plaintext letter.

```
Keyword:     KEYWO RDKEY WORDK EYWOR DKEYW ORDKE YWORD KEYW
Ciphertext:  sxuwg kkofc ohfid mkagz wgeqp vvzyv qpcww sqco
Plaintext:   ITWAS THEBE STOFT IMESI TWAST HEWOR STOFT IMES
```

Clearly the strength of the Vigenère cipher is against frequency analysis. A simple look at any plaintext message and the corresponding cipher proves the point. And letter from the keyword picks 1 of the 26 possible solutions.

## The Kasiski Method for Solving the Vigenère Cipher
It took nearly 300 years to develop a solution to the Vigenère cipher. The breakthrough came in 1863 thanks to a Prussian major, Kasiski. The Kasiski method involved three steps:

1. **Finding patterns.** In order to use this method, one needs to identify all repeated patterns in the ciphertext. Clearly for plaintext to be enciphered twice, the key needs to go through a number of rotations. Any pattern over three characters is certainly not accidental.
2. **Factoring distances between repeated bigrams.** Kasiski's method suggests that the distance between the repeated patterns must be a multiple of the keyword length. For each instance, we write down starting positions and then compute the distance between each successive position:

| Bigram | Location | Distance | Factors |
|--------|----------|----------|---------|
| wg | 20 | 20-3=17 | 1, 17 |
| co | 37 | 37-9=28 | 1, 2, 7 |
| qp | 30 | 30-23=7 | 1, 7 |

3. **Interpretation.** Factoring distances between bigrams helps in interpreting or narrowing down the search for a keyword, which can then be used to decrypt the plaintext message. In our example above the common factors are 1 and 7. Clearly there is less likelihood of a 1-character keyword. This narrows down our task of figuring out the keyword (Note: the keyword used in the above example is the 7-character-long KEYWORD).

There are other kinds of substitution ciphers, which are not discussed here. These are topics of discussion for a more advanced text on cryptography. Readers interested in these techniques may find discussions on *One-Time Pads, Random Number Sequences, Vernam Ciphers,* useful and interesting.

### *Transpositions*
Often referred to as permutation, the intent behind transpositions is to introduce diffusion in the decryption process. So a transposition entails rearranging letters in the message. By diffusing the information across the ciphertext, it becomes difficult to decrypt the messages. **Columnar transposition** is perhaps the simplest form of transposition. In this case characters of plaintext are rearranged into columns. As an example, the message IT WAS THE BEST OF TIMES IT WAS THE WORST OF TIMES would be written as:

```
I T W A S
T H E B E
S T O F T
I M E S I
T W A S T
H E W O R
S T O F T
I M E S X
```

The ciphertext for this message would be:

```
itsi thsi thtm wetm weoe
awoe abfs sofs seti trtx
```

Note the X in the last column. An infrequent letter is usually used to fill in the short column. In columnar transpositions, output characters cannot be produced unless the complete message has been read. So there is a natural delay with this algorithm. For this reason this method is not entirely appropriate for long messages.

**Using Digrams for Cryptanalysis**

In any language there are certain letters that have a high frequency of appearing together. These are referred to as **digrams**. For example, the 10 most common digrams in the English language are: EN, RE, ER, NT, TH, ON, IN, TE, AN, and OR. The 10 most common trigrams are: ENT, ION, AND, ING, IVE, TIO, FOR, OUR, THI, ONE. It may seem that encryption using transposition would be rather simple to analyze and break; however, since plaintext is largely left intact, it becomes rather difficult to decrypt. A lot is left to human judgment rather than algorithms and statistical procedures.

In decrypting transpositions, the first step is to calculate letter frequencies. This is done by breaking text into columns. We know that two different strings of letters in ciphertext will represent adjacent plaintext letters. The task at hand then is to find out where in the ciphertext a pair of adjacent columns is. The ends of the columns have also to be found. The process of finding this is laborious and involves an extensive comparison of strings in the ciphertext. A block of ciphertext is compared with characters successively farther away for the block. For instance, if the first block is between the first character and the seventh character, then comparison is done with the second character and the eighth. This process is carried out for the complete ciphertext. As the analysis is conducted, note is taken of two issues. The first relates to an emergent pattern that might exist. This is usually interpreted by assessing the frequency of digrams for common English. Second, any chance occurrences of patterns need to be eliminated. It is possible to do this by looking at variances in digram frequencies. Let us look at the example used above by considering a block of seven.

Considering the comparisons shown in Table 3.3, some emerging digrams could be identified, *viz.* IS, IT, IE. Following this, the relative frequency is calculated. There are standard relative frequencies available for various digrams in the English language. Consequently, the mean and standard deviation of each list of frequencies can be calculated. A high mean would imply that some digrams are likely and a low standard deviation suggests that all digrams are likely and the mean was not raised artificially. Finally, the matches of fragments of ciphertext are extended.

**Table 3.3.** Comparisons

| i | s |
|---|---|
| t | t |
| s | h |
| i | t |
| t | m |
| h | w |
| i | e |
| t |   |
|   | m |
|   | w |
|   | e |
|   | o |
|   | e |
|   | a |

|   | s |
|---|---|
| i | t |
| t | h |
| s | t |
| i | m |
| t | w |
| h | e |
| i | t |
|   | m |
|   | w |
|   | e |
|   | o |
|   | e |
|   | a |

|   | s |
|---|---|
|   | t |
| i | h |
| t | t |
| s | m |
| i | w |
| t | e |
| h | t |
| i | m |
|   | w |
|   | e |
|   | o |
|   | e |
|   | a |

|   | s |
|---|---|
|   | t |
|   | h |
| i | t |
| t | m |
| s | w |
| i | e |
| t | t |
| h | m |
| i | w |
|   | e |
|   | o |
|   | e |
|   | a |

|   | s |
|---|---|
|   | t |
|   | h |
|   | t |
| i | m |
| t | w |
| s | e |
| i | t |
| t | m |
| h | w |
| i | e |
|   | o |
|   | e |
|   | a |

|   | s |
|---|---|
|   | t |
|   | h |
|   | t |
|   | m |
| i | w |
| t | e |
| s | t |
| i | m |
| t | w |
| h | e |
| i | o |
|   | e |
|   | a |

## Conventional Encryption Algorithms

**Stream ciphers** convert one symbol of plaintext into a symbol of ciphertext. Stream ciphers are relatively fast compared with block ciphers. However, some block ciphers working in certain modes can effectively operate as stream ciphers. Stream ciphers are developed out of a specialist cipher, the Vernam cipher. Typical examples include the RC4 and the Software Optimized Encryption Algorithm (SEAL).

Clearly there are some advantages to stream ciphers. The speed with which the transformation can be carried out is a key benefit. Each symbol gets encrypted immediately upon being read. As a result there is very little chance of error propagation. Even if there is an error, it affects only that character. However, since each symbol is enciphered separately, it also contains the complete set of information. Anyone trying to decrypt can pretty much do so by analyzing individual characters. Stream ciphers are also susceptible to malicious insertions and modifications. It is possible to do so since each symbol is enciphered separately. A person who decrypts a given message can easily generate a synthetic message and transmit it and yet make it appear as genuine.

**Block ciphers** convert a *group* (fixed length) block of plaintext into ciphertext through the use of a secret key. Decryption is done through reverse transformation through the same key. Most block ciphers use a block size of 64 bits. One type of transposition used in block ciphers is *columnar*. In this transposition the complete message is translated as one block. There is usually no relationship between block size and character size. So block ciphers work on blocks of text to produce blocks of cipher.

Block ciphers can be applied in an iterative manner, thus having several rounds of encryption. Clearly this improves the level of security. Each iteration typically applies a subkey (derived from the original key) for a special function. This additional computing requirement has an impact on the speed at which encryption can be managed. Ultimately a balance needs to be established between the level of security, speed, and the appropriateness of the method. There are various kinds of block ciphers, which include DES, IDEA,

SAFER, Blowfish, and Skipjack. Skipjack is used in the US National Security Agency (NSA) Clipper chip.

## Data Encryption Standard

Data Encryption Standard (DES) is the formal description of the Data Encryption Algorithm. Initially developed by IBM, it was later adopted by the US government in 1977. The algorithm forms the basis for automatic teller machines and is widely used to secure financial data. DES uses techniques based on the work of Horst Fiestel.[1]

DES input is a block of 64 bits. An initial permutation of 64 data bits is done. Only 56 bits of the 64-bit key are used. The reduction is done by dropping bits 8, 16, 24, ... 64. The bits are assumed to be parity bits and carry no information. Next, the 64 permuted bits are divided into two halves—left and right. Each of these is of 32 bits. The key is shifted right by a number of bits and then permuted. It is then combined with the right half and then combined with the left half. This combination results in a new right half, the old right becomes the new left. This process, known as a cycle, is repeated 16 times. Once all the cycles are completed, there is one last permutation, which is the reverse of the initial permutation. The cycle of substitutions and permutations in DES is shown in Figure 3.2.

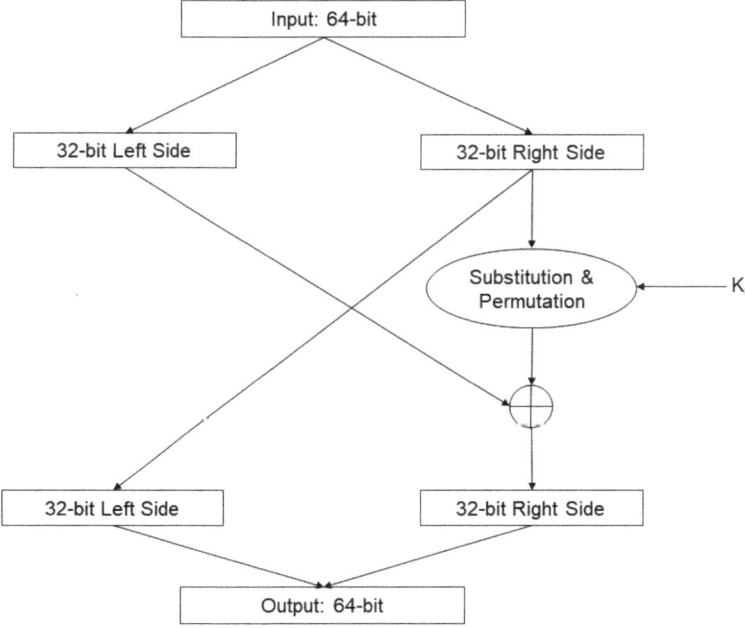

**Figure 3.2.** A simplified Data Encryption Algorithm showing a single cycle

---

1 DES is an important standard and its details are worthy of being understood. In this section we have provided an overview A detailed discussion is beyond the scope of this chapter. However, a good description and overview can be found in Coppersmith, D. (1994), "The Data Encryption Standard (DES) and its strengths against attacks," *IBM Journal of Research and Development* 38(3): 243–250; Schneier, B. (1996), *Applied cryptography* (New York: Wiley).

In each cycle of the data encryption algorithm, there are four different operations that take place. First, the right side of the initial permutation is expanded from 32 bits to 48 bits. Second, a key is applied to the right side. Third, a substitution is applied and results condensed to 32 bits. Fourth, permutation and combination with left side is undertaken to generate a new right side. This process is shown in Figure 3.3.

The expansion from 32 bits is through a permutation. This helps in making the two halves of the ciphertext comparable to the key and provides a result that is longer than the original. This is later compressed. The condensation of the key from 64 bits to 56 is another important operation. This is achieved by deleting every 8th bit. They key is split into two 28-bit halves, which are then shifted left by a specific number of digits. Then the halves are brought together. This results in 56 bits. Forty-eight of these bits are permuted to be used as a single key. The results of the key are moved into a table where six bits of data are replaced by four bits through a substitution process. This is commonly referred to as the S-box. In the next stage, 48 of the 56 bits are extracted through permutations. This is referred to as the P-box. A total of 16 substitutions and permutations complete the algorithm.

Ever since the National Security Agency adopted DES, it has been marred by controversy. The agency never made the logic behind S- and P-boxes public. There have been concerns of certain trapdoors being embedded in the DES algorithm so that the NSA could use an easy and covert means to decrypt. There were also concerns about the reliability of designs. Concerns were also raised about sufficiency of 16 iterations. However, numerous experiments have shown that only eight iterations are sufficient. The length of the key has also been an issue of concern. Although the original key used by IBM was 128 bits, DES uses only a 56-bit-long key.

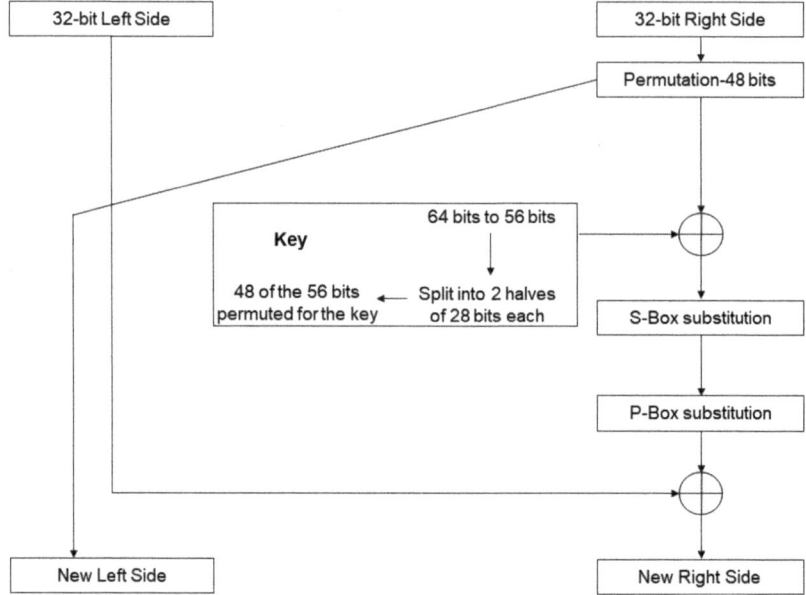

**Figure 3.3.** Details of a given cycle

## IDEA

Besides DES, there are other encryption algorithms. The International Data Encryption Algorithm (IDEA) emerged from Europe through the work of researchers Xuejia Lai and James Massey in Zurich. IDEA is an iterative block cipher that uses 128-bit keys in eight rounds. This results in a higher level of security than DES. The major weakness of IDEA is that a number of weak keys have to be excluded. DES on the other hand has four weak and 12 semi-weak keys. Given that the total number of keys in IDEA is substantially greater at two 128, it leaves only two 77 keys to choose from.

IDEA is widely available throughout the world. It is considered to be extremely secure, particularly for analytic attacks. Brute force attacks generally don't work since with a 128-bit key, the number of tests has to be significantly high. Even allowing for weak keys, IDEA is far more secure than DES. Things have now changed because of parallel and distributed processing.

## CAST

CAST is named for its designers, Carlisle Adams and Stafford Tavares. The algorithm was developed while they were working for Nortel. The algorithm is 64-bit Feistel cipher that uses 16 rounds. Keys up to 128 bits are allowed. CAST-256, a variant, uses keys of up to 256 bits. Pretty Good Privacy (PGP) and many IBM and Microsoft products use CAST.

## AES

The Advanced Encryption Standard (AES) is intended to replace DES. It is based on the work of Joan Daemen and Vincent Rijmen of Belgium. The algorithm, named Rijndael, is currently undergoing extensive trials and evaluation. It appears to be extremely secure and there are hopes that it will be used in a wide range of applications, including smart cards.

What emerges from the discussion in previous sections is that key length is an important factor in determining the level of security. Clearly the 56-bit keys used in DES are not secure. But, neither are the conventional padlocks. There is no doubt that it's important to balance security with cost, time, sensitivity of data/communication, among other elements, when security considerations are being weighed. The system developed needs to consider the level of security relative to the expected life of an application and the increased speed of the computers. It is increasingly becoming easier to process longer keys. Software publishers also have a responsibility to make public their cryptographic elements for public scrutiny. In many ways this helps in building trust.

# Asymmetric Encryption

Asymmetric encryption was proposed by Diffie and Hillman [3], who observed that the process could be used in reverse to produce a digital signature. The primary goal was not the confidentiality of the message but to authenticate the sender and to guarantee the integrity of the message. The contents of the message, the plaintext portion, remain in plaintext format (see Figure 3.4). The digital signature portion of the message is a mathematical digest of the message that has been encrypted using the sender's private key. The relationship observed by Diffie and Hillman was that anything encrypted using the public key can

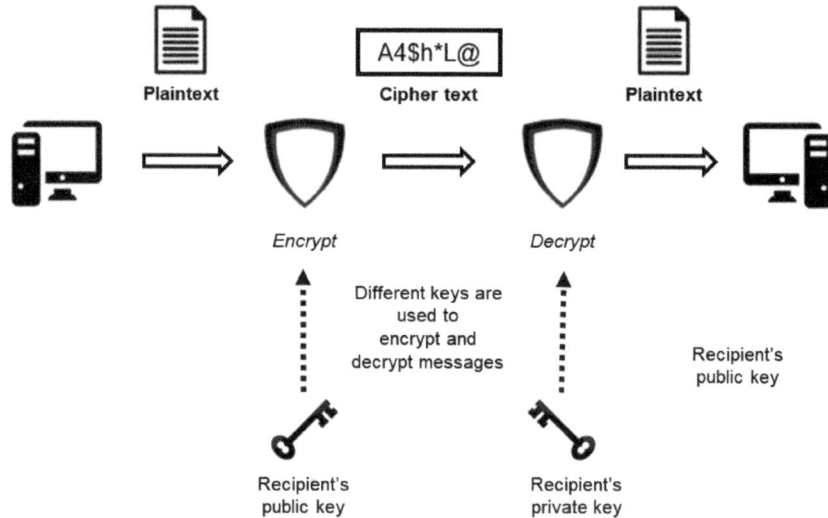

**Figure 3.4.** Asymmetric encryption

be decrypted using the private key and anything encrypted using the private key could be decrypted using the public key. Since the private key and its associated password are under the control of only one individual this allows for authentication that that person and only that person could have originated the message.

The integrity or inalterability of the contents of digitally signed messages comes about through the "hashing" process. The hashing process as it relates to digital signatures is quite different from the hashing process used to convert a key filed to store an address in a database environment. A cryptographic hash function such as SHA-1 or MD4/MD5 is a one-way process that produces a fixed-length digest of the original plaintext document.

One of the most important features of a cryptographic hash function is its resistance to collisions [1]. Since the digest is a fixed length, 128 bits for MD5 and 160 for SHA-1, there is a probability that more than one message will map to the same digest. The larger the digest of the hash function the lower the probability of a collision occurring. A hash function is analyzed as being weakly resistant when, given one message, it is not possible to find the second with the same hash. It is strongly resistant if it is possible to find two messages with the same hash. Hash functions operate on blocks of contiguous bits and are exceptionally sensitive to any change in the ordering or the value of the bits. This sensitivity is where the integrity feature is derived.

The digital signing process starts with a plaintext file. Using one of the cryptographic hash functions a hash of this file is calculated. The hash or message digest is then encrypted using the sender's private key. The plaintext file and the encrypted hash aka digital signature are then concatenated together and transmitted to the receiver. Upon receipt, the two parts of the message, the plaintext file and the digital signature, are separated and the recipient then runs the same hash algorithm against the plaintext file. The encrypted hash is decrypted using the sender's public key. The two hashes are compared. If they match the

## Asymmetric Encryption • 57

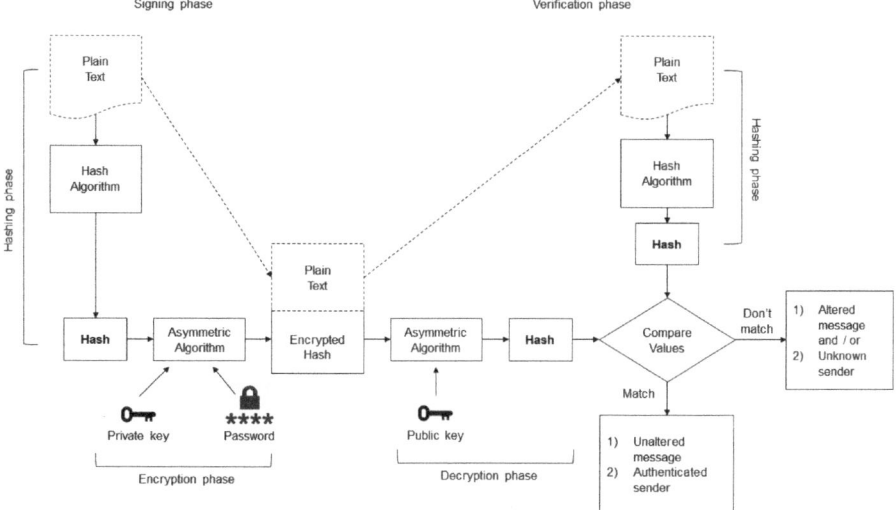

**Figure 3.5.** The digital signature process

recipient knows that the file has not been altered and the sender has been authenticated. Figure 3.5 graphically depicts the digital signature process.

## Authentication of the Sender

The authentication of the identity of the sender requires verification by a third party as to the identity of the sender. The requirement for authentication comes about because any individual can generate a key pair with any name associated with the keys. Authentication is the process of associating the object key with an individual. This is accomplished according to one of two methodologies. The PGP (pretty good privacy) model provides for authentication through a process known as a web-of-trust. The business world uses a hierarchical structure that has been standardized as X.509.

The web-of-trust authentication model is based on a decentralized transitory trust principle. If an individual, I1, knows for a fact that a second individual I2 has generated a key pair K2, then I1 signs the key K2 with their key K1. When a third individual, I3, receives the key K2, and recognizes I1 through their signing and trusts I1, trust is then transferred to I2 and I3 can safely assume that the key K2 belongs to the individual who is claiming ownership. The number of signatures or vouchers to the identity of the key and its creator may continue to increase until every possible recipient knows one of the individuals who have vouched for the authenticity of the key. This model works very well in small communities or environments where a central authority is impractical or inadvisable.

The X.509 structure is based on a hierarchical model where there is one ultimately trusted endorser, root certificate authority. The "root" transfers its endorsement through a series of sub-endorsers that will finally authenticate the key as belonging to the stated individual. Figure 3.6 shows this hierarchical structure. In the business environment, the

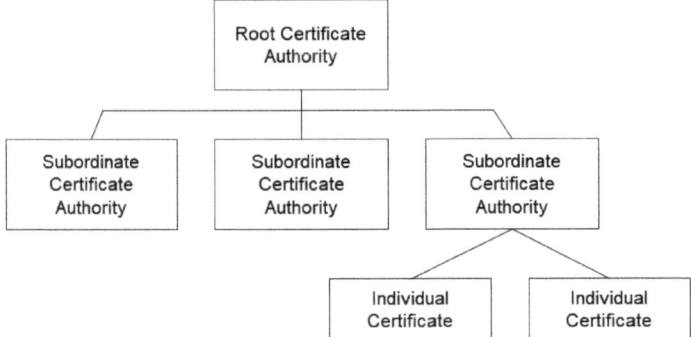

**Figure 3.6.** Hierarchical X.509 certificate structure

root certificate is held by either the corporate headquarters or a third party organization that specializes in this area such as VeriSign (www.verisign.com). For two individuals from different organizations to conduct business, they would need to arrive at a common certification scheme. This could be accomplished by exchanging root certificates or by having both root certificates signed by a trusted third party.

In the X.509 environment each key has a single endorsement, that of the authority immediately superior to it. The root signs it own key. The certificate chain is the certificate of the individual who signed a document and all of the certificates that signed that individual's certificate and subordinate certificates back to the root certificate. This chain establishes the authenticity of the individual. Extensive discussion of the X.509 format and certification schemes can be found in Atreya et al. [1].

## RSA

Any discussion of asymmetric encryption would be incomplete without the mention of the RSA encryption method. Previous sections introduced various kinds of ciphers, but in this section we will exclusively focus on the RSA method. RSA are the initials of the three inventors of this method: Rivest, Shamir, and Adleman.

The RSA encryption method is based on a rather simple logic. It is rather easy to multiply numbers, especially if computers are used, but is very difficult to factor the numbers. If one were to multiply 34537 and 99991, the result is calculated manually or with a computer to be 3453389167. However, if one were given the number 3453389167, it is difficult to find the factors manually. A computer will however use all possible combinations. The logic used by the computer would be to check for something that is of the size of the square root of the number that has to be factored.

As the size of the digits increases computing factors also becomes difficult, unless the number is a prime. If the number is a prime, it cannot be factored. The RSA algorithm chooses two prime numbers $p$ and $q$. Multiplying them makes a number $N$, where $N = pq$. Next $e$ is chosen, which is relatively prime to $(p-1)*(q-1)$. $e$ is usually a prime that is larger than $(p-1)$ or $(q-1)$. Then we compute $d$, which is the inverse of $e$ mod $n$.

A user will freely distribute *e* and *n*, but keep *d* secret. It may be noted that although *N* is known and is a product of very large prime numbers (over 100 digits) it is not feasible to determine *p* and *q*. Neither is it possible to derive the private key *d* from *e*.

# Future of Encryption

Researchers are and will always be in a constant search of new technologies and innovations to enhance security features. Encryption is no different. With the current encryption techniques there are several problems and since the organizations that need to secure their data are increasing exponentially, enhancements to current techniques is very important.

## Quantum Cryptography

The standard techniques used for exchanging the keys are inefficient. The RSA1024, which was commonly used, is now broken and is no longer considered safe as per NIST standards. There are however a few algorithms like RSA2048 that are still approved, but it won't be for long: as computers are becoming faster and bigger, these algorithms will be broken soon.

Researchers are constantly working on creating new methods to help improve the software-based key interchange mechanism using a new technology called post-quantum cryptography. These methods are projected to be effective even after powerful quantum computers become highly advanced. These methods are based on an improvable assertion that certain numerical algorithms are difficult to predict and reverse.

Quantum cryptography is based on the laws that govern quantum physics. It is a method used to transmit secret keys over long distances. Since the laws are governed by quantum physics, this method uses photons or light for transmission.

The risk of intercepting the key/message is still present; however, the probability of recreating or copying the data is reduced. Data can be copied but not perfectly accurately. This method is argued to be "provably secure" as comparison of measurements of the properties of a fraction of these photons shows that no interception is happening and hence the keys are securely transferred. Since this method is used just for key transmission and not cryptography, some researchers refer to it as "Quantum Key Distribution" (QKD) instead of Quantum Cryptography.

The capability of this method to support no-copying or the no-cloning transfers makes this the technology of the future. QKD is a technology new to the US; however, it has been already implemented in several banks and government agencies in Europe. This method is especially popular in Switzerland. However, the method lacks commercial acceptance in the United States. Research in this new technology is pushing the distance over which these quantum signals can be sent securely and accurately. Unused optical fibers laid by the telecommunication companies, which are at par with the laboratory standards, have been used for trails and the quantum signals were transferred accurately to up to 300 kilometers. The practical systems however are limited to 100 kilometers.

An architecture including a secure node acting as a bridge between sequential QKD systems will help extend the practical range of this emerging technology and will in turn allow keys to be transmitted over wide networks making large-scale implementations practical and achievable.

Many nations are interested in moving towards a technology that is not hackable and where the data security is high. Though no technology is ever foolproof and eventually all technologies become vulnerable, QKD is the best feasible solution currently available. It does have a few challenges, but continued improvements and innovation in this field will help tackle the misses and make it fit for use.

# Blockchains

While a large portion of the population is at least tangentially familiar with Bitcoin and the concept of a decentralized digital currency, they may be much less familiar with the underlying technology that enables Bitcoin—a technology known as "blockchain." The lack of awareness is understandable, as Bitcoin was the first real implementation of blockchain technology, invented by a mysterious developer known only as Satoshi Nakamoto, the official creator of Bitcoin, and thus of the blockchain technology that underpins it. The developer is only known by name and nothing more: as "Satoshi Nakamoto" is a pseudonym, the true identity or identities behind it are still unknown to this day. In the introduction to the seminal work published in 2008, *Bitcoin: A Peer-to-Peer Electronic Cash System*, Satoshi Nakamoto describes the backbone of blockchain technology: "The network timestamps transactions by hashing them into an ongoing chain of hash-based proof-of-work, forming a record that cannot be changed without redoing the proof-of-work" [5]. The transactions are hashed into a single block, and all blocks make up the chain, hence the name "blockchain." In the portion of the paper that follows, the author provides a more detailed and thorough explanation of this process, and how it applies to the blockchain and Bitcoin in general. While Nakamoto was the first to successfully implement blockchain technology and make it a practical solution, they were not the first to attempt to solve what is known as the double-spending problem, which is the purpose of blockchain in Bitcoin. The term "double-spending" dates as far back as 2007, when it was discussed by Osipkov et. al (2007) [6] where they state, "One major attack on electronic currency is double-spending, where a user may spend an electronic coin more than once. Unless the merchant accepting the coin verifies each coin immediately, double-spending poses a significant threat." This dilemma is at the heart of the motivation for blockchain technology, and in creating a hash-based decentralized ledger system, Nakamoto solved the problem of double-spending in the Bitcoin system. Nakamoto's development of blockchain technology to solve the problem of double-spending is highly relevant in modern society and in the years since publication; others have developed non-Bitcoin uses of blockchain technology. For example, the concept of "smart contracts" as described by Kariappa Bheemaiah [2]: "Smart contracts are programs that encode certain conditions and outcomes. When a transaction between 2 parties occurs, the program can verify if the product/service has been sent by the supplier. Only after verification is the sum transmitted to the suppliers account." With the advent of the post-Bitcoin blockchain landscape in 2014, referred to as "Blockchain 2.0," new technologies are being rapidly developed which take advantage of blockchain to underpin a more secure transaction ledger.

As noted above, blockchains were created to prevent the problem of double-spending within cryptocurrency. However, in implementation it is essentially a distributed ledger system, wherein participants work to build blocks on the chain by hashing individual

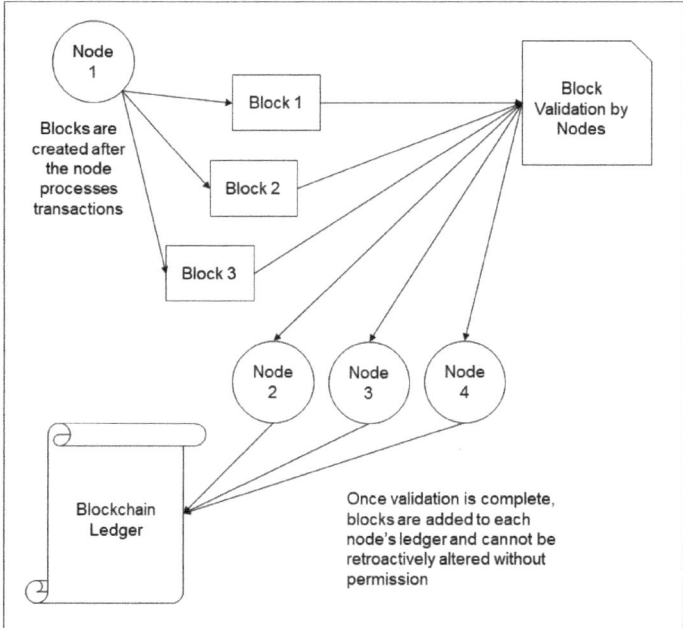

**Figure 3.7.** Blockchain creation and validation process

transactions with the chain acting as the ledger. As a blockchain is a decentralized distributed ledger, it means that multiple participants involved in building the blocks can hold a copy of the ledger. In technical terms, every participant in the blockchain that holds a copy of the ledger is known as a "node." As all nodes hold a copy of the ledger, it means the ledger is decentralized and does not exist in a single location, which prevents a central authority from altering the ledger in any way as all nodes must "agree to" any additions to the ledger. Therefore, the most important role the nodes have is to add blocks to the chain, which is done by processing transactions (of which blocks are composed) through a one-way cryptographic hash function that cannot be altered retroactively [4]. This means that a blockchain then acts as an open and distributed ledger that records transactions between various parties in an efficient, verifiable, and permanent manner. These types of secure transactions can be very appealing to any organization interested in maintaining information in a secure, yet efficient manner such as medical or banking records. A simple illustration of blockchain can be seen in Figure 3.7.

## Applications of Blockchain Technology

It is difficult to discuss real-world applications of blockchain technology without discussing Bitcoin, as it is the foremost use of blockchain technology. Generally, the predominant application of blockchain technology has been cryptocurrency, with Bitcoin being the most popular and well known. Bitcoin can be envisioned as the first wave of cryptocurrency which inspired the invention of many new forms of cryptocurrency, such as Altcoin, Dogecoin, Ethereum, and Litecoin. All of these operate in a manner similar to Bitcoin, with a decentralized blockchain serving as their backbone. However, a novel application of the blockchain that is now being pioneered is its use in the creation of "smart contracts." The term "smart contract" is somewhat broad. There are three components that help understand the concept:

1. The program is recorded "on" the blockchain, giving it permanence and censorship resistance.
2. The program can control blockchain assets, which means, for example, it can store and transfer cryptocurrency.
3. The program is executable by the blockchain. This means that no one can interfere with the operation.

## Alternative Blockchains

The large majority of what has been discussed so far has been related to "Blockchain 1.0," which pertains to Bitcoin-related blockchain. As previously mentioned, we are in what is currently known as the post-Bitcoin era (post-2014), which is now referred to as "Blockchain 2.0," or the post-Bitcoin blockchain. In some instances, these new alternative blockchains stay within the realm of finance, yet new applications outside of concepts such as "smart contracts" have arisen. For example, an Israeli ride-sharing company known as La'Zooz has adopted blockchain technology for their own use, giving rise to practical alternative blockchain applications in daily business practices. At La'Zooz, drivers give rides in exchange for Zooz tokens, which are a type of cryptocurrency derived from the blockchain. Drivers can later exchange those Zooz tokens for rides as passengers. While La'Zooz is a relatively small player in the ride-sharing industry, financial giants such as JP Morgan Chase have taken notice of the blockchain and are developing their own blockchain-based systems. Chase is developing "Quorom" in collaboration with Ethereum, a cryptocurrency company whose blockchain operates in a manner similar to Bitcoin. Interestingly, nodes on Quorom's blockchain must be properly vetted (by a central authority) before they are allowed onto the chain, and this is a departure from Bitcoin protocol, wherein nodes are allowed to interact on the blockchain without the permission of a higher authority. This departure means that Quorom is a centralized blockchain, as opposed to the decentralized blockchain found at the heart of Blockchain 1.0. Quorom represents a closed for-profit private development in Blockchain technology, as only those directly involved at JP Morgan Chase can contribute to its development. This is in contrast to projects like Counterparty, an open-source peer-to-peer platform. Counterparty rides atop the Bitcoin blockchain. This means that all Counterparty transactions occur on the Bitcoin blockchain and its primary usage thus far has been in the development of peer-to-peer smart contracts. However, unlike many other Blockchain 2.0 developments, Counterparty's code is open-source, meaning it is available for anyone to download and improve (counterparty.io).

### In Brief

- Cryptography incorporates within itself methods and techniques to ensure secrecy and authenticity of message transmissions.
- Cryptanalysis is the process of breaking in to decipher the plaintext or the key.
- Our aim in any encryption process is to transform any computer material and communicate it safely and securely, whether it be ASCII characters or binary data. Encryption can be carried out in two forms: **substitution** and **transposition**.
- Often referred to as permutation, the intent behind transpositions is to introduce confusion in the decryption process.
- **Stream ciphers** convert one symbol of plaintext into a symbol of ciphertext.
- **Block ciphers** convert a *group* (fixed length) block of plaintext into ciphertext through the use of a secret key.
- **Quantum cryptography** is the new emergent form of security mechanisms.
- **Blockhain technologies** have taken encryption technologies to new heights. We now see the emergence of smart contracts and centralized blockchains.

## Questions and Exercises

**Discussion questions.** These are based on a few topics from the chapter and are intentionally designed for a difference of opinion. These questions can best be used in a classroom or a seminar setting.

1. Clearly encryption is essential in ensuring secrecy of communication. Identify characteristics of encryption that make it rather impossible to decrypt.
2. Suppose the language of the plaintext message is not English. Enumerate steps that you would follow in decrypting the message.
3. Map an encryption plan for a typical organizational network. What kinds of encryptions would you use and how would you go about the implementation process?

**Exercise.** Scan the popular press for available encryption tools. Draw a conceptual map of how these could be used in the context of a typical retail organization. Possible retail stores that could be used: bookstore (both online and brick and mortar); department store; grocery store.

### Short Questions

1. The science of _____ seeks to ensure that the messages transmitted are kept confidential, maintain their integrity, and are available to the right people at the right time.
2. The field of _____ includes methods and techniques to ensure secrecy and authenticity of message transmissions.

3. The range of methods used for breaking the encrypted messages is referred to as _____.

4. Once a document has been encrypted it is referred to as _____ text.

5. A _____ text document is any document in its native format.

6. The _____ algorithm is designed to produce a ciphertext document that cannot be returned to its plaintext form without the use of the algorithm and the associated key(s).

7. In _____ encryption, a single key is used to encrypt and decrypt a document.

8. It is the _____ that holds the means to decrypt, and therefore it becomes important to establish a secure channel for sending and receiving it.

9. Ciphers that use the same key for both encrypting and decrypting plaintext are referred to as _____ ciphers.

10. Ciphers using a different key to encrypt and decrypt the plaintext are termed as _____ ciphers.

11. A brute force attack where the opponent will typically undertake a range of statistical analysis on the text in order to understand the inherent patterns is called a _____ text attack.

12. An attack that utilizes information regarding the placement of text such as in the header of an accounting document or a disclaimer statement is referred to as a _____ text attack.

13. Encryption can be carried out in two forms: _____ and _____.

14. In any language there are certain letters that have a high frequency of appearing together. These are referred to as _____.

15. Ciphers which generally convert one symbol of plaintext at a time into a symbol of ciphertext are referred to as _____ ciphers.

16. Ciphers that convert a *group* (fixed length) block of plaintext into ciphertext through the use of a secret key are referred to as _____ ciphers.

17. Initially developed by IBM, _____ was later adopted by the US government in 1977. Hint: It inputs a block of 64 bits, but only uses 56 bits in the encryption process.

18. A cryptographic _____ function such as SHA-1 or MD4/MD5 is a one-way process that produces a fixed-length digest of the original plaintext document.

# Case Study: The PGP Attack

While man-in-the-middle attacks are nothing new, several cryptography experts have recently demonstrated a weakness in the popular email encryption program PGP. The experts worked with a graduate student to demonstrate an attack which enables an attacker to decode an encrypted mail message if the victim falls for a simple social engineering ploy.

The attack would begin with an encrypted message sent by person A intended for person B, but instead the message is intercepted by person C. Person C then launches a chosen ciphertext attack by sending a known encrypted message to person B. If person B has their email program set to automatically decrypt the message or decides to decrypt it anyway, they will see only a garbled message. If that person then adds a reply, and includes part of the garbled message, the attacker can then decipher the required key to decrypt the original message from person A.

The attack was tested against two of the more popular PGP implementations, PGP 2.6.2 and GnuPG, and was found to be 100% effective if file compression was not enabled. Both programs have the ability to compress data by default before encrypting it which can thwart the attack. A paper was published by Bruce Schneier, chief technology officer of Counterpane Internet Security Inc.; Jonathan Katz, an assistant professor of computer science at the University of Maryland; and Kahil Jallad, a graduate student working with Katz at the University of Maryland. It was hoped that the disclosure would prompt changes in the open-source software and commercial versions to enhance its ability to thwart attacks, and to educate users to look for chosen ciphertext attacks in general.

PGP is by far the world's most well-known email encryption software and has been a favorite since Phil Zimmermann first invented PGP in 1991, and it has become the most widely used email encryption software ever since its inception. While numerous attacks have been tried, none have succeeded in actually breaking the algorithm to date. But with the power of computers growing exponentially, cracking this or even more modern algorithms is only a matter of time.

1. What can be done to increase the time required to break an encryption algorithm?
2. What is often the trade-off when using more complex algorithms?
3. Phil Zimmermann had to face considerable resistance from the government before being allowed to distribute PGP. What were their concerns, and why did they finally allow its eventual release?
4. Think of other social engineering schemes that might be employed in an effort to intercept encrypted messages.

Source: "PGP Attack Leaves Mail Vulnerable," *eWeek*, August 12, 2002.

# Case Study: Visa Goes Blockchain

Visa has just announced a preview of Visa B2B Connect, a new platform that Visa is developing with blockchain startup Chain to give financial institutions a simple, fast, and secure way to process business-to-business payments globally.

Visa is working to build Visa B2B Connect using Chain Core, an enterprise blockchain infrastructure that facilitates financial transactions on scalable, private blockchain networks. Building on this technology, Visa is developing a new near real-time transaction system designed for the exchange of high-value international payments between participating banks on behalf of their corporate clients. Managed by Visa end-to-end, Visa B2B Connect will facilitate a consistent process to manage settlement through Visa's standard practices.

> "The time has never been better for the global business community to take advantage of new payment technologies and improve some of the most fundamental processes needed to run their businesses," said Jim McCarthy, executive vice president, innovation and strategic partnerships, Visa Inc. "We are developing our new solution to give our financial institution partners an efficient, transparent way for payments to be made across the world."
>
> "This is an exciting milestone in our partnership with Visa," said Adam Ludwin, chief executive officer of Chain. "We are privileged to support Visa's efforts to enhance the service it provides to its clients and shape the future of international commerce with this Blockchain-enabled innovation—streamlining business payments among financial institutions and their customers around the world."

With Visa B2B Connect, Visa aims to significantly improve the way international B2B payments are made today by offering clear costs, improved delivery time, and visibility into the transaction process—ultimately reducing the investment and resources required by banks and their corporate clients to send and receive business payments.

Visa B2B Connect, which Visa plans to pilot in 2017, is designed to improve B2B payments by providing a system that is:

- **Predictable and transparent:** Banks and their corporate clients receive near real-time notification and finality of payment.
- **Secure:** Signed and cryptographically linked transactions are designed to ensure an immutable system of record.
- **Trusted:** All parties in the network are known participants on a permissioned private blockchain architecture that is operated by Visa.

Visa is a global payments technology company that connects consumers, businesses, financial institutions, and governments in more than 200 countries and territories to fast, secure, and reliable electronic payments. We operate one of the world's most advanced processing networks—VisaNet—that is capable of handling more than 65,000 transaction messages a second, with fraud protection for consumers and assured payment for merchants. Visa is not a bank and does not issue cards, extend credit, or set rates and fees for consumers. Visa's innovations, however, enable its financial institution customers to offer consumers more choices: pay now with debit, pay ahead with prepaid, or pay later with credit products.

Chain (wwwchain.com) is a technology company that partners with leading organizations to build, deploy, and operate blockchain networks that enable breakthrough financial products and services. Chain is the author of the Chain Protocol, which powers the award-winning Chain Core blockchain platform. Chain was founded in 2014 and has raised over $40 million in funding from Khosla Ventures, RRE Ventures, and strategic partners including Capital One, Citigroup, Fiserv, Nasdaq, Orange, and Visa. Chain is headquartered in San Francisco, CA.

1. Unprecedented transparency of transactions makes financiers uneasy. How do you think Visa will be able to successfully balance secrecy and transparency in the context of blockchain adoption?
2. There is an argument that if data were replicated across all banks using some sort of a shared settlement system, such as the one Visa aspires to have in place, it potentially becomes cumbersome. Discuss.
3. How would blockchain adoption enable security in the transfer of funds?

Reproduced with permission from Richard Kastelein. Richard is founder of *Blockchain News* and The Hackitarians Foundation, http://www.the-blockchain.com. The news item was also featured in various other publications, including the *Wall Street Journal*, October 22, 2016.

# References

1. Atreya, M., et al. 2002. *Digital signatures*. New York: RSA Press.
2. Bheemaiah, K. 2015. Block Chain 2.0: The renaissance of money. *Wired*, Feb. 17.
3. Diffie, W and M. Hellman. 1976. New directions in cryptography. *IEEE Transactions on Information Theory* 22: 644–654.
4. Iansiti, M. and K. Lakhani. 2017. The truth about Blockchain. *Harvard Business Review* 95(1): 119–127.
5. Nakamoto, S. 2012. *Bitcoin: A peer-to-peer electronic cash system*. http://www.bitcoin.org/bitcoin.pdf.
6. Osipkov, I., et al. 2007. Combating double-spending using cooperative P2P systems. In *Proceedings of the 27th International Conference on Distributed Computing Systems, ICDCS'07*. Washington, DC: IEEE.

# CHAPTER 4

# Network Security*

> I think that hackers...are the most interesting and effective body of intellectuals since the framers of the US constitution....No other group that I know of has set out to liberate a technology and succeeded. They not only did so against the active disinterest of corporate America to adopt their style in the end.
>
> —Stewart Brand,
> Founder, *Whole Earth Catalogue*

Joe Dawson encountered a new problem. His network administrator, Steve, had walked in the other day and declared that he was being paid far less than the market and that there was no reason for him to continue in his role. Sure Steel Inc.'s networks were dependent on this one person, whose departure would pose a new challenges and risks to the company. Joe had asked Steve to come back next week to discuss details. Essentially, Joe wanted some time to think about the challenges. He remembered advice given by his MITRE[1] friend Randy:

1. Ensure that everybody in the company knows that your network administrator is leaving.
2. All physical and electronic access needs to be terminated.
3. In the future, ensure that all employees sign a computer use and misuse policy.

Randy had also suggested that it is usually best to have an enterprise implementation of public key infrastructure that supports access to all resources. Randy had pointed out the benefits of this include the ability to revoke an employee's key when he or she decides to leave.

All these steps would be useful in the future, but Joe had a more immediate problem. How could he somehow keep Steve for the time being and yet develop a policy of some kind to deal with such issues in the future? When Joe had first embarked on to the cyber security journey, he had been introduced to a book by Matt Bishop. "Computer Security: Art and Science" (Addison-Wesley). He had then thought about the "art" aspects of cyber

---

* This chapter was prepared by Michael Sosnkowski.

1 MITRE is a private, not-for-profit corporation that operates FFRDCs—federally funded research and development centers. If you've ever flown in a jet or used GPS, you've benefited from technology with roots in an FFRDC (see www.mitre.org).

security, "where there any?" He had though. However after having read Matt Bishop's book, Joe had begun to feel that the majority of security issues were technical in nature. But the latest challenge faced by him was not technical. This was a human resource management issue. That evening as Joe drove home from work, his thoughts wandered into issues related to the nature and scope of security. Every single time Joe had felt that he had come to grips with the problem, a new set of issues had emerged. If it was managing access to systems, he was forced to consider structures of responsibility. If it was secure design, he had to understand formal models. Now it was network security, but he had to deal with human resource management issues. One thing was certain: he needed to ensure that security went beyond technical and socioorganizational measures. Perhaps management of security was sociotechnical in nature.

Joe remembered some of the debates on this topic area during his time at the university. At that time there was a lot of hype about sociotechnical systems. In particular he remembered the work of Eric Trist from the early 1950s at the Tavistock Institute. While studying the English coal-mining industry, where mechanization had actually decreased worker productivity, Trist had proposed that systems have both technical and social/human aspects, which are tightly bound and interconnected. He had argued that it was the interconnections more than the individual elements that determined system performance. This was an interesting argument and had some connection with his efforts to ensure security at Sure Steel. The argument also resonated with what Enid Mumford had said in her book, *Systems Design* [3]. Mumford had argued, "Designing and implementing socio-technical systems is never going to be easy. It requires the enthusiasm and involvement of management, lower level employees and trade unions. It also requires time, training, information, good administration and skill."

By this time, Joe had reached home. "I am sure Steve is going to be okay even if he leaves," he said aloud. Steve was, after all, an ethical man. Joe made a cup of coffee and went to his computer to check his email. Randy had sent Joe an email. He had cut and pasted a quote from *CIO* magazine. The original was from Gary Bateman, VP for IT at Wabash National Corporation. It read as follows:

> I'm reminded of a story where a professor is discussing ethics with one of his students. The professor posed the question to his student, "If you were presented with the opportunity to cheat on a test for a million dollars, would you do so." The young student pondered the question for a minute and then rationalized his answer by saying that in this situation, because a very large amount of money is involved for such a small indiscreet act, he would accept the payoff. The professor then asked if he would commit the same act for one dollar. The student, highly offended answered, "Of course not. Professor, what kind of man do you think I am?" The professor wisely answered, "Young man, we have already established that. We are only trying to establish the price."

*So very true*, Joe thought. He indeed had to learn more about the computer networks to appreciate what could or could not go wrong.

# TCP/UDP/IP Protocol Architecture

Networking technologies can be viewed from an information systems security professional's perspective as the signaling between two computers using a "Morse code"–like system of dots and dashes. These dots and dashes are used to symbolize letters, which are then grouped into words and strung together as sentences. The sentences become the instruction codes between client and server computers for data transmissions and system actions. As an example, imagine two people each with flashlights on separate hills at night. As they start to flash on and off to each other, sending their Morse coded messages, it would be necessary that both people had agreed to the meaning of the amount of flashes on and off to stand for the letters. This agreement of what the on/off flashing means is called "protocol" of how to behave and what to expect. The medium they use to transmit the Morse code would be in the form of light photons from the blinking flashlights. If they were shouting to one another, the medium would be the air carrying the sound. The start and end of the communication would need some sort of special blinking pattern to denote the separation of the sentences or conversations. In data networking, this is known as the "framing" of the data, as it provides the border between the start and end of the conversation.

Modern-day networking is likewise based on the same standards needed for the Morse code flashlight users: a physical medium such as radio waves, fiber optic or copper wire cables, to transmit the blinks; an agreed upon start/stop signaling from each other, which is called "framing," for the word collection of ones and zeros (010111011); and "protocols" referring to network commands on how to format the conversation between two devices. The network equipment used for the transmission of data are simply specialized computers that are designed for the accepting, storing, and retransmission of data signals. As such, like any computer they are made up of memory, static storage, CPUs, operating systems, and network interface cards, and are therefore vulnerable to similar attacks and compromises.

There are two types of network devices used to move data traffic: routers and switches. Routers interconnect to dissimilar devices such as wide area networks (WAN), the Internet, and legacy network technology in the organization. Routers are considered the "traffic cops" for the flow of information inside and outside of the network, and to accomplish this, they have signaling rules called "protocols" on what to do for network failure for rerouting of traffic. Switches are used closest to the end user of the network and provide a simple structure for moving data at extremely high speeds within a building or local area network (LAN). The flow of traffic is from a user's client computer, to a switch, to a router, and then out to the Internet or to a local data center to access the services needed. Routers and switches are limited in the security controls. To augment the security of a network, firewalls and intrusion detections or prevention systems (IDS/IPS) are inserted directly into the flow of the client traffic to and from the server, for the implementation of security policies and as logical and physical blocks to traffic.

The OSI and TCP/IP models structure the protocols used for signaling and the separation of function in a network. This is done by dividing the network signaling into logical divisions or "layers," each of which is responsible for an aspect of the signaling or, in other words, the transmission of data across the network. The security professional's interest in network structures lies in defining the points that allow access for investigation, interruption, and attacks. The lowest layer is that of the physical cables, radio, or microwave mediums.

As a reminder, all signaling is either an "on" or "off" time sequenced sending of signals from one point to another, much like that of the two people with flashlights described earlier. The "on" becomes the "1" of a binary computer code, and the "off" is represented by the "0." Fiber optic cables use this on/off pulsating of light photons to represent the zeros and ones, copper wire cables use voltage levels of high/low, and radio and microwave systems use the peaks and trough of the wave.

The physical layer known as "Layer 1" in the OSI model can be intercepted and compromised by copying the signaling methods to a secondary source by inserting an "optical splitter" known as a TAP for fiber optics, rogue access points for wireless, or the copping of network transmission from one router/switch port to another electronically for copper wire and fiber connections.

The grouping of the off/on (or in other words "0,1") is called "framing." The second layer up on the network signaling structure is called "Layer 2" and contains the broad start and stop signaling used to alert the switches to the beginning and ending of a transmission of data. This layer is nonhuman readable and is used by the switch and a computer's network interface card (NIC), which plugs directly into the switch to format the application data for use on the network. The term "MAC address" or "media access control" is the method for defining the data to be moved on a switch's forwarding from one switch port to another until it reaches the destination computer or router for further forwarding. The term "Virtual Local Area Network" (VLAN) is used, since the switch can impose a starting "tag" number onto the MAC frame, which is used to differentiate one group of computers from another, even though they both share the same physical medium. From a security point of view, if the transmitting switch is compromised and controlled, then the data frames can have their VLAN tag changed and/or copied, allowing for logically separate network traffic to be blended and therefore exposed to unauthorized users.

Wireless LAN installations also use MAC addresses, and the security controls used are based on allowing specific client MAC addresses to connect (or not) to the network. On a wireless network (or wired network as well), this can be compromised by falsely using an allowed MAC address and thereby accessing the wireless network as a legitimate user to do malicious activities. Please note that Layer 2 uses the term "datagram" for the unit of framed 1's and 0's. The third layer "Layer 3" is the network transport system signaling in which logical addresses are used. In Layer 3 the computer's network interface card's MAC address is used by Layer 3 to create a one-to-one relation to a logical address called an Internet Protocol Address, or more simply, the "IP address." The IP address has become the standard addressing format for internal and Internet-based communications and as such allows for abstraction to apply security control to the flow of traffic as it transfers on the router, security system, or IDS, but not by a switch, as the switch is only considering MAC addresses. The method of compromise can occur similarly as that at Layer 2, with the false numbering of a computer with a known, permitted IP address. The term "packet" is used in place of the Layer 2 "datagram" with Layer 3, as it incorporates the IP header and the data into one unit term.

The final network division is that of Layer 4 and is correlated with the various processes and services a server/computer provided to other computers. All the basic service functions such as file transfer (TCP port 20/21), secure shell (TCP port 22), and web services (TCP port 80/443) are used as the public facing connection point or "socket" into a server to

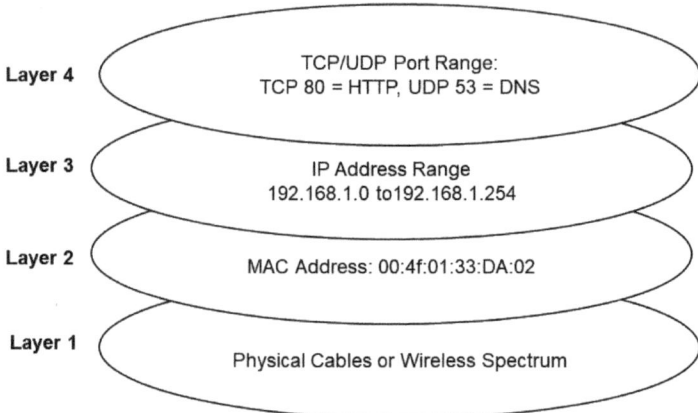

**Figure 4.1.** Network signaling and the layers

transmit the instructions and responses for use. TCP/UDP ports are software registers that are "listening" for incoming requests to connect and are related to specific server functions such as file sharing, websites, remote administration access, and so on. Figure 4.1 shows the graphic representation of the layers.

The original creators of the IP and TCP protocols did not include security constructions in the makeup of the version four (IPv4) IP address design scheme. The focus at the time was to create the ability to transmit files, provide remote administration sessions, and have automatic rerouting during a large network disturbance. Therefore the creators assumed the communications would likely be done over secured telecommunication lines between secure facilities such as military and government institutions. Because of these assumptions, the concepts of encryption and secure authentication were not the focus, but rather the assurance of data delivery over a continental telephone infrastructure. The original IPv4 address scheme was built with the goal to ensure forwarding of the packets over bandwidth-constrained transmissions. Terms such as "fragment," "header checksum," and "Internet header length" (IHL) are used to ensure the packets can be assembled by the receiving computer into a data stream for the processor to work on (see Figure 4.2).

Many of the network correction scheme sections used in IPv4 were eliminated in the new version IPv6, since newer telecommunication technologies removed these constraints. At the same time, IPv6 had its size of addressable IP host numbers increased from IPV4's $2^{32}$ (about 4.3 billion) to IPv6's $2^{128}$ (about 340 trillion), to allow for the anticipated growth in new endpoint devices such as mobile, personal, and other similar devices (see Figure 4.3).

The importance of IPv4 and IPv6 for the security professional is that these protocols were envisioned to be end-to-end addressing schemes without devices such as IDS, firewalls, and proxies in the middle of the transmission, which could modify the IP address. As such, with end-to-end connectivity there would also exist a visibility of who was connecting to what and the ability to record, prevent, or permit this connection from occurring as well as identify the speakers involved. However, in place of the originally envisioned one-to-one connections, the heavy use of network address translations (NATs) systems such as

| Bits 0 - 15 | | | Bits 16 - 32 | |
|---|---|---|---|---|
| Version | IHL | Type of Service | Total Length | |
| Identification | | | Flags | Fragment Offset |
| Time to Live | | Protocol | Header Checksum | |
| Source Address (i.e. 192.168.1.33) | | | | |
| Destination address (i.e. 10.22.0.99) | | | | |
| Options | | | | |

**Figure 4.2.** Original IPv4 address scheme

| Bits 0 - 15 | | Bits 16 - 32 | |
|---|---|---|---|
| Version | Traffic Class | Flow Label | |
| Payload Length | | Next Header | Hop Limit |
| Source Address (i.e.2001::001) | | | |
| Destination address (i.e. 2001:db8::8:9000:437B) | | | |

**Figure 4.3.** IPv6 address scheme

routers, proxies, load balancers, and firewalls allowed for the implementation of many-to-one addresses. The result was the obscuring of the speakers in the communications, as well as the ability to reduce the need for more Internet routable IP address for each client and server. As a result, while the ideal security design for the professional is to have every device given a unique IP address for ease of identification and tracking, the reality is that with IPv4 the move to IPv6 has been slowed due to the success of network address translation devices. Further, even with the full implementation of IPv6 and the ability of each device to communicate directly and knowingly with another devices, this would create the loss of anonymity, in addition to opening a vast vector for attack.

The IP addressing of server, workstations, and mobile device systems allows for inter-communications at a machine level; however, for human use it is easier to remember an alphabetical name rather than a series of numbers. Therefore the Domain Name Service (DNS) is used as a distributed database for locating a computer's IP address number and its corresponding "name," commonly used by people like "xyzcompany." The DNS architecture uses a limited set of internet holding servers known as the "root" domain servers, which now house the ".com," ".org," ".gov," and so on. The next level down called the parent and child domains. This allows for the hierarchical naming conventions to be quickly referenced and located for use (see Figure 4.4).

For example, the website "www.newcar.xyzcompany.com" is found in the following manner: The client computer asks for the full name of its local DNS server. The local DNS sever in turn asks for or "queries" the root domain server for the location of "xyzcompany.com." The root DNS replies with the IP address of xyzcompany's DNS server. Again, the local DNS server queries xyzcompany's DNS server for "www.newcar" and receives a response back from xyzcompany's DNS server with the IP address of the web server hosting "www.newcar.xyzcompany.com." The importance of this system for the security professional is

74 • Chapter 4 / Network Security

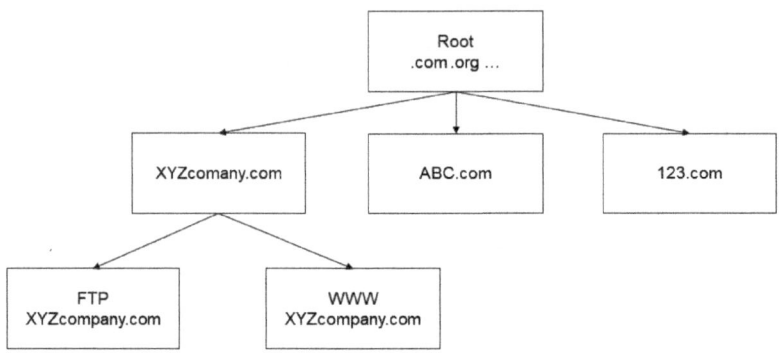

**Figure 4.4.** The DNS scheme

the need for availability of DNS resources, not only for the client but for all servers and devices needing to talk with each other either on or off the local network. A disruption of this service in the form of a distributed denial of service (DDoS); manipulation of DNS records, known as "DNS poisoning"; and the subsequent redirection to false websites and resources can result in a major loss in service as well as data integrity for client data. Currently there are efforts to increase communication encryption between clients and DNS servers, as well as efforts to enhance DNS to DNS server transfer of information. However, due to the complexity involved, this initiative has not been completed.

Once the identification of the IP addresses of the various servers and client endpoints are known and their location listed with DNS, the next step is to ensure the data packets have mechanisms independent of the infrastructure to recover from the loss of service. The use of agreed upon methods of transporting and accessing computer services like web or file exchanges is called a "protocol," much like the etiquette of good behavior used in polite society. The Transport Control Protocol (TCP) and User Datagram Protocol (UDP) provide the methods of adjustment needed to recover from such outages of connections. The TCP concept of "sessions" is used for the alignment of the conversation between the client and servers for access to resources. Briefly, a client will initiate what is known as a "three-way" handshake, or a request, response, and acknowledgement

Technically this is known as the Sync (SYN), Sync/Acknowledge (SYN/ACK), and acknowledgement (ACK). From this point on, the conversation will flow from client to server and server to client on specific numeric number called "ports," which allow for a computer processor to accept data on its memory register. Typical ports used by TCP are 80 for HTTP, 443 for HTTPS, and 22 for Secure Shell. The default standard list of defined system service ports ranges from 0 to 1024, with number ports above 1024 to 65535 for dynamic use (see Figure 4.5). This is important in security when reviewing the various attack types possible. A "SYN" attack is one in which a malicious client or multiple clients start the three-way handshake but never complete the process, thereby exhausting the server's memory areas used for new network connections.

In contrast, the UDP protocol does not set up a "session" with a three-way-handshake but rather trusts the underlying infrastructure to provide recovery, should packets be lost. The benefit is that the start and end of UPS connections is very quick and traditionally

| Bits 0 - 15 | | | | | | | | Bits 16 - 32 |
|---|---|---|---|---|---|---|---|---|
| Source Port ( i.e. 1024 – 65535) | | | | | | | | Destination Port (i.e. 80,53,...) |
| Sequence Number | | | | | | | | |
| Acknowledgment Number | | | | | | | | |
| Data Offset | Reserved | CWR | ECE | ACK | PSH | RST | SYN | FIN | Window Sized |
| Checksum | | | | | | | | Urgent Pointer |
| Options | | | | | | | | Padding |
| DATA | | | | | | | | |

**Figure 4.5.** System server ports

has been used in voice and video streaming data, where speed and low latency are of the highest importance. From the perspective of the firewall, the UDP traffic is passed quickly, and the ability of controlling the "state" or the point of a session with TCP "SYN, SYN/ACK, ACK" is lost, as UDP does not have an known "state" for their connections. For this reason, applications can use this stateless aspect of the UDP protocol to "wrap" their data and bypass the control of the "stateful" inspections of firewalls. One production example of this is Google's Chrome web browser, which uses the "QUIC" protocol to improve speed and security for the user but diminishes visibility for the security team. Finally, when an IP addresses is combined with TCP/UDP ports, the union is called a "socket." The visualization is that of a client connecting into a server much like an electrical cord or data patch cable connect into a wall "socket."

The term "tunneling" is used when one protocol is encased within another, as can be shown by the use of TCP/IP being "tunneled" across an Ethernet by the exchange of the header of each switch connection. More commonly, the term "VPN tunneling" appears in security literature as "encryption of the data packets." The virtual private network (VPN) is an example of devices in which regular data packets using IP addresses with TCP or UDP enter in one of the VPN's network interfaces, become encrypted (scrambled "1s" and "0s"), and then exit onto the network to be moved to the remote VPN server for decryption. The point to remember here is that the encrypted packet moves over the same network as unencrypted data traffic, and only the term "tunnel" is used to denote the inability of network security devices to read the network packets without the decryption keys.

Three main terms for divisions of network infrastructure are LAN, WAN, and Internet. LAN refers to a "local area network," which is visualized as a home or office space and up to a large building or confined building infrastructure, and is only accessible to users within the wiring or wireless signal range. WAN, or "wide area network," is a connection of LANs via telecommunication-company-owned data links, which may consist of lease lines such as copper phone lines, fiber, or leased metro-Ethernet networks. Lastly, the Internet can be thought of as a global distributed interconnection of LANs and data links consisting of public and private fiber optic cables and publicly accessible DNS services. The relevance of these terms from a security perspective lies in the LAN, WAN, and Internet peering points, or connection points in which security devices are placed. These devices are known

| Bits 0 - 15 | Bits 16 - 32 |
|---|---|
| Source Port ( i.e. 1024 – 65535) | Destination Port (i.e. 80,53,..) |
| 16 bit UDP Length | 16 bit UDP Checksum |
| Acknowledgment Number ||
| DATA ||

**Figure 4.6.** The UDP protocol

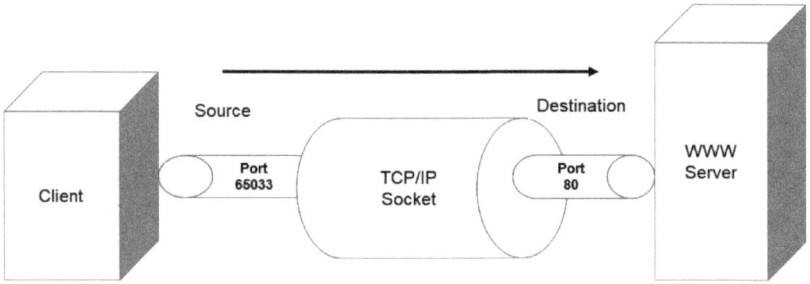

**Figure 4.7.** The socket arrangement

as "middle boxes" and are known as taps, packet captures, firewall, proxies, and other filters that are placed. The value this brings to security is as point of entry, control, and policy, as well as forensics and compliance to standards. In conclusion, by knowing these networking terms, the security professional is in the position to understand the various methods hackers and malicious software use to penetrate and overwhelm networks and eventually computer systems.

## Middleware Devices

As single networks are connected to external networks and the Internet, concerns of security are elevated and companies rely on computer systems, which are in the "middle" of the client-to-server traffic. This prevents unauthorized access to and from the internal network. A firewall is considered the first line of defense in protecting private information and denying access from intruders to secure systems on the internal network. Firewalls are devices, either software- or hardware-based or a combination of both, used to enforce network security policies. The most popular defenses for networks is the placement of stateful network firewalls, which limit the range of IP address and ports allowed in or out of a network's perimeter. The word "stateful" refers to the TCP three-way-handshake status or "state" of the TCP/IP session as traffic is crossing the firewall. Stateful firewalls can detect and eliminate the issues caused when malicious users send spoofed sessions to servers and thereby attempt to breach the network access of the server with memory depletion via buffer overflows, when too many connection packets are sent in a short period of time. Firewalls

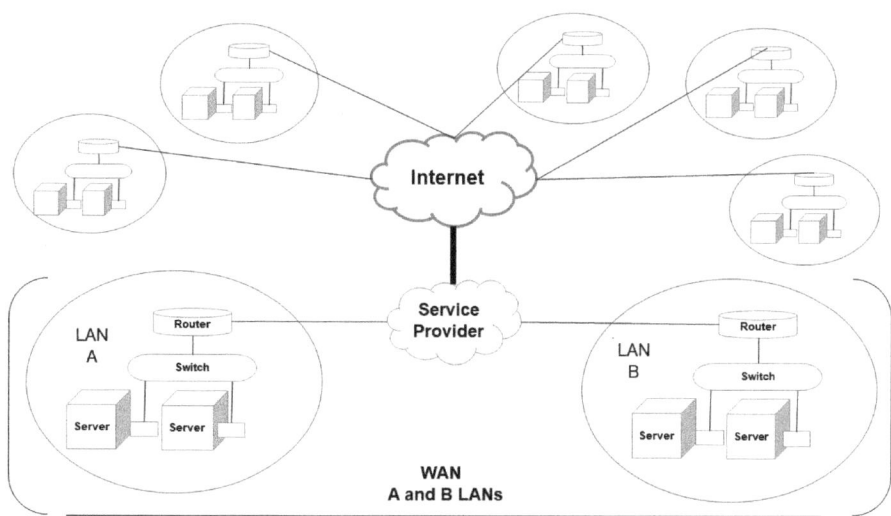

**Figure 4.8.** Network structures

can also be implemented on host computers in the form of A nonstateful access list, which allows access based on various ranges of address and ports as well.

A firewall provides different tasks such as:

- Enforce network policies on the incoming packets by looking at the Internet and transport layer data.
- Control traffic by defining zones such as DMZs (demilitarized zones), internal network, and perimeter networks.
- Serve as an audit point for all traffic to and from the internal network.

Firewalls can be configured to take specific action on the incoming and outgoing packets based on different types of network rules, such as rule-based packet filtering, port filtering/forwarding, and NAT (network address translation). Rule-based packet filtering is a packet filtering process used by the firewall device to block or allow network traffic based on a set of rules defined by the administrator. The rules are applied by the firewall to all the packets, thereby denying potentially dangerous traffic patterns and allowing only authorized access to specific systems. Common examples of packet filters are pf for Unix and iptables or ipchains for Linux operating systems.

Port forwarding or filtering is the process by which firewall devices can forward the traffic addressed to a certain port on a specific machine to another. This method allows a certain type of traffic to be sent to only specific systems, as defined by the firewall rule. For example, forwarding port 80 using firewall rules or at the router will send all browser-based packets to the web server on the internal network and nowhere else. NAT is a technique in which source and destination IP addresses are rewritten as they pass through the firewall or router. This serves the dual purpose of hiding the internal IP addresses of critical systems, as well as allowing multiple hosts on a private internal LAN to access the Internet using a single public IP address.

It is also a common practice to establish different zones within the internal network based on different levels of trust. For instance, highly secure data stored on database systems should be in a well-trusted zone where only authorized traffic is permitted access. The firewall is configured with high security level to specifically monitor traffic to this zone. A DMZ (demilitarized zone) sits between an organization's internal network and the Internet. The DMZ contains host systems to provide services to the external network (such as a web service hosting the company's website) while protecting the internal network from possible intrusions into these hosts. The host systems are vulnerable to attack and therefore constantly monitored by the firewall. However, a compromised system in the DMZ does not pose a threat to other internal systems, as they are separated from the rest of the machines in the internal network. Connectivity to the DMZ is allowed from the internal network, but no access is allowed from the DMZ to the internal network.

As a result of the usefulness of firewalls, for many decades the restrictions imposed by source and destination IP address and TCP/UDP port were adequate to provide the first layer of defenses in network security. The value was their ability to apply a security policy across all the business segments of an institution as a means to objectively and authoritatively limit access in and out of a network. However, these limitations placed burdens on the utility of the network, and as a result users found various methods of bypassing perimeter firewalls. One example is the initiation of TCP sessions from within the interior LAN network to proxy sites, which then allows the user to return home but still "tunnel" back to their work desktop via the existing proxy session. Other methods such as the use of encryption to avoid deeper packet inspection or the use of UDP based connections like that used in the Google "QUIC" protocol remove much of the benefit of a stateful firewall logging packets for further analysis. To combat these issues, next generation firewalls incorporate more deep packet inspections. Such inspections are seen with intrusion detections, or with proxy servers with the ability to decrypt and reencrypt packets in real-time while logging the usage. One of the best aspects of firewalls or proxies as the first line of demarcation between internal and external is the use of bulk filtering of traffic to and from the internal network. The filtering lessens the burden on traffic processing by computers as well as removes the noise seen on security device looking for suspicious traffic.

## VPN and Proxy

Virtual private networks (VPN) are often seen as the solution to insecure network communication over untrusted networks such as the Internet or internally for sensitive transactions such as PCI money activities. As stated earlier, the VPN technology relies on network-based devices or software on the host computer that individually encrypt the data packets prior to transmission and post receipt. In order for the packets to be decrypted, either a proxy device is needed that has the encryption "keys" or access to the VPN devices is needed to copy the decrypted cleartext traffic to another monitoring system. Most commercially available VPN servers have a 1 gigabit network connection on the inside and outside network, which can be scaled by adding more appliances, but with additional cost for the hardware and licensing. The issues come when VPN is seen as a security tool beyond its original use of temporary remote access to secure systems. If used in place of secure segments and firewalls, the VPN server becomes a point of failure for the institution or at least a bottleneck of bandwidth for data transfer, due to limitations of the interfaces of the VPN server connecting to the network. The value of a VPN system is its use within the context of a multilayer method of defense.

## IDS, IPS, SEIMs

Once the traffic from a VPN or via a firewall has entered the interior network and the traffic is no longer encrypted, then further security concerns can be addressed. One tool that allows for the automatic real-time analysis of high speed network traffic is an intrusion detection server (IDS) or the intrusion prevention servers (IPS), which are stationed behind the perimeter firewall to review and filter the traffic as it is forwarded to servers and the interior of an organization. The focus of the IDS is to detect and alert a security team of potential issues, while an IPS uses a protection mechanism to block automatically suspicious traffic. Both IDS and IPS use signature-based analysis of the data inside of the network packets or, in other words, the pattern matching of the data to known malicious software used by hackers. In addition, the use of heuristics behavior of traffic may be used to determine if an intruder is stealthily cracking into the organization. Security information and event management (SIEMs) has the ability to review massive log and network traffic flow data and then produce correlations to indicated potential attacks and breach attempts. There are multiple commercial and open-source versions of IDS/IPS, like Security Onion's open-source product or Palo Alto Network's next generation firewall. These devices require the packets to be unencrypted to gain the full benefit of the inspections. The value brought is that the system is able to review hundreds of thousands of connections and sort out potential breaches, as well as implement controls to block traffic, should a real-time attack occur. While event triggers can be included to interject blocks and resets of selected TCP sessions, the real value is the correlation of various traffic patterns to discern potential stealth attacks and other longitudinal comprises.

# Types of Network Attacks

Network attacks come in different forms, and their security will remain a major concern as long as there is connectivity of some sort with other systems. Attacks are primarily attempts to forge, steal, or gain access to systems by manipulating, sniffing, or redirecting data transmitted across the network. A majority of the attacks on a network occur by taking advantage of vulnerabilities of the operating systems and by exploiting inherent weaknesses of the Internet, transport, and application layers of the TCP/IP protocol architecture that have not been secured by the network administrator. Attackers come up with innovative and powerful methods to gain access to weakly secured networks and unpatched operating systems. To be able to secure a network and the computers on the network, it is important to know how attacks are launched on a network. In most cases, an attacker follows a sequence of steps to launch an attack or to identify a potential vulnerability that can be exploited. The two main steps involved in attacking a network are reconnaissance and scanning.

## Reconnaissance

This is the information gathering step where the intruder tries to gather as much information about the network and the target computer as possible. The attacker seeks to perform an unobtrusive information gathering process without raising alarms about his activity. On a network, this involves collecting data regarding network settings such as subnet ids, router configurations, host names, DNS server information, and security level settings. Corporate web servers, DNS servers, SMTP mail servers, and wireless access points are often targets of this form of information inquiry used by the attacker. Once sufficient information is available regarding the network, the attacker seeks target devices or computers as the next step in the information gathering phase. On collecting this information, the intruder starts the probe for identifying the target operating system. This is a critical step, as each operating system has its own unique set of known vulnerabilities that can be exploited. Identifying and fingerprinting the operating systems of the target computers make it easy for the attacker to focus on known vulnerabilities of the operating system that the system owner might have forgotten or failed to fix. Information pertaining to usernames, unprotected network sharable folders, and unencrypted password policies are easily detected in this phase of the information gathering process.

## Scanning

After the reconnaissance phase, the hacker is armed with enough information to start the scanning phase, while being cautious not to raise an alarm. Scanning is done in different ways and usually is aimed at networks, ports, and hosts. Network scanning involves the attacker sending probing packets to the identified network-specific devices such as routers, DNS servers, and wireless access points to check and gain information about their configuration settings. For example, a compromised DNS server will provide a great deal of information about a company's servers and host systems. Many firms use a block of IP addresses that are statically assigned for servers. In addition, most companies rely on Dynamic Host Configuration Protocol (DHCP) to automatically assign an IP address from a predefined range of IP addresses to the client computers, such as desktops, laptops, and PDAs. Access to this sort of information provides vital data to the hacker that will help him fixate his attack on target computers of interest that are worthy of the effort.

Host scanning provides the hacker with information regarding the vulnerabilities of the target host system. The attacker uses different tools to connect to the target host and probe the targeted machine to check if any known vulnerabilities (such as common configuration errors and default configuration and other well-known system weaknesses) specific to the operating system are present that can be exploited.

Most common break-ins exploit specific services that are running with default configuration settings and are left unattended. Using port scanning, the attacker can know the kinds of services that are running on the targeted hosts. This helps the hacker attack vulnerabilities that are specific to the services running on the host. For instance, finger is a Unix program that returns information about the user who owns a particular email. On some other systems, finger returns additional user information such as the user's full name, address, and phone number, assuming this information is stored on the system. Finger runs on port 79, and unless the port has been turned off, a hacker who can access this port on a Unix system that stores all company information can easily gather valuable information without having administrative privileges for the system. With all this information in hand, the attacker can then proceed to launch a full-fledged application or operating-system-based attack or a network-based attack.

## Packet Sniffing

Besides the common operating-system-based attacks, the network itself is a major cause of concern for security. Some common network-based attack techniques include sniffing, IP address spoofing, session hijacking, and port scanning. Sniffing techniques are a double-edged sword, since they can be used for the general good as well as for potentially negative outcomes. On one hand, they can be used to detect network faults, while on the other, those with a malicious intent can use such acts to sniff sensitive data (such as passwords for email and website accounts) without the owner being aware of the deed. In essence, a discussion on sniffing also highlights the importance of encrypting data that is transmitted across the network.

Packet sniffing is done by using programs called "packet sniffers" that operate on the data link layer of the TCP/IP protocol architecture to gather all network traffic. Packet sniffers can thus be viewed as devices that plug into the network via a computer's interface card (or network card) and eavesdrop on the network traffic. They capture the binary data on the network and translate it into human readable form. This functionality of packet sniffers is used by network security administrators to monitor network faults (such as a rogue computer sending out too many ARP packets) or by an attacker to probe for critical data sent across the network.

One important point related to sniffing is that the sniffer software can be used by an attacker to sniff on a network only if access to the network has been gained. In other words, the probing software has to be on the same network from where the data is captured. For example, if John and Emily are engaged in an Internet chat session that Jane wants to overhear, she cannot do so unless she has access to the path that the data travels during the chat session. To make this possible, hackers usually gain access to the host computer and install Trojan software (for spying), and thus the sniffer itself is on the same wire as the users. To successfully sniff all packets on a LAN, the end systems have to be connected to a hub and not a switch. A hub echoes packets from each port to every other port, thereby making them easily accessible to a sniffer program running on any machine connected to the hub. A packet sniffer attached to any network card on the LAN can thus run in a promiscuous mode, silently watching all packets and logging the data. However, if the computers on the LAN are connected via a switch instead of a hub, then things are different. The switch does not echo all packets to every other port, and therefore the sniffer program cannot read all the packets. Switched LAN configuration thus provides good sniffer protection. In this case, the approach used by the hacker is to trick the switch into exhibiting behavior like a hub. The attacker can flood the switch with an ARP request (a protocol that is used to find a host's MAC address from its IP address), which causes the switch to echo the packets to all other ports or redirect traffic to the sniffer system.

Many commonly used Internet applications like POP mail, SMTP, FTP, and chat messengers send data (and passwords!) in clear text. Sniffing can easily gather all the unencrypted data sent by the services running on the host system and store them in a file that can be analyzed by the hackers at their convenience. In order to ensure protection against sniffing, some important, yet simple, steps can be taken to prevent unsolicited eavesdropping. For instance:

1. Check with the email service provider to see if the email client can be configured to support encrypted logins. The email server has to support this feature in order for the client to allow encrypted login.
2. Even when using encrypted logins, the email messages are still transmitted in clear text. Use encryption (such as PGP at www.pgpi.org) for added security for the message content of the emails.
3. Consumers who shop online on a regular basis should make it a habit to verify that all credit card and banking transactions are conducted only on websites that support SSL or S-HTTP. It is recommended that credit card information not be given to untrustworthy websites or websites that fail to provide optional information such as contact information or fax or phone numbers.
4. Remote connections to servers using telnet should be avoided. SSH (secure shell) should be used instead of telnet so that traffic is always encrypted.
5. If possible, change the network to a switch rather than a hub on a LAN.

## IP Address Spoofing

IP address spoofing is a form of attack that takes advantage of security weaknesses in the TCP/IP protocol architecture. This form of attack is used by hackers to hide their identity and to gain access by exploiting trust between host systems. In IP spoofing, the attacker forges the source IP address information in every IP packet with a different address to make it appear that the packet was sent by a different computer. IP spoofing is mainly used to defeat network security and firewall rules that rely on IP-address-based authentication and access control. By changing the source IP address information in the packets, the hacker remains anonymous and the target machine is incapable of correctly determining the identity of the attacker. This form of attack is also used to exploit IP-based trust relationships between networks or computers. It is common on some corporate networks to have internal systems allow a user to login based on a trusted connection from an allowed IP address without the use of a login ID and password. If the intruder has gathered enough information about the trust relationships, the attacker can then gain access to the target system without authentication by forging the source IP address to make it appear as if the packet is originating from the trusted system.

Packet filtering is one form of defense to prevent IP address spoofing. Ingress filtering is the filtering technique that can be used to block packets from outside the network with a source address inside the network. Ingress filtering is implemented at the gateway to a network, router, or the firewall.

The ease with which the source IP address in a packet can be masked together with the ability to make easy sequence number predictions also leads to other common forms of attacks such as man-in-the-middle and denial-of-service. A man-in-the-middle (MITM) attack is when the hacker is able to intercept messages between the communicating systems and modify the messages without the two parties being aware of it. The attacker can control the flow of information and read, eliminate, or alter the information that is transmitted between the two end systems.

Although public-key cryptography was devised as a means to allow users to communicate securely, man-in-the-middle attacks can still be launched to intercept the transmitted messages. Therefore digitally trusted keys such as a certificate authority (CA) assigned by a trusted third party are preferred to public-key cryptography, to correctly endorse the two communicating parties and secure their transactions.

## Flooding

One of the most difficult attacks to defend against, the denial-of-service (DoS) attack, is also based on IP address spoofing. In this case, the hacker is not particularly concerned about stealing information or manipulating data on a target computer. The malicious intent is to create inconvenience through vandalism in the form of disrupting communication by consuming bandwidth and resources. DoS attack relies on malformed messages directed at a target system, with the intention of flooding the victim with as many packets as possible in a short duration of time. The attacker uses a series of malformed packets directed at the victim's host computer, while the host computer tries to respond to each packet by completing the TCP handshake and transaction, causing excessive usage of the host CPU resources or even causing the target system to crash. DoS attacks take different forms.

SYN flooding is a type of DoS attack where the attacker sends a long series of SYN TCP segments (synchronized messages) to the target system, forging multiple TCP connections. The target machine is forced to respond with a SYN-ACK to each of the incoming packets before the connection can be established. However, the attacking system will skip the sending of the last ACK (acknowledge) message before the final connection is established. The target host will wait for this last ACK message from the requesting client, which is never sent! A half-connection of this sort causes the target system to allocate resources in the hope of fulfilling the connection request from the client. Flooding the target with SYN packets brings the target host to a crawl. Barely being able to keep up with the incoming requests for TCP connection, the target system now starts denying connection requests from legitimate users, which ultimately results in a system crash if other operating system functions are starved of valuable CPU resources.

Smurf attack is another denial-of-service attack that uses spoofed broadcast ping messages to flood a target system. Here the perpetrator uses a long stream of ping packets (ICMP echo) that is broadcast to all IP addresses within a network. The packets are spoofed with the source IP address of the intended target system. ICMP echo is a core protocol supported by the TCP/IP protocol architecture and is commonly used by the ping tool to determine whether a host is reachable and the time it takes for the packet to get to and from the host. Since each ICMP echo request message receives an echo response message, all host systems that received the broadcast ping packet will reply back to the source IP address—in this case the spoofed IP address of the victim. A single broadcast echo request now results in large volumes of echo response that will flood the victim host. On a multiaccess broadcast network, the response echo directed to the target victim easily falls into hundreds and hundreds of echo responses. Firewall rules can be set to specifically drop ping broadcasts, and newer routers can be configured to stop smurf attacks.

Another very popular form of DoS attack is the distributed DoS (DDoS). In this case, multiple compromised host systems participate in attacking a single target or target site, all sending IP-address-spoofed packets to the same destination system. DDoS is highly effective due to the distributed nature of the attack. Since multiple compromised source systems are used to launch a DDoS attack, its makes it difficult to block the traffic and even more difficult to trace the attacker. Ideally, to protect computers against a DoS attack, outgoing packets from a network should also be filtered. Egress filtering can be used at the firewall or the gateway to the Internet to drop packets from inside the network with a source address that is not from the internal subnet. This prevents the attacker within the network from performing IP address spoofing to launch attacks against external machines.

## Operating System Attacks

Poorly managed computers with unpatched operating systems and badly designed applications are targets for culprits looking to steal, copy, or manipulate data in the form of an operating-system-based attack. Once the hacker has identified the operating system, the services running on the host and applications that use ports to communicate with other computers, an attack can be undertaken in many different ways. Some common operating-system-based attacks are discussed as follows. It should be noted that routers and switches also can suffer from such computer system attacks, as they are structurally the same as general purpose computers.

## Stack-Based Buffer Overflow

A stack is a data structure that works on the principle of last-in, first-out (LIFO) and is commonly used by operating systems to store important instructions relating to the different processes running on a computer. Applications commonly use data buffers to store input data, which goes into the program stack. Properly written applications need to check the length of the input data to ensure that it is not longer than the allocated buffer space, but this is frequently overlooked, especially by novice programmers. If the application was poorly written and makes bad use of the stack, an attacker can write a program that causes the stack to overflow, leading to alteration of the application's execution. For example, an overflowed stack will give the attacker the ability to force the application that caused the overflow to spawn a command shell, which can then be used to execute commands on the target system.

Generally, buffer overflows are caused by careless programming and are common in programs such as C and C++, which are limited in run-time checking and automated memory allocation. Stack-based buffer overflow attacks can be avoided by ensuring that only extensively tested applications from reliable software developers or vendors are installed on a host system. If the organization relies on in-house developed software, steps should be taken to ensure that only safe source code libraries are used on the development computers. Using programming languages that perform automated memory management such as Java or Microsoft .NET also ensures that the host computers are less vulnerable to buffer overflow attacks.

## Password

Passwords are the most commonly used form of authentication of a user to a computer system. Password attacks are also the most commonly used mode of attack against an operating system. In many cases, the default password settings are left unchanged, and this is common knowledge that can be easily used to break into a computer. Password attacks are also undertaken by guessing, by dictionary attacks, or through the use of brute force cracking. It is not surprising that passwords can often be guessed fairly easily, since many users tend to use weak passwords, usually relating to who they are. It is common to see users having blank passwords, the word "password," their pet's name or children's names, or their birthplace. Needless to say, such passwords can be easily guessed by a determined cracker. Indeed, guessing has emerged as the most successful method of password cracking.

Dictionary attacks also exploit the tendency of people to use weak passwords that are slight modifications of dictionary words. Password-cracking programs can encrypt each word in the dictionary and simple modifications of each word, including reversing a word, and check them against the system to see if they match. This is simple and feasible because the attack software can be automated and run in a clandestine mode without the user even knowing about it. Guessing and dictionary attacks together have consistently been shown to be the most effective way to hack into computer systems.

Brute force attack is the last resort and involves trying all possible random combinations of letters, numbers, and special characters (punctuations, etc.). This is computationally intense and most unlikely to succeed unless the password is too small. However, brute force attacks might be effective against a poorly designed encryption algorithm.

The best method to prevent password-based hacks is to ensure that the users comply with strong password requirements. Using a good encryption algorithm or hashing algorithm, in conjunction with a minimum password length that is not short, are proven ways to keep attackers at bay. In a corporate environment, password cracking can be prevented by using a well-designed and well-implemented security policy that eliminates easily guessable words. Some guidelines for password policies are as follows:

- They should have a minimum of eight characters.
- They must not contain a username or part of full names.
- They must contain characters from at least three of the four following classes:
  - English uppercase letters A, B, C,...Z
  - English lowercase letters a, b, c,...z
  - Westernized Arabic numerals 0, 1, 2,...9
  - Non-alphanumeric (special characters like $, #, @ symbols), punctuation marks, and other symbols
- They should have expiration dates.
- Passwords cannot be reused.

Although there are alternatives to password-based authentication (such as Kerberos, which relies on tickets, or those based on certificates), further research is necessary for them to become an industry standard. Until then, it is imperative that passwords are properly chosen and policies regarding password settings are strictly enforced to prevent a potential system exploit.

## Web Applications

These have become a highly sophisticated means to acquire access to personal information as more and more software applications are now becoming web-based. Several techniques are used in this form of attack. Account harvesting methods such as phishing and pharming and poorly implemented web applications are common culprits in this form of attack.

Phishing can be defined as the fraudulent means to acquire sensitive personal information such as usernames, passwords, and credit card details through deceptive solicitations using the name of businesses with a web presence. The attacker, posing as a trustworthy source, seeks information from the victims by sending an official-looking email, instant message, and so on, disguising a real need for sensitive information from the user. This is a form of social-engineering attack, where confidential information is obtained by manipulating legitimate users or tricking them to do so against accepted social norms and policies.

Popular targets are users of online banking services and online payment services such as PayPal, online auction sites such as eBay, and popular online consumer shopping websites. Phishers usually work by sending out email spam to a large number of potential victims, directing the user to a web page that appears to belong to the actual website but instead forwards the victims' information to the phisher. The email messages are aptly worded with a subject and message that is intended to make the recipient take immediate actions either by going to the website link (URL) provided or by replying directly to the email. A common approach is to inform the recipients that their account has been deactivated and that to fix the issue they need to provide the correct username and password information. The convenient link provided in the email takes the recipient to a fake website that appears to be from a trustworthy source, and once the user information is entered, the data is forwarded to the attacker.

URL spoofing is a common way to redirect a user to a website that looks authentic. For example, http://www.paypal-secure-login.com might appear to be a reliable domain name associated with the popular online payment service at http://www.paypal.com. In reality, this website might be a spoof, with templates that look identical to the actual PayPal website. Users who enter their login information to this fake website are essentially providing the phisher with the actual login data that they can use to take over the account from the actual website.

Besides URL spoofing, it is also common to provide misspelled URLs or subdomains—for example, http://www.yourbankname.com.spamdomain.net. The user is easily fooled into believing that she is actually interacting with her bank website when in reality all her activity is being tracked by the person who set up the spoof website. Also common are website addresses that contain the @ symbol, similar to http://www.google.com@members.aol.com. Although it seems like a URL to the popular search engine website, the page is actually using a member name called "www.google.com" to login to a server named "members.aol.com." Although such a user does not exist in most cases, the first part of the link looks legitimate, and unless users are cautious, they are redirected to alternative websites where their information is collected and subsequently misused. Figure 4.9 shows an actual email

**Figure 4.9.** Example of a suspect email

received by the author, directing him to visit a particular site. (Note the author did not even have such a condition!) Clicking on the website, though, results in a different message and directs the user to enter Facebook login credentials.

Pharming is a more advanced form of website-based attack, where a DNS server is compromised and the attacker is able to redirect traffic to a popular website to another alternative website, where user login information is then collected. DNS servers are responsible for resolving Internet website names to the IP address of the server that hosts the website, and a pharming attack changes the IP address related to a website to an alternative IP address that is owned temporarily by the attacker. In this type of attack, the alternative website and IP address is maintained by the hacker only for a very short time, since violation of this sort gets noticed very quickly.

However, due to the popularity of websites that are the targets of this form of attack, a short time span usually provides the attacker with large volumes of user login information, most of which will provide detailed credit card information, addresses, and phone numbers of authentic users of the actual website. In addition, the actual website holder may have no means to easily detect those accounts that have been compromised due to randomness of access to the website and the unbounded geographic locations from where the site can be reached.

Web application session hijacking is used by the attacker to take advantage of the weakness in the implementation of a website. Innovative entrepreneurs seeking wealth by taking advantage of the web to reach the masses often set up websites quickly to sell goods and services. Transaction data to and from Internet-based web applications are targets of prying eyes, also with the intent to capitalize on the vulnerabilities of poorly implemented websites. Any monetary or credit card transaction that is not secured via encryption can be easily sniffed by a watching attacker. Sniffing is the process of capturing each packet of data and eventually retrieving valuable information from these packets. Internet eavesdroppers use an assortment of tools to quickly capture relevant and vital data of interest if transactions are not encrypted using protocols designed to securely transmit private data and documents. Secure Socket Layer (SSL) and S-HTTP (Secure HTTP, also known as HTTPS) are examples of protocols that support encryption before data is transmitted via the Internet. Whereas SSL creates a secure connection between the user and the server, over which any amount of data can be sent securely, S-HTTP is designed to transmit data associated with individual web pages securely.

Besides encryption protocols, session tracking mechanisms are used by websites, to ensure privacy by forcing timeouts based on inactive intervals of usage. Improperly implemented session tracking mechanisms can be used by an attacker to hijack the session of a legitimate user. In order to exploit this vulnerability, an attacker establishes a session with the web server by logging in. Once logged in, the attacker tries to determine the session ID of a legitimate user and then change his session ID to a value currently assigned to the actual user. The application is now made to believe that the attacker's session belongs to the legitimate user, and using this exploit, the attacker can do anything a legitimate user can do on the website.

Web application–based attacks have become a serious concern to users, especially since these attacks do not require the attacker to gain direct access to the end user's computer. In the United States, the federal anti-phishing bill (the Anti-Phishing Act of 2005) introduced to Congress proposes that criminals charged with bogus websites and spam emails intended to defraud consumers could be fined up to $250,000 and serve a jail term up to five years. Most software vendors recognize the seriousness of the problem and have joined in efforts to crack down on phishing, pharming, and email spamming.

# New Trends in Network Security

In this section we discuss the new and emerging trends in network security. Some of these are bound to have a significant impact in years to come. Virtualized services and micro-segmentation are transforming the way network security is understood.

## Hypervisor and Virtualized Server Infrastructure

The exponential growth of the information age has provided rapid growth in services such as data housing and archiving, processing, and application interface for the consumers of the information. To provide this access, the traditional data center has become the end point for the collection, processing, and distribution of such services, which in turn has required the growth of physical hardware and housing.

The traditional data center infrastructure (Figure 4.10) is composed of a physical server with an operating system (Windows or Unix/Linux) connecting to the network by its network interface card (NIC) via a physical cable such as a fiber or copper wire, called a "patch cable," to a network switch. From the switch other cables of fiber interconnect into larger switches and then to a "router," which is the main traffic control center. Finally, firewalls, IDS, and proxy servers are selectively used to secure the data centers as traffic enters and leaves. Of note is that the previously discussed use of virtual local area networks (VLANs) are paired with a single range of IP addresses called a subnet. Within the "subnet/vlan," there is no security control available independently, then, that place on the individual servers by the server administrators in the form of access control permissions.

Virtualization of servers (Figure 4.11) was developed to counter the large capital expenditures of duplicating physical servers and networks. A virtualized server infrastructure uses the same conceptual structure as the traditional data center design described previously.

The difference comes from the use of software on specialized equipment called a "hypervisor," which provide software versions for the network cabling, memory, processors, storage, switches, and routers. The virtual server's O/S is loaded onto the hypervisor, which is called "bare metal," as there is only a small Linux kernel operating system on the hypervisor that allows for booting, remote control, and deployment of the virtual server operating system. Due to the capacity of the hypervisor, multiple servers of differentiating operating systems and types like web or files can be placed on the same hypervisor hardware. The virtual

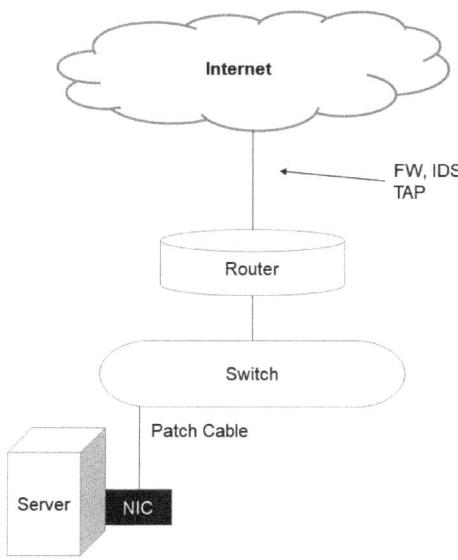

**Figure 4.10.** Traditional data center infrastructure

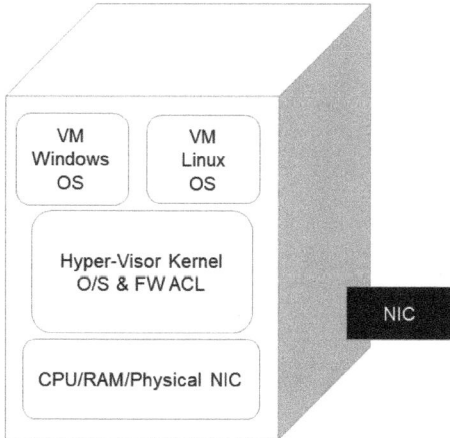

**Figure 4.11.** Virtualized server

server's operating system sends data as it would on a real server, but in place of a physical network card, a virtual network card is defined in the hypervisor kernel and passes the traffic to other virtual servers or out the hypervisors physical NIC to the rest of the network.

The challenge then becomes where to interject security devices into such a virtualized environment. The kernel on the hypervisor is limited in capacity, so additional software can be loaded in the larger user space environment where a hosted operating system would be a stage, and therefore instantiate a virtual appliance on the hypervisor to provide antivirus, IDE, and firewall capabilities. Similarly, built-in capabilities of a virtualized access control list provided by the software run in the kernel space between the hypervisor and virtual servers, which can be used to filter dynamically over many host servers access. This then extends the control from one physical hosting server to a large cluster of physical hosts. Such a distributed network infrastructure limits where a physical antivirus (AV), IDE, or firewall can be inserted; however, leveraging a virtualized appliance within the hypervisor helps security professionals insert network controls and thereby enforce policy on server-to-server as well as client-to-server interactions.

## Micro-segmentation to Secure Virtualized Network Traffic

In data-networking terms, the flow of network connections is split into four directions: North-South and East-West (Figure 4.12). This terminology is a reference to the origin and destination of the connection. When a user on the Internet connects to an internal server through a firewall, this is referred to as "North-South," since the user is coming from above (the Internet) down to a server (internal). When a server connects to another server in the same internal network, it is referred to as "East-West," since the traffic moves from side to side in a network diagram.

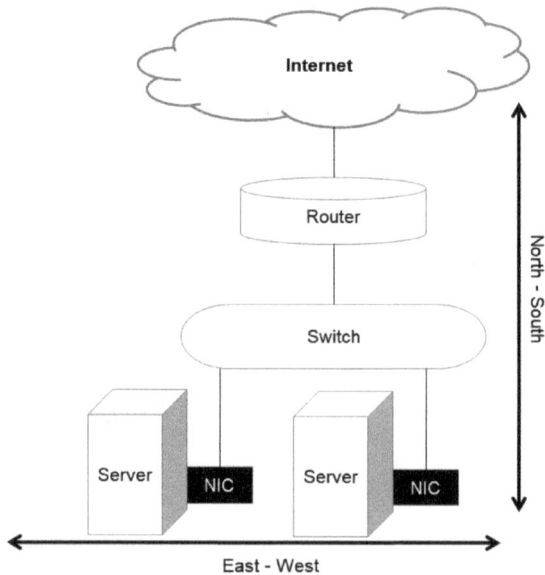

**Figure 4.12.** Four directions of network connections

For the security professional, the traditional focus has been the implementation of control points using firewalls, ACLs, and IDS at the North-South flow of traffics, since the unknown clients on the Internet pose the greatest threat. However, should a server be compromised from an external source and then used as launching point to attack adjacent servers on the same switching infrastructure, this is known as an East-West traffic attack (Figure 4.13).

The issue is that the servers within the same IP subnet address space do not have any protection, as there is no intra-vlan or intra-subnet security filters other than those applied by the individual system administrators as host based access control list. From a security management perspective, this is problematic, as the cooperation of the server administrator will need to be gained, as well as procedures and audits agreed upon to ensure there is consistency with the security policy and its application. For this reason, many organizations look to the use of firewalls to ensure independent controls with more central points of implementation. One traditional method of control is to divide the functionality of the services into a "three tiered" system and provide the public connecting servers (known as "public facing") to be behind their own firewall/load balancing/IDS device. The typical web application is built upon a web, application, and database server system. Each of these servers is separate from each other, with individual subnets and firewalls that limit the type of communication. Client request traffic only interacts with the web server tier, which in turn connects to the application server to request an action. The application server then connects to the database server, which holds the information and replies back to the web server. Finally, the web server responds to the initial client request. In this manner, the client only has direct access to the limited abilities of the web server—without connectivity to the applications or database servers—thus limiting the attack vector of the server system as a whole.

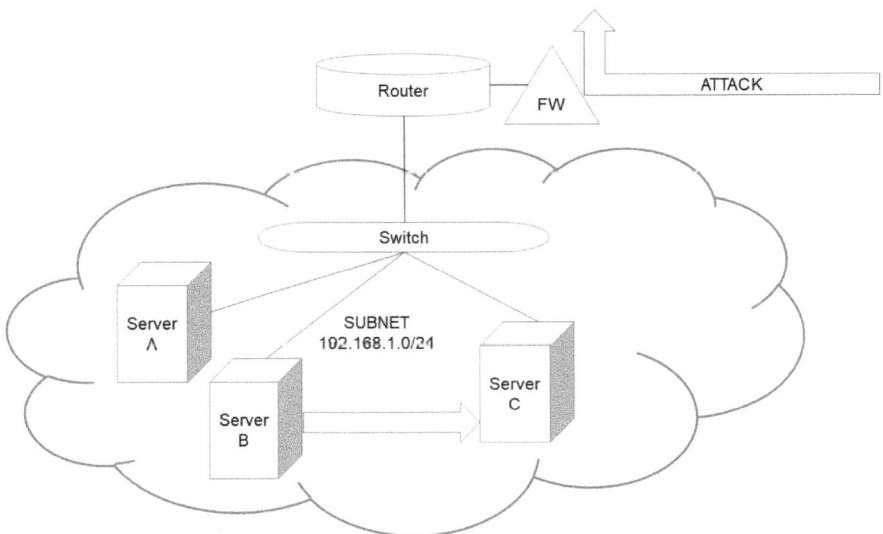

**Figure 4.13.** East-West traffic attack

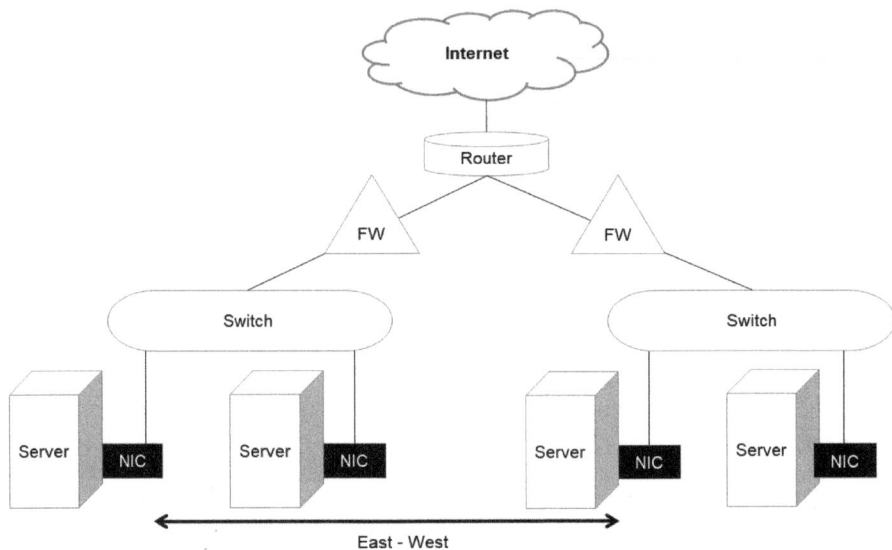

**Figure 4.14.** A possible secure configuration

One enhancement that can be done with server traffic is that of positioning a physical firewall as the next hop connection from the server's subnet to the rest of the network. This method allows for the stateful inspection of inbound and outbound connections to be controlled, as well as a single point of control. In addition, another method to help limit the exposure to other servers from an East-West attack is that of limiting the size of the subnet to support 2, 6, or 14 hosts, thus reducing the vector of attack (Figure 4.14).

Securing East-West traffic was historically difficult, due to the need to make small subnets or implement hardware-based control of the switch ports per connection. With the use of virtualization and the implementation of ACLs between the hypervisor kernel and the hosted virtual O/S, it becomes feasible to control East-West at a much larger scale. The concept is a reiteration of a server-based access control list that filters which external systems are allowed to connect to the server but requires that each individual system administrator coordinate and maintain each individual server's security controls. The security virtualization is known as "micro-segmentation," since the access control list has been pushed down to the virtual servers beyond the limit of the subnet's gateway. By approaching centralized security filtering in this manner, it removes the administrative burden from the server administrators and allows for the implementation of a consistent security policy being applied directly to server-to-server communication, independent of the reliance on multiple server administrators to implement "host"-based access control lists (see Google patents, in particular [4]).

## Cloud-Based Network Security

East-West traffic has increased in usage and importance from a security perspective with the introduction of "cloud"-based server infrastructure (see [1]). The term "cloud" comes from the usage of the cloud network diagram symbol to represent all connections from the internal network, across the Internet, to a remote network. As more and more services were hosted in remote data centers for email, websites, and more, the symbol of a cloud was used to denote the Internet or WAN network. This "cloud" symbol in turn became associated with services based in remote locations, such as large pay-as-you-use data centers—hence "cloud" computing. The growth of third-party hosting applications for enterprise and consumers, such as email, social media, word processing, and file storage, has brought about the connotation that the "cloud" itself is a destination for the technologies discussed and is used for processing, storing, and distributing services.

The growth and development of high-speed wired and cellular wireless networks has allowed cloud-based services to be instantly functional, since the slowness of connection speeds decreased. This explosion of usage means that a user's phone or workstation application may request a web page, and on the cloud data center infrastructure and corresponding East-West traffic may flow from authentications servers to application servers, databases, and even files systems for a single web page request. The advent of the Internet of Things, wherein devices will participate in machine-to-machine communication and uploading of information for big data mining, is also expected to magnify the amount of East-West traffic as well. A cloud data center internally is a vast collection of inexpensive generic servers and switches interconnected with network cabling and attached to large storage arrays. In general, the infrastructure is intended to provide for the massive redundancy of the hardware, as well as redundancy of software applications and file storage. To this end, the security setup does not implement discrete physical firewalls, intrusion prevention, or routers but rather virtualizes the functionality into the software instance running, as is done in with micro-segmentation, explained earlier. Access for clients, management, and server-to-server communication to the virtual servers, firewall, and so on is done via an application programming interface (API) call from software. This allows for the implementation of scripting technologies to automate large deployments, configuration changes, and dynamic reallocation of resources, without the need to physically touch the servers. For the security professional who is working for an entity that has services hosted in the "cloud," there are limits to what is auditable and attainable in specifying the cloud-based security design.

At this point in time, all of the major cloud providers are able to be compliant with the major security standards for both physical and logical security. However, the boundary between the structural setup of the cloud provider and that of the client hosted system or applications will still require review and monitoring from the security professional. At this time, cloud providers do not provide independent monitoring and logging to the detail, which may be required by customers' audit reports. Therefore, with cloud installations, the customer would install their own SIEMs, syslog, and other monitoring virtual appliances. (For a discussion on attack modelling in SIEM systems, see [2].)

## In Brief

- An understanding of TCP/IP protocol architecture is essential if security of networks is to be assured.
- A firewall is a barrier that separates sensitive components from danger. Firewalls can have both software and hardware components. Ensure that your systems are protected with a firewall.
- Viruses are a significant source of security threat. Good housekeeping practices ensure that systems remain protected from viruses and Trojan horses.
- Good sense in ensuring confidentiality of information from accidental and intentional disclosure not only ensures privacy but also protects the information resources available on corporate networks.

## Discussion Questions

These questions are based on a few topics from the chapter and are intentionally designed for a difference of opinion. They can best be used in a classroom or seminar setting.

1. If you are a network administrator and have suspected some hacking activities, suggest steps you would take to investigate the break-in.
2. Create a statement to be distributed to various corporate employees as to what they should do when they receive an email attachment.
3. Differentiate between targeted attacks and target of opportunity attacks.

## Exercise

Imagine that your computer has been infected with adware and you are in a virtualized network. You systematically follow steps to remove the adware, but now other servers are showing the same adware infection. Discuss what may be wrong and what further steps you can possibly take to ensure complete eradication.

## Short Questions

1. Frames use 48-bit _____ addresses to identify the source and destination stations within a network.
2. Thirty-two-bit _____ addresses of the source and destination station are added to the packets in a process called encapsulation.
3. Which Transport layer standard that runs on top of IP networks has no effective error recovery service and is commonly used for broadcasting messages over the network?

4. A _____ is considered the first line of defense in protecting private information and denying access by intruders to a secure system on the internal network.

5. What technique serves the dual purpose of hiding the internal IP addresses of critical systems, as well as allowing multiple hosts on a private internal LAN to access the Internet using a single public IP address?

6. Most common break-ins exploit specific services that are running with _____ configuration settings and are left unattended.

7. What technique can attackers use to identify the kinds of services that are running on the targeted hosts?

8. What type of attack is the most commonly used mode of attack against an operating system?

9. An advanced form of website-based attack where a DNS server is compromised and the attacker is able to redirect traffic of a popular website to another alternative website, where user login information is collected, is called _____.

10. A packet sniffer attached to any network card on the LAN can run in a _____ mode, silently watching all packets and logging the data.

11. A(n) _____ attack relies on malformed messages directed at a target system, with the intention of flooding the victim with as many packets as possible in a short duration of time.

12. A(n) _____ _____ attack uses multiple compromised host systems to participate in attacking a single target or target site, all sending IP address spoofed packets to the same destination system.

13. Computer users should ensure that folders are made network sharable only on a need basis and are _____ whenever they are not required.

14. From a security perspective, it is important that not all user accounts are made a member of the _____ group.

15. An account _____ policy option disables user accounts after a set number of failed login attempts.

# Case Study: The Distributed Denial of Service Attack

A recently distributed denial of service attack (DDoS) against a large DNS service provider showed the weaknesses inherent in the Internet domain name system. On October 21, 2016, during the early morning hours, users on the East Coast found difficulty in connecting with Amazon.com, Wired.com, the *New York Times*, and other websites. The issues stemmed from the inability to lookup the IP addresses of these websites from the large DNS infrastructure company "DYN," located in New Hampshire. DNS is the phonebook for the Internet, allowing computers to change a human readable name such as "Amazon.com," seen in the web browser, to a machine-readable IP address, which in turn is used to connect to the remote server by the local computer. Without the ability to lookup names and convert them to routable IP addresses, the human user would have to enter in the IP address by hand number by number, and thus destroy the usability of the Internet. This attack had three waves: the first at 7 a.m. EST, then early noon, and again at 4 p.m. EST. What was of interest was the size of the waves of attacks numbering in the tens of millions of devices sending numerous connections as well as the sources—webcams and DVRs. What had been unleashed by malicious actors was a botnet virus that targeted the Internet of Things devices (i.e., home security systems webcams, DVRs, and other "things"), which the owners left set to their default passwords. Once the botnet virus, called "Mirai," had spread throughout home user's network, all was ready for the call to attack by the command and control server against any target the hackers wanted. The "DYN" attack and similar outages brings into focus three areas of concern to the security professional: legacy protocols (DNS) that were not designed with security as an integral aspect; the lack of accountability of service providers who transport malware and attack traffic; and finally, lack of responsibility by manufacturers and users to secure devices that have the potential to massively compromise our daily life.

The case is based on Newman, L. H. (2016, October 21). What We Know About Friday's Massive East Coast Internet Outage. Retrieved November 22, 2016, from https://www.wired.com/2016/10/internet-outage-ddos-dns-dyn/.

1. Describe a layered security approach that would prevent such a DDoS attack.
2. What measure could have allowed earlier detection of such an attack from the service provider and home networks?

## References

1. Balus, F.S., et al., *System and method providing distributed virtual routing and switching (DVRS)*. 2016, Google Patents.
2. Kotenko, I. and A. Chechulin. *Common framework for attack modeling and security evaluation in siem systems*. in *Green Computing and Communications (GreenCom), 2012 IEEE international conference on*. 2012. IEEE.
3. Mumford, E., *Systems design - ethical tools for ethical change*. 1996, London: Macmillan Press Ltd.
4. Nordenstam, Y. and J. Tjäder, *System for traffic data evaluation of real network with dynamic routing utilizing virtual network modelling*. 2002, Google Patents.

# PART II

# FORMAL ASPECTS OF INFORMATION SYSTEMS SECURITY

# CHAPTER 5

# Planning for Information System Security

*"Speak English!" said the Eaglet. "I don't know the meaning of half those long words, and what's more, I don't believe you do either!"*

*The Red Queen said, "Now, here, it takes all the running you can do to keep in the same place. If you want to get somewhere else, you must run at least twice as fast as that!"*

—Lewis Carroll,
*Alice in Wonderland*

Over the past several months Joe Dawson had really immersed himself in the subject of security. However, the more he read and talked to the experts, the more confused he became. Clearly, Joe felt, something was not right. After all, managing security should not be that difficult. In terms of his own business, Joe had really worked hard to be at a point where his business was rather successful. So, managing security should not be that tough. He was after all qualified and competent. But on most occasions Joe got lost in the mumbo jumbo of terminology, which made it quite impossible to make sense of anything.

As Joe considered the complexities of security and its implementation in his organization, he was reminded of a book he had read while in the MBA program—*The Deadline* by Tom DeMarco. This was an interesting book that presented principles of project management in a succinct manner. In particular, Joe remembered something about processes for undertaking software development work. He reached out for the book and began flipping through the pages. Aah! It was there on page 115. There were four bullet points on developing processes:

- Model your hunches about the processes that get work done.
- Use the models in peer interaction to communicate and refine thinking about how the process works.
- Use the models to simulate results.
- Tune the models against actual results.

Wasn't the advice given by DeMarco so very true for any organization, any implementation? Joe thought for a moment. Clearly security management was about identifying the right kind of process and ensuring that it works. This means that he had to think proactively about security, plan for it, and have the right process in place. If the business process had been sorted out, wouldn't that result in a high-integrity operation? Wasn't high integrity a cornerstone of good security? It all seemed to fall into place. Maybe he had started at the wrong place by focusing on the technological solutions, Joe thought. Maybe security was not about technology at all. At this point Joe was interrupted by a phone call.

It was Steve, who lived four houses down the lane. Both Steve and Joe were Lakers fans. After discussing Lakers strategies to get back the key position player, Steve asked, "Do you have a wireless router?" "Yes," said Joe. "I think I am picking up your signal. You need to secure it." After all, understanding technology was important, Joe thought instantly. Joe did not know how to fix the problem. And Steve volunteered to come over and help Joe out.

Although the wireless router problem got resolved, Joe was still uncomfortable with strategizing about security at SureSteel. Should he sit down with his technology folks and write a security policy? Should he simply let the policy emerge in a few years? How was he going to deal with other companies and assure that his systems were good enough? Was there any need to do so? These were all very true and genuine questions. Joe understood the nature and significance of these questions. He had been formally trained to appreciate and deal with these issues. Joe knew that these issues and concerns were indeed the building blocks of the strategy process. Henry Mintzberg had written a wonderful article in *California Management Review* in 1987, which Joe remembered and knew was still relevant. Mintzberg had conceptualized strategy in terms of five Ps—plans, ploys, patterns, positions, perspectives.

Strategy as a *plan* is some sort of a consciously intended course of action. Joe knew that any security plan he initiated at SureSteel was going to be formally articulated. Strategy could also be a *ploy*. In this case specific "maneuvers" to outwit opponents trying to impregnate SureSteel systems would have to be developed. Joe remembered his computer geek friend Randy, who had said that maintaining security of systems is like a "Doberman awaiting intruders." In many ways, Joe thought, a ploy is a deterrent strategy.

Joe was also responsive to Mintzberg's conceptualization of strategy as a *pattern*. Clearly it is virtually impossible for SureSteel to identify and establish countermeasures for all possible threats. What Joe had to do was identify patterns that might exist as the organization went about doing its daily business. The dominant patterns that might emerge would form the basis for any further learning and strategizing that Joe might be involved in.

In his pursuit to achieve a good strategic vision for SureSteel, Joe did not want to create security plans that would hinder the job of his employees. After all, security is a key enabler for running a business smoothly. Such a conception suggests that strategy is a *position*—a position between the organization and the context, and between the day-to-day activities of the company and its environment. Clearly a *perspective* has to be developed. A strategic security perspective would allow for a security culture to be developed, allowing all employees to think alike in terms of maintaining security.

Such thinking was helping Joe to consider multiple facets of security. All he needed was a means to articulate and structure the thinking.

Formal IS security is about creating organizational structures and processes to ensure security and integrity. Since organizing is essentially an information handling activity, it is important to ensure that the proper responsibility structures are created and sustained, integrity of the roles is maintained, and adequate business processes are created and their integrity established. Furthermore, an overarching strategy and policy needs to be established. Such a policy ensures that the organization and its activities stay on course.

Various IS security academics and practitioners have identified the need to understand formal IS security issues. The call for establishing organizationally grounded security policies has been made by numerous researchers and practitioners. One of the earlier papers to make such a call, and published in *Computers & Security*, was in 1982. Entitled "Developing a Computer Security and Control Strategy," the paper focused on establishing appropriate security strategies. The author, William Perry, argues that a computer security and control strategy is a function of establishing rules for accessibility of data, processes for sharing business systems, and adequate system development practices and processes. Perry also identifies other issues such as competence of people, data interdependence rules, etc. [15]. The arguments proposed in his article are indeed relevant even today.

More than two decades later, in another interesting paper published in *Computers & Security*, Basie von Solms and Rossouw von Solms presented the 10 deadly sins of information system security management [18]. Central to their argument is the importance of structures, processes, governance, and policy. The authors note that information system security is a business issue, and security problems cannot be dealt with by just adopting a technical perspective. Therefore although it's important to establish system access criteria and technical means to secure systems, these will perhaps not work or will fail if adequate organizational structures have not been put in place. A summary of the 10 deadly sins postulated by Solms and Solms appears in Table 5.1.

The 10 deadly sins identified by Solms and Solms essentially suggest four classes of formal IS security issues. These include:

- Security strategy and policy—development of a security strategy and policy that would determine the manner in which administrative aspects of IS security are managed.
- Responsibility and authority structures—a definition of organizational structures and how subordinates report to superiors. Such a definition helps in establishing access rules to systems.
- Business processes—defining the formal information flows in the organization. Information flows have to match the business processes in order to ensure integrity of the operations.
- Roles and skills—identifying and retaining the right kind of people in organizations is as important as defining the security policy, structures, and processes.

**Table 5.1.** 10 deadly sins of IS security management [18]

| Deadly sins of information system security |
|---|
| 1. Not realizing that information security is a corporate governance responsibility (the buck stops right at the top) |
| 2. Not realizing that information security is a business issue and not a technical issue |
| 3. Not realizing the fact that information security governance is a multi-dimensional discipline (information security governance is a complex issue, and there is no silver bullet or single "off the shelf" solution) |
| 4. Not realizing that an information security plan must be based on identified risks |
| 5. Not realizing (and leveraging) the important role of international best practices for information security management |
| 6. Not realizing that a corporate information security policy is absolutely essential |
| 7. Not realizing that information security compliance enforcement and monitoring is absolutely essential |
| 8. Not realizing that a proper information security governance structure (organization) is absolutely essential |
| 9. Not realizing the core importance of information security awareness amongst users |
| 10. Not empowering information security managers with the infrastructure, tools, and supporting mechanisms to properly perform their responsibilities |

# Formal IS Security Dimensions

There can possibly be a number of security dimensions in the formal parts of an organization. These are identified and discussed below.

## Responsibility and Authority Structures

When responsibility and authority structures are ill-defined or not defined at all, it results in a breakdown of the formal controls systems. Clearly adopting adequate structures will go a long way in establishing good management practices and will set the scene for effective computer crime management. The notion of structures of responsibility goes beyond the narrowly focused concerns of specifying an appropriate organizational structure. Although important, exclusive focus on organizational structure issues tends to skew the emphasis towards formal specification. In a 1996 paper, Backhouse and Dhillon [3] introduced the concept of structures of responsibility to the information system security literature. They suggest that responsibility structures provide a means to understand the manner in which responsible agents are identified within the context of the formal and informal organizational environments. The most important element of interpreting structures of responsibility is the ability to understand the underlying patterns of behavior.

Backhouse and Dhillon evaluate issues and concerns related to interpreting structures of responsibility. Their inherent argument is that the organizational structures are manifestations of the roles and reporting structures of organizational members. It is for this reason that any organizational structure should be modeled on the basis of understanding the communication necessary for achieving a coordinated action. After all, an organizational structure is a means to achieve a coordinated outcome. With respect to security, Backhouse and Dhillon argue that understanding the communication flows and identifying places where the communication might break down allows an assessment of IS security. They further argue that usually security problems are a consequence of communication breakdowns and lack of understanding of behaviors of various stakeholders. The structures of responsibility provide a means to understand the manner in which responsible agents are identified; the formal and informal environments in which they exist; the influences they are subjected to; the range of conduct open to them; the manner in which they signify the occurrence of events; the communications they enter into; and, above all, the underlying patterns of behavior.

---

### Box 5.1. Data Stewardship

One way in which responsibility and authority structures have been manifested in organizations is through the role of a data steward. Many companies now have this role as part of their organizational structure.

The data steward role is primarily responsible for data quality. Although uncommon in smaller organizations, most large companies are recognizing the importance of data quality and ensuring that someone in the company is in charge. Although there is a movement in the industry that emphasizes the importance of data quality, there are many organizations that essentially dwell within the "find-and-fix" paradigm, usually undertaken after a project has been completed. A typical corporate data stewardship role may have one data steward assigned to each major data subject area. Such individuals may not carry the title of a data steward, but may nevertheless perform the functions typical of maintaining data quality.

Typically, data stewards manage data assets to improve reusability, accessibility, and quality. They may be involved in data naming standards and developing consistent data definitions for documenting business rules. So, in one sense a role such as that of a data steward manifests itself by enforcing organizational hierarchy and business rules. This is an important aspect of this role since it ensures integrity of the operations.

Jonathan Geiger in an article in *Teradata Magazine* (September 2008), summarizes the requirements of a data steward as follows:

- **Business knowledge:** Data stewards must understand the business direction, processes, rules, requirements, and deficiencies.
- **Business-area respect:** They need to influence business decisions and gain business-area commitments.
- **Analysis:** When faced with multiple options, they must examine situations from many angles.
- **Facilitation and negotiation:** They must facilitate the proponents of conflicting viewpoints to arrive at a mutually satisfactory solution.
- **Communication:** Stewards need to effectively convey the business rules and definitions and promote them with the business areas as well.

## Organizational Buy-In

The effectiveness of the security policy is a function of the level of support it has from an organization's executive leadership. Although this may sound obvious, it is indeed the most challenging task. A related challenge is that of educating the employees. It is easier to harden the operating systems and undertake virus scans than it is to communicate security policy tenets to various stakeholders.

There is a two-fold need for **executive leadership buy-in**. First, it assures staff buy-in. When the executive leadership visibly validates the security policy and procedures, it becomes easier to "sell" the program to organizational staff members. If however the executive leadership does not support the policies and procedures, it becomes difficult, if not impossible, to convince the rest of the organization to adopt the security policy. Second, executive leadership buy-in ensures funding for a comprehensive IS security program.

Support from the **IT department** for the security policy and procedures is also essential. Consensus needs to be reached regarding the best practices to protect enterprise information assets. There usually are more than one means to establish such protective mechanisms. While it may be important to acknowledge the importance of different approaches, it is equally important to identify the best possible way to achieve security objectives. If the debates on the best possible course of action continue, it can then become detrimental to the overall success of the security policy. If departments themselves cannot agree on the best possible course of action, then support from non-technical staff becomes difficult as well.

**User support** is another important ingredient. User support resides in the people throughout the organization and represents a critical functional layer that could be rather useful in the overall defense strategy. A strategy of "locks and keys" becomes inadequate if people inside the organization open those locks (i.e. subvert the controls). Once the organizational shortcomings have been identified, the next step is to establish an education and training program.

The National Institute of Standards and Technology (NIST) in their document NIST 800-14 ("Generally Accepted Principles and Practices for Securing Information Technology Systems") prescribes the following seven steps to be followed for effective security training:

1. **Identify program scope, goals, and objectives**. The scope of the program should provide training to all types of people who interact with IT systems. Since users need training which relates directly to their use of particular systems, a large organization-wide program needs to be supplemented by more system-specific programs.
2. **Identify training staff**. It is important that trainers have sufficient knowledge of security issues, principles, and techniques. It is also vital that they know how to communicate information and ideas effectively.
3. **Identify target audiences**. Not everyone needs the same degree or type of security information to do their jobs. A computer security awareness and training program that distinguishes between groups of people, presents only the information needed by the particular audience, and omits irrelevant information, will have the best results.

4. **Motivate management and employees.** To successfully implement an awareness and training program, it is important to gain the support of management and employees. Consider using motivational techniques to show management and employees how their participation in a security and awareness program will benefit the organization.
5. **Administer the program.** Several important considerations for administering the program include visibility, selection of appropriate training methods, topics, materials, and presentation techniques.
6. **Maintain the program.** Efforts should be made to keep abreast of changes in computer technology and security requirements. A training program that meets an organization's needs today may become ineffective when the organization starts to use a new application or changes its environment, such as by connecting to the Internet.
7. **Evaluate the program.** An evaluation should attempt to ascertain how much information is retained, to what extent security procedures are being followed, and general attitudes toward security.

## Security Policy

It goes without saying that a proper security policy needs to be in place. Numerous security problems have been attributed to the lack of a security policy. Possible vulnerabilities related to security policies occur at three levels—policy development, policy implementation, policy reinterpretation. Vulnerabilities at the **policy development** level exist because of a flawed orientation in understanding the range of actual threats that might exist. As will be discussed later in the chapter, security policy formulation that does not consider the organizational vision or is developed in a vacuum often results in it not being adopted. Such policies cause more harm than good. Clearly organizations tend to have a false sense of security due to the policy, thus resulting in ignoring or bypassing the most obvious controls.

Some fundamental issues that could possibly be considered in good security policy formulation include:

1. The strategic direction of the company needs to be incorporated at both micro and macro levels.
2. Clarification of the strategic agenda sets the stage for developing the security model. Such a model identifies the relationship between the business areas and the security policies for that business area.
3. The security policies determine the processes and techniques required to provide the security but not the technology.
4. The implementation of security policies entails the development of procedures to implement the techniques defined in the security policies. The implementation stage defines the nature and scope of the technology to be used.
5. Following the implementation there is a constant need to monitor the security processes and techniques. This enables checks to be made to ascertain effectiveness at three levels: policy, procedure, and implementation. In particular, an assessment

is made of the uptake of the security policies; implementation of procedures; detection of breach of procedures. Monitoring also includes assessment and reassessment to ensure that procedures match the original requirements.
6. A response policy is also an integral part of a good security policy. It pre-empts a security failure and determines the impact of a failure at policy, procedure, implementation, or monitoring levels. It is essentially the security breach risk register.
7. Finally, a program is needed to establish procedures and practices for educating and making all stakeholders aware of the importance of security. Staff and users also need to be trained on methods to identify new threats. In the current changing business environment new vulnerabilities constantly keep emerging and it's important to have the requisite competence to identify and manage them.

An important aspect of the security model is the layered approach. One cannot begin working on any layer without having taken certain prerequisite steps. The design of formal IS security can best be illustrated as layers, shown in Figure 5.1.

To summarize, at a formal level we consider IS security to be realized through maintaining good structures of responsibility and authority. Organizational buy-in and ownership of security by top management are the key success factors in ensuring security. Finally, the importance of security policies cannot be underestimated. A general framework for conceptualizing security policies incorporates three interrelated considerations:

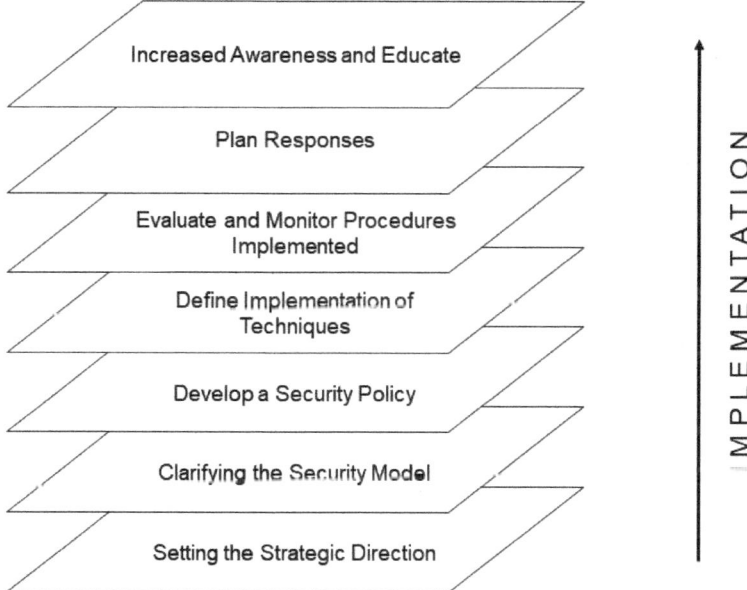

**Figure 5.1.** Layers in designing formal IS security

- Organizational considerations related to structures of responsibility for information system security
- Ensuring organizational buy-in for the information system security program
- Establishing security plans and policies and relating them to the organizational vision

## Security Strategy Levels

There is often confusion between the various terms—strategy, policy, programs, and operating procedures. The term *strategy* is used to refer to managerial processes such as planning and control, defining the mission and purpose, identification of resources, critical success factors, etc. Corporate strategy has been considered as the primary means to cope with the environmental changes that an organization faces. It is often considered as a *set of policies* which guides the scope and direction of an organization. However, there is much confusion between what is designated as a policy and what as strategy. Ansoff [1] (p. 114) traces the origin of the term *strategy* to "military art, where it is a broad, rather vaguely defined, 'grand' concept of a military campaign for *application* of large-scale forces against the enemy." In business management practices, the term *policy* was in use long before *strategy* but the two are often used interchangeably, despite having very different meanings. In practice, a policy refers to a contingent decision.[1] Therefore, implementing a policy can be delegated, while for implementing a strategy executive judgment is required. The term *program* is generally used for a time-phased action sequence that guides and coordinates various operations. If any action is repetitive and the outcome is predetermined, the term *standing operating procedure* is used.

In the realm of IS security, it is important to differentiate between these terms since there is much confusion. Clearly there needs to be a strategy as to what needs to be done with respect to security. Such a strategy should determine the policies and procedures. However, in practice rarely is a strategy for security created. Most emphasis is placed on policies, implementation of which is generally relegated to the lowest levels, and it is simply assumed that most people will follow the policy that is created.

If we accept that secure information systems enable the smooth running of an enterprise, then what determines the ability of a firm to protect its resources? There are two routes (Figure 5.2). Either a firm considers security as a strategic issue and hence operates in an environment designed to maintain consistency and coherence in its business objectives, or, a firm may position itself such that it gains advantage in terms of the risks afforded by the environment. This has traditionally been achieved by performing a risk analysis. This demarcation identifies two levels of a strategy within an organization: the corporate and the business level.

At a corporate level the security strategy determines key decisions regarding investment, divestment, diversification, and integration of computing resources in line with other business objectives. The primary concern here is to take decisions regarding the nature and

---

[1] The *Oxford English Dictionary* defines policy as: "prudent conduct, sagacity; course or general plan of action (to be) adopted by government, party, person, etc." In business terms "policy" denotes specific responses to specific repetitive situations. Typical examples of such usage are: "educational refund policy," "policy for evaluating inventories," etc.

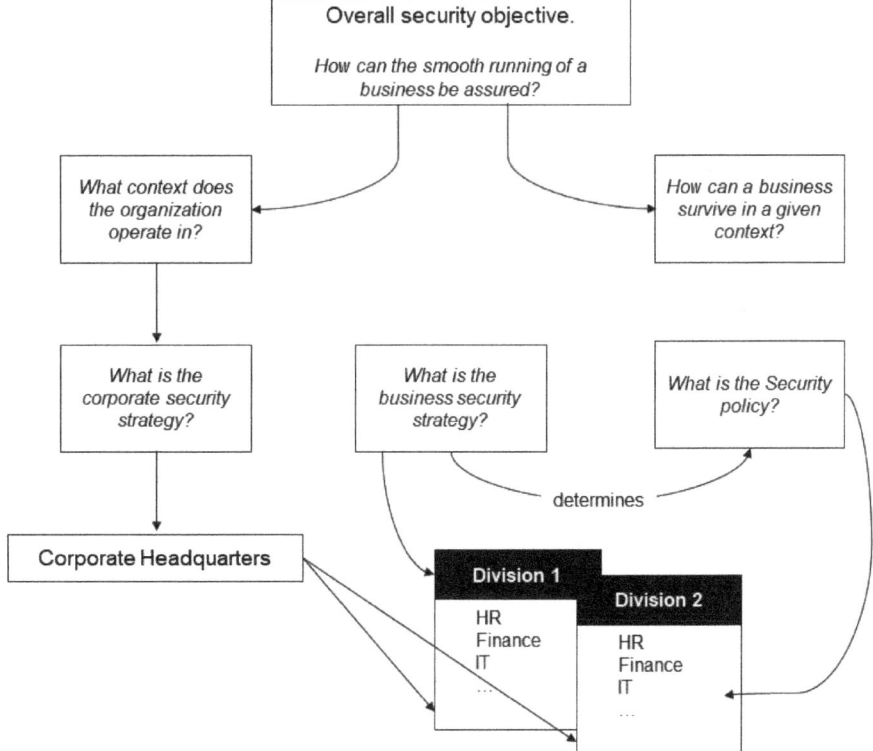

**Figure 5.2.** Strategy levels

scope of computerization. At a business level, the security strategy looks into the threats and weaknesses of the IT infrastructure. In the security literature many of these issues have been studied under the banner of risk analysis. The manner in which risk analysis is conducted is a subject of much debate, as are the implementation aspects. While a business security strategy defines the overall approach to gain advantage from the environment, the detailed deployment of the procedures at the operational level is an issue of concern for functional strategies (i.e. the security policy). These functional strategies may either specifically target major organizational activities such as marketing, legal, personnel, finance, or may be more generic and consider all administrative elements.

Most of the existing research into security considers that policies are the *sine qua non* of well-managed secure organizations. However, it has been contended that "good managers don't make policy decisions" ([19]; p. 32). This avoids the danger of managers being trapped in arbitrating disputes arising out of stated policies rather than moving the organization forward. This does not mean that organizations should not have any security policies sketching out specific procedures. Rather, the emphasis should be on developing a broad security vision that brings the issue of security to center stage and binds it to the organizational objectives. Traditionally, security policies have ignored the development of such a vision, and instead a rationalistic approach has been taken which assumes either a condition of partial ignorance or a condition of risk and uncertainty. Partial ignorance occurs when alternatives cannot be arranged and examined in advance. A condition of risk presents alternatives that are known along with their probabilities. Under uncertainty, alternatives may be known but not the probabilities. Such a viewpoint forces us to measure the probability of occurrence of events. Policies formulated on this basis lack consistency with the organizational purpose.

# Classes of Security Decisions in Firms

One of the problems, as discussed in the previous section, relates to relegating IS security decisions to the operational levels of the firm. Inadvertently, this results in lack of ownership by the top management. In most cases this means that the senior management adopts a "hands off" attitude. Such an attitude might work if a firm has little dependence on IT systems. In the current business environment this is rarely the case though. For this reason it is prudent to differentiate between different classes of IS security decisions and to award adequate importance to each of the classes at the relevant levels. Different classes of IS security decisions are presented in Table 5.2 and discussed in the paragraphs below. It should however be noted that these classes are not mutually exclusive. There is obviously a fair amount of overlap, with the core purpose of the security decisions relating to *configuring and directing the resource conversion process so as to optimize attainment of objectives*. Clearly the main objective with respect to security is to create an environment where there is no scope for abusing the systems and processes.

## Strategic Decisions

One of the fundamental problems with respect to security is for a firm to choose the right kind of an environment to function in. Strategic security issues, therefore, relate to where the firm chooses to do its business. If a given firm chooses to set up headquarters in a war-ravaged environment, clearly there will be increased threat to physical security. Or, if a firm chooses to be headquartered in an environment where bribery is rampart, it increases the chances that company executives will engage in unethical acts, which at times may result in subverting existing control structures. Strategic decisions for security can also relate to the nature and scope of a firm's relationship to other firms and the contexts within which it might choose to operate. For instance, if a firm chooses to integrate its enterprise systems with a US-based firm, it clearly will have to ensure compliance with corporate governance principles as mandated by the Sarbanes-Oxley Act of 2001. Furthermore, any change to an existing business process will have legal implications for either of the partners.

Allocation of resources among competing needs therefore becomes a critical problem in terms of strategizing about security. Apart from high-level corporate governance and firm location issues, which no doubt are important, issues such as return on investment in security products and services become important. IT directors will have to ask some very fundamental questions, such as: *Are investments in security products and services paying off?* Addressing this issue would have a range of implications for success in ensuring security. Today many managers are asking this question [6]. Indeed, over the past few years investments in security have been going up, but so have the number and range of security breaches. This would mean that perhaps the security mechanisms are not working. Or maybe the

**Table 5.2.** IS Security decision classes

|  | **Strategic** | **Administrative** | **Operational** |
|---|---|---|---|
| Problem | To select an environment that ensures the smooth running of the business | To create adequate structures and processes to realize adequate information handling | To optimize work patterns for efficiency gains |
| Nature of the problem | Allocation of resources among competing needs | Organization, structuring, and realization | Ensuring business process integrity Scheduling resource application Supervision and control |
| Key decisions | Setting security objectives and goals Resource allocation for security strategy Infrastructure expansion strategy Research and development for future operations | Organizational: Structure of information flows; authority and responsibility structures Structure of resource conversions: Establishing high-integrity business processes Resource acquisition: Financing security operations; return on security investments; facility management | Identifying operating objectives and goals Costing security initiatives Operational control strategies Policies and operating procedures for various functions |
| Key characteristic | Decisions generally centralized Generally partial ignorance of actual operations and challenges Non-repetitive decisions | Balancing conflicting demands of strategy and operations Conflicts between individual and group objectives Decisions generally triggered by strategic or operating problems | Decentralized decisions Known risks Repetitive problems and decisions Suboptimization because of inherent complexity |

security investments are being made in the wrong places. It could also be that the benefit of a security investment is intangible and that it is rather difficult to link a tangible investment in security to a tangible benefit. After all, most security-related investments are triggered by fear, uncertainty, and doubt [16].

Whatever may be the reasons for lack of security investment payoff, it is important that the key decisions about security objectives be identified. Indeed, this is where the problem with security payoffs resides. While many organizations have engaged in identifying security issues and have created relevant security policies, there is a clear mismatch between what the policy mandates and what is done in practice. Researchers have termed this as a gap in "espoused theory" and "theory-in-use" [13]. Espoused theories are the actions that people write, while theories-in-use are what people actually do. Theories-in-use therefore have different degrees of effectiveness, which are learned.

Espoused theories and theories-in-use are part of the double loop learning concept (see Figure 5.3), which creates a mindset that consciously seeks out security problems, in order to resolve them. The double loop mindset results in changing the underlying governing variables, policies, and assumptions of either the individual or the organization. Fiol and Lyles [10] classify higher-level organization learning as a double loop process, yielding organizational characteristics such as acceptance of non-routine managerial behavior, insightfulness, and heuristics behavior. In contrast, the single loop mindset ignores any security contradictions. One reason is that the blindness is designed by the mental program that keeps us unaware. We are blind to the counterproductive features of our security actions. This blindness is mostly about the production of an action, rather than the consequences of the actions. That is why we sometimes truly do not know how we let something happen. Thus, organizations exhibiting single loop security exhibit minimal, if any, security contradictions in their underlying governing values, variables, policies, or assumptions, and the mindset that Fiol and Lyles classify as lower-level organization learning yields organizational characteristics such as rules and routine.

When using the double loop learning security framework (Figure 5.3), assumptions underlying current espoused theories and theories-in-use are questioned and hypotheses about their behavior are tested publicly. The double loop is significantly different from the inquiry characteristics of single loop learning. To begin with, the organization must become aware of the security conflict. It must identify the actions that have produced unexpected outcomes—a mismatch (error), a surprise. They must reflect upon the surprise to the point where they become aware that they cannot deal with it adequately by doing better what they already know how to do. They must become aware that they cannot correct the error by using the established security controls more efficiently under the existing conditions. It is important to discover what conflict is causing the error and then undertake the inquiry that resolves the security conflict. In such a process, the restructured governing variables become inscribed in the espoused theories. Consequently, this allows the espoused theories and theories-in-use to become congruent and thus more susceptible to effective security realization.

In summary, the proposed double loop security design (Figure 5.4) has four basic steps: (1) discovery of espoused and theory-in-use; (2) bringing these two into congruence, inventing new governing variables; (3) generation of new actions; and (4) generalization of consequences into an organizational match.

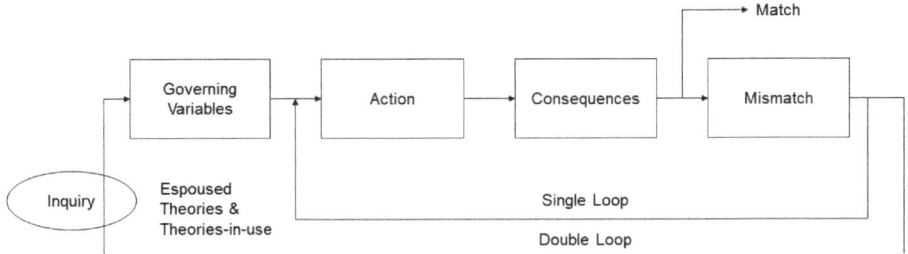

**Figure 5.3.** Double loop learning

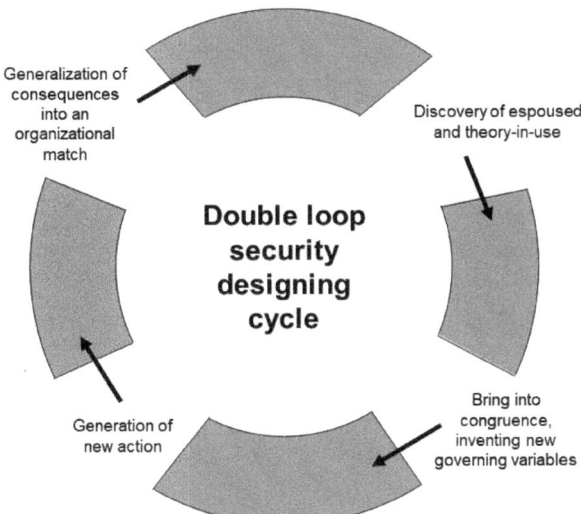

**Figure 5.4.** Double loop security design process

## Administrative Decisions

An understanding of a range of strategic aspects of IS security is clearly an important aspect. Equally important, if not more, is an understanding of structures and processes that should be created to adequately deal with information handling. Inability to properly design structures and processes can be a major reason why many security breaches take place. Usually, design of structures and processes is considered to be beyond the realm of traditional IS security. However, as stated previously, structures and processes are increasingly becoming more central to planning and organizing for security (*cf.* consequences of the Sarbanes-Oxley Act).

One of the key decisions with respect to structures and processes relates to responsibility and authority. It goes without saying that any organization needs to have in place a process for doing things. If the substantive task at hand is *order fulfillment* for example, then a business process needs to be created that identifies a range of activities that will be undertaken when the first order comes in. Each of the activities will have information flows associated with it. It is prudent to not only map all the information flows, but also undertake consistency and integrity checks. Obviously redundancy has to be taken care of. Traditionally, mapping of information flows and integrity checks have been done whenever new computer-based systems have been developed. This activity has taken the form of drawing data flows, establishing entity relationships, etc. Use of data flows and entity relationships is not (and should not be) restricted to design and development of new IT solutions. In fact, it is an important task that all organizations should undertake.

While creating processes, associated organizational structures also need to be designed. At the confluence of the processes and structures reside the responsibilities and authorities. In large organizations it is rather difficult to balance the process and structure aspects. As a consequence, responsibilities and authorities never get defined properly. Even if they are, delineation of responsibilities and authorities is invariably not undertaken. Subsequently, when computer-based systems are developed, ill-defined responsibilities and authorities often get reflected in the system. There are two reasons for this. First, the individuals who are usually interviewed by the analysts to assess system requirements occupy roles that have been ill defined. Second, even though the system developed imposes certain structures of responsibility, the mix of business processes and structures is not geared to deal with it.

The issue at hand has been well discussed and studied by a number of researchers and practitioners alike. The cases of Daiwa and Barings Bank have been well researched [11-12, 17]. Dual responsibility and authority structures and their subsequent abuse by Nick Lesson of Barings Bank is an excellent case in point, as is the 2008 Société Générale fiasco because of fraudulent transactions created by Jérôme Kerviel. As has been well documented, Lesson was able to subvert the controls because the structures had not been well defined in the latest round of changes. A similar situation brought about the demise of Daiwa Bank and Kidder Peabody [8].

Decisions related to formal administrative controls deal with establishing adequate business structures and processes so as to maintain high-integrity data flow and the general conduct of the business. Establishing adequate processes also ensures compliance with regulatory bodies, organizational rules and policies. Therefore, it goes without saying, good business processes and structures ensure the safe running of the business and prevent crime from taking place. Clearly, mature organizations have well-established and institutionalized processes and newer enterprises have to engage in the process of innovation and institutionalization. To a large extent high-integrity processes are a consequence of adequate planning and policy implementation.

Some aspects that Dhillon and Moores [8] recommend as immediate steps to ensure that proper responsibilities and authorities are established include:

- Setting standards for proper business conduct
- Monitoring employees to detect deviations from standards
- Implementing risk management procedures to reduce the opportunities for things to go wrong
- Implementing rigorous employee training, instituting individual accountability for misconduct

Another aspect related to administrative decisions is that of facilities management. While most of the organizational security resources get directed to protecting the logical aspects, little consideration is given to the physical aspects and general facilities management. In a study of security infrastructure at a UK-based local authority, Dhillon [5] found that there were absolutely no physical access controls to the server rooms. This was in spite of a significant thrust made toward security. In presenting the findings of the study Dhillon observed:

> The hub of the IT department of the Local Council is the Networking and Information Centre. This Centre has a Help Desk for the convenience of the users. At present there is no physical access control to prevent unauthorised access to the Help Desk area and to the Networking and Information Centre. The file servers and network monitoring equipment remains unprotected even when the area is unoccupied. In fact the file servers and network monitoring equipment throughout the Council should be kept in physically secure environments, preferably locked in a cabinet or secure area. This would prevent theft or deliberate damage to the hardware, application software or data on the network. Access to the Help Desk area can typically be restricted by a keypad. The auditors had identified these basic security gaps, but concrete actions are still awaited.

Such behavior on the part of organizations is indeed very common. Lack of consideration of the basic hygiene and simplest controls is often overlooked. To some extent this can be tied back to issues of—Who is responsible? Who has authority? Who is accountable?

## Operational Decisions

In most firms there are a myriad of operational problems that require immediate attention. To a large extent such problems can be managed if initial design of work patterns and activities is done with care. This would ensure significant efficiency gains. However, such detailed work flow analysis and review is rarely done. As a consequence, small operational problems end up affecting the administrative and strategic levels as well. Since there is little flexibility and authority in the hands of operational staff, any problem automatically becomes an issue for the top management. The volume of such problems is usually great, essentially because of the need for daily supervision and control.

Although staff at the operational level cannot and should not be given authority to "tweak" the business processes, it is prudent for the higher management to take some key decisions related to identifying operational goals and objectives. If the goals and objectives are clarified, it pretty much sets the stage for establishing operational control strategies and policies and procedures for various functional divisions. Careful planning and establishing proper checks and balances are perhaps the cheapest of the operational level security practices. Once the design for various procedures has been adequately undertaken, it helps in identifying the range of relevant security initiatives.

The premise on which operation decisions are taken is based on classic probability theory. Most of the risks that the operations of the business might be subjected to are usually known. Hence, there is usually a good idea of the cost associated with the risks. Therefore, it becomes relatively easy to calculate the level of overall risk given the following equation:

$$R = P \times C$$

where R is the risk, P the probability of occurrence of an event, and C the cost if the event were to take place.

## Prioritizing Decisions

The balance between strategic and operating decisions is to a large extent determined by a firm's environment. However, there is a need to identify a broad range of objectives, both strategic and operational. Dhillon and Torkzadeh [9] undertook an extensive study of values of managers in a broad spectrum of firms. Their findings identified 25 classes of objectives for IS security. Dhillon and Torkzadeh concluded that although it is possible to classify the objectives into fundamental and means, it is rather difficult to rank them. Although tools and techniques, such as Analytical Hierarchical Modeling, are available to rank the objectives, any ranking would still be context specific. Figure 5.5 presents a network of means and fundamental objectives.

The fundamental objectives are ultimately the ones that any organization should aspire to achieve. These are also the high-level objectives that should form the basis for developing any security policy. Failure to do so will result in policies that do not necessarily relate to organizational reality. As a consequence, there is confusion in the means of achieving the security objectives. For instance, there are always calls for increasing awareness among organizational members. However, in practice there may be a lack of proper communication channels. Furthermore, responsibility and accountability structures may not exist. This results in awareness programs becoming virtually ineffective.

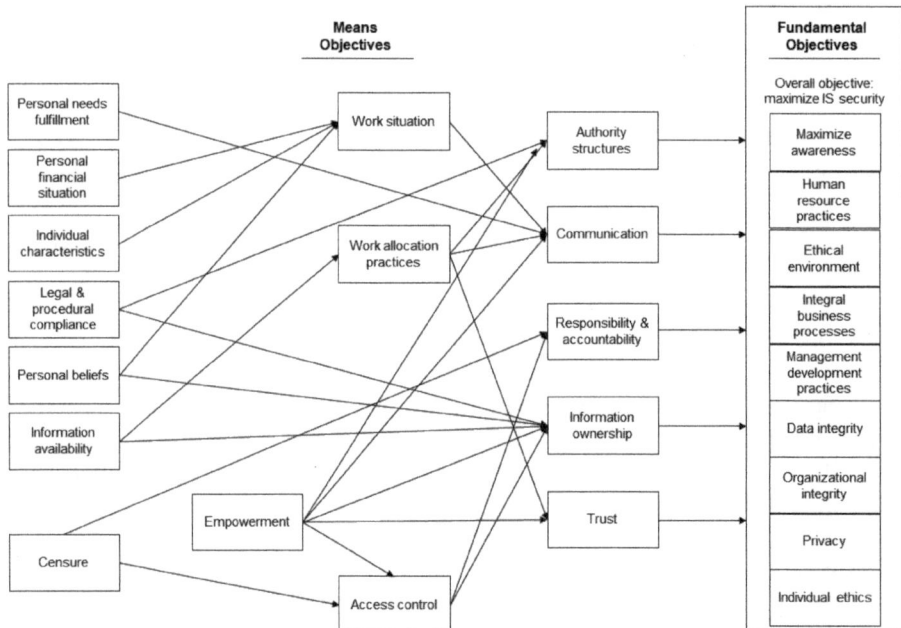

**Figure 5.5.** A network of IS Security means and fundamental objectives

Similarly, it may be difficult to realize any of the fundamental objectives if the appropriate means of achieving them have not been clarified. Data integrity, for example, cannot be maintained if ownership of information has not been worked out. Maintaining integrity of business processes is a function of adequate responsibility and accountability structures. Both means and fundamental objectives are a means to ensure that the espoused theory of IS security and the theory-in-use are congruent. In many ways, properly identifying and following the objectives ensures that double loop learning takes place.

## Security Planning Process

The importance of planning for IS security cannot be overstated. Clearly, there is a need to systematically identify and address a range of performance gaps. Such gaps might exist in setting objectives or in establishing mechanisms for their implementation. Whatever the nature and scope of the performance gaps, there is a need to establish a method that will guide project managers, systems developers, and security analysts in ensuring that proper security is built into the organization. Usually, the primary challenge in any IS security plan is that of stakeholder involvement. This is a rather critical aspect of any security strategy. Past research has shown that a lack of understanding of what the stakeholders want is perhaps a main reason why various security policies do not get accepted in organizations.

A useful way to conceptualize security is to use Peter Checkland's Soft System Methodology (SSM) [4]. One of the core tenets of SSM is to think of the ideal situation (*systems thinking*) and the real-world situation (*real-world thinking*). This thinking is always done by people actually involved in the work situation. In the first instance, a rich picture of the problem situation, with all its complexity, is developed. Perceptions of all stakeholders are captured. This is followed by building models of the real situation. Due consideration is given to the organizational goals and objectives. The conceptual models are then compared with the problem situation. This results in developing feasible and desirable changes that might be necessary. The various steps of SSM are used in an iterative manner. Their application is not necessarily sequential. Rather, it is important to go back to particular situations since it helps in developing clarity of an uncertain and an ambiguous situation.

SSM has been extensively used in numerous fields as a means of understanding problematic situations. Application of SSM to management of IS security has not been extensively undertaken, however. One exception is the doctoral work completed by Helen Armstrong [2] in 1999. Based on SSM Armstrong developed the Orion security strategy, which was subsequently used to manage IS security in a health care environment. The Orion model offers an alternative way to plan and manage IS security as opposed to highly structured approaches. The focus is shifted away from situations where exclusive reliance is placed on security specialists to one where users are made aware of security issues. This helps users themselves to identify a range of security controls. The result of this is that users feel responsible for IS security in their given work area. This results in developing a holistic view of security.

The situation mandated by Orion shifts the emphasis away from a security expert who might be dictating a range of protective measures. Rather, the role evolves into one of an advisor to the user and management, who feel responsible for IS security. This reduces the possibility of end user resistance to IS security mechanisms and even the implementation of inappropriate security controls. Usually, it is the end users in the organization who know of all the vulnerabilities in business processes and systems. They are also the ones who know the best methods to "plug the holes." It, therefore, makes sense to work with the users to ensure security.

As has been discussed elsewhere in this book, traditional IS security mechanisms have largely been confined to technical and risk attributes of IS security. The Orion method is one means of shifting the focus away from an exclusively technical or a risk-dominated security approach. The method details steps to study the current and ideal security situations and identify any gaps. Appropriate measures to fill the gaps are then suggested. This allows for security to be integrated into the organizational mindset. This helps in integrating security thinking into all organizational activities rather than it being considered a separate activity.

Involving users in the Orion method has two distinct advantages:

1. It affords an opportunity to tap into the knowledge of the users. Since staff members are usually engaged in day-to-day operations, they understand the range of activities that come together to achieve a given process objective. Tapping into this knowledge domain is important since technical security experts will never be able to understand the depth of the problem situations and business processes as well as the people engaged in the activities on a day-to-day basis.
2. It also implicitly helps in increasing awareness of the range of security issues among coworkers. Since employees are involved in uncovering the security vulnerabilities and identifying protection mechanisms, they become more inclined to follow the security measures that might be implemented. As Armstrong [2] notes, "One of the keys to the success of the Orion approach is the marrying of people with information security and organizational expertise to build a holistic consciousness integrated into the organizational mindset" (p. 343).

## Orion Strategy Process Overview

A high-level representation of the Orion security strategy process appears in Figure 5.6. The Orion strategy process is conceptualized at two planes of reality:

1. Level 1: This is the physical world, where all actions and processes can be seen and measured.
2. Level 2: This is the abstract or the conceptual level. Idealized processes and work situations exist at this level.

Level 2 allows participants to think beyond the confines of the physical reality and engage in creative thinking. In the Orion strategy process diagram, oval shapes denote activities. The boundary is denoted by a solid line. The boundary, usually, encompasses the main activities, separating the main area being considered. The range of inputs could include legal and regulatory requirements, policies put forward by the board of directors, inputs from other organizations/systems, etc. Outputs are the well-defined security requirements, action plans, and reporting structures. The Orion strategy process has seven steps in all.

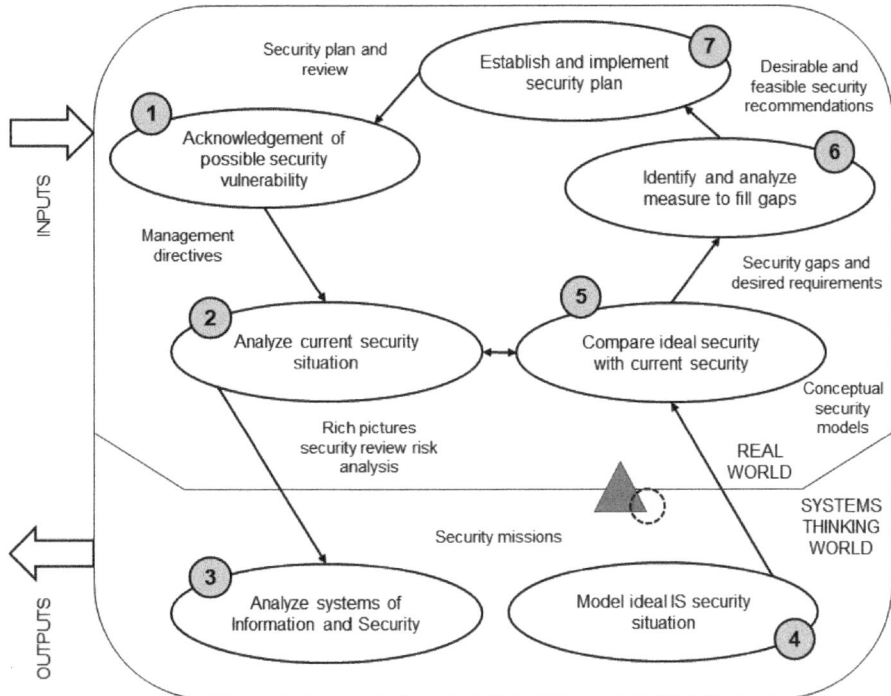

**Figure 5.6.** A high-level view of the Orion strategy (reproduced from Armstrong [2])

These are numbered in Figure 5.6 for reference purposes. Although it is recommended that all activities are undertaken in a sequential manner at least to begin with, following each of the steps in a "cause-effect" mode is not mandated by the process. In the remainder of this section, the details of the Orion strategy process are discussed.

*Activity 1: Acknowledgement of possible security vulnerability.* This activity involves the collection of perceptions of the problem situation. Multiple stakeholders are interviewed and their perception of the situation is recorded. No analysis is undertaken *per se*. This not only helps in understanding the range of opinions about security, but is also a stepping stone for building consensus.

*Activity 2: Identify risks and current security situation.* A detailed picture of the current situation is drawn. Particular attention is given to the existing structures and processes. The structure is the physical layout of the organization, formal reporting structures, responsibilities, authority structures, formal and informal communication channels. Softer power issues are also mapped as research has shown that these do have a bearing on the management of IS security [7]. Process is looked at in terms of typical input, processing, output, and feedback mechanisms. This involves considering basic activities related to deciding to do something, doing it, monitoring the activities as progress is made, impact of external factors, and evaluating outcomes. The result of this stage is a detailed description of the situation. Usually, a lot of pictures are drawn, security reports are reviewed, and outcomes of traditional risk analysis studied.

*Activity 3: Identifying the ideal security situation.* At this stage hypotheses concerning the nature and scope of improvements are developed. These are then discussed with the concerned stakeholders to identify both "feasible" and "desirable" options. In particular, this involves developing a high-level definition of systems of doing things and the related security—both technical and procedural. It is important to note that a "system" should not necessarily be viewed in terms of a technical artifact. Rather, a system, as discussed in earlier parts of this book, has both formal and informal aspects. Activity 3 is rooted in the ideal world. Here we detach ourselves from the real world and think of ideal types and conceptualize ideal practices.

*Activity 4: Model ideal information systems security.* This stage represents the conceptual modeling step in the process. All activities necessary to achieve the agreed-upon transformation are considered and a model of the ideal security situation developed. This involves analyzing the systems of information and defining important characteristics. The security features should match the ideal types defined in Activity 3. An important step in Activity 4 is monitoring the operational system. In particular three sub-activities are undertaken:

1. Measures of performance are defined. This generally relates to assessing the efficacy (does it work), efficiency (how much work completed given consumed resources), and effectiveness (are goals being met). Other metrics besides efficacy, efficiency, and effectiveness may also be used.
2. Activities are monitored in accordance with the defined metrics.
3. Control actions are taken, where outcomes of the metrics are assessed in order to determine and execute actions.

*Activity 5: Comparison of ideal with current.* At this stage the conceptual models built earlier are compared with the real-world expression. The comparison at this stage may lead to multiple reiterations of activities 3 and 4. Prior to any comparison, however, it's important to define the end point of Activity 4. There is a natural tendency to constantly engage in conceptual model building. However, it is always a good idea to move rather quickly to Activity 5 and then return to Activity 4 in an iterative manner. This not only helps in building better conceptual models, but also enables undertaking an exhaustive comparison. Comparison, as suggested in Activity 5, is an important step of the Orion strategy process. There are four particular ways in which the comparison can be done.

1. Conceptual model as a base for structured questioning. This is usually done when the real-world situation is significantly different from the one depicted in the conceptual model. The conceptual model helps in opening up a range of questions that are systematically asked to understand aspects of the real-world situation.
2. Comparing history with model prediction. In this method the sequence of events in the past are reconstructed and then comparison is done to understand what had happened in producing it and what would have happened if the relevant

conceptual model was actually implemented. This helps in defining the meaning of the models, allowing for a satisfactory comparison.
3. General overall comparison. This comparison relates to discussing the "Whats" and "Hows." The basic question addressed relates to defining features that might be different from present reality and why. In Activity 5, the comparison is undertaken alongside the expression of the problem situation expressed in Activity 2.
4. Model overlay. In this method there is a direct overlay of the two models—real world and conceptual. The differences in the two models become the source of discussions for any change.

*Activity 6: Identify and analyze measures to fill gaps.* This stage involves a review of the desired solution space. The wider context of the problem domain is reviewed for possible alternative solutions. The source of this review is a function of the solution that is sought. If the intent is to identify devices, then vendors are approached. If the procedures need to be redesigned, then compliance consultants need to be brought in (at least in the US where the Sarbanes-Oxley Act is mandating such compliance). It is important to note that at this stage no alternatives are dismissed. All are reviewed and adequately analyzed.

*Activity 7: Establish and implement security plan.* Recommendations developed in Activity 6 are considered and solutions formulated. An implementation plan is devised. Detailed tasks are identified. Criteria to subsequently measure success are also established. At this stage, integration of security into overall systems and information flows is also considered. It is important at this stage to ensure that the means used to establish security are appropriate and do not conflict with the other controls. Resources used for implementing security are then calculated and adequately allocated. Such resources would include people, skills, time, equipment, among others. On completion of implementation and training, the success is reviewed in light of the original objectives so that further learning can be achieved.

# IS Security Planning Principles

It should be clear from the discussion so far that organizations need to develop a security vision that ties corporate plans with the tactical security policy issues. There are numerous cases where although information technology has been considered as a strategic resource, little effort has been made to address the security concerns. Even where security implications have been thought of, a narrow technical perspective has been considered. Such a perspective hinders progress in establishing good security. So what are the fundamental principles that organizations need to have in place if proper IS security strategies and plans are to be realized? The remaining part of this chapter discusses the principles.

## *Principles*

In furthering our understanding of security policies, we should be able to study the security policy formulation process from the perspective of people in an organization, thus allowing us to avoid causal and mechanistic explanations. By adopting a human perspective, we tend to focus on the human behavioral aspects. Security policy formulation is therefore not a set of discrete steps rationally envisaged by the top management, but an emergent process that develops by understanding the subjective world of human experiences. Mintzberg [14] contrasts such "emergent strategies" with the conventional "deliberate strategies" by using two images of *planning* and *crafting*:

> Imagine someone planning strategy. What likely springs to mind is an image of orderly thinking: a senior manager, or a group of them, sitting in an office formulating courses of action that everyone else will implement on schedule. The keynote is reason—rational control, the systematic analysis of competitors and markets, or company strengths and weaknesses, the combination of these analyses produces clear, explicit, full-blown strategies.
>
> Now imagine someone crafting strategy. A wholly different image likely results, as different from planning as craft is from mechanization. Craft invokes traditional skill, dedication, perfection through the mastery of detail. What springs to mind is not so much thinking and reason as involvement, a feeling of intimacy and harmony with the materials at hand, developed through long experience and commitment. Formulation and implementation merge into a fluid process of learning through which creative strategies emerge. (p. 66)

This does not necessarily mean that systematic analysis has no role in the strategy process, rather the converse is true. Without any kind of an analysis, strategy formulation at the top management level is likely to be chaotic. Therefore, a proper balance between *crafting* and *planning* is needed. Figure 5.7, hence, is not a rationalist and a sequential guide to security planning, but only highlights some of the key phases in the information system security

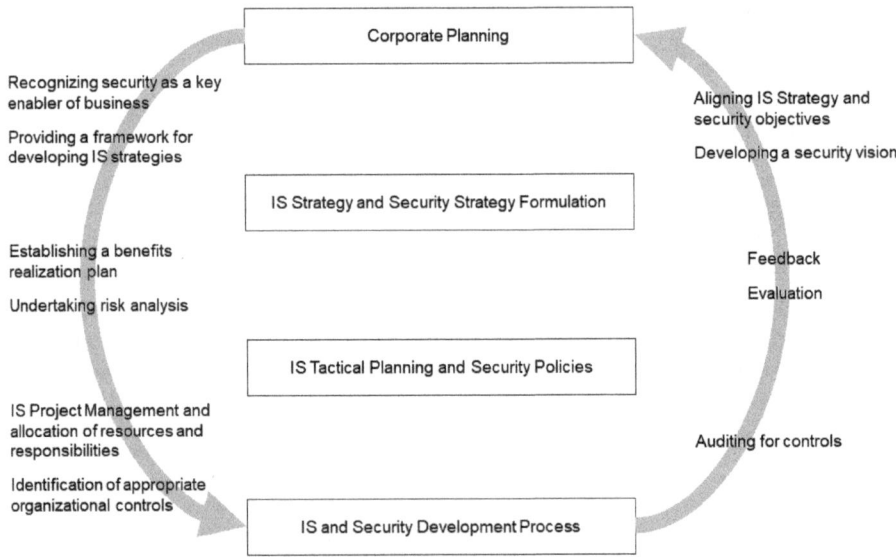

**Figure 5.7.** IS security planning process framework

planning process. Underlying this process is a set of principles which would help analysts to develop secure environments. These are:

1. **A well-conceived corporate plan establishes a basis for developing a security vision.** A corporate plan emerging from the experiences of those involved and the relevant analytical processes forms the basis for developing secure environments. A coherent plan should have as its objective a concern for the smooth running of the business. Typical examples of incoherence in corporate planning are seen in a number of real-life situations. The divergence of IT and business objectives in most companies and the mismatch between corporate and departmental objectives illustrate this point. Hence, an important ingredient of any corporate plan is a proper organizational and a contextual analysis. In terms of security it is worthwhile analyzing the cultural consequences of organizational actions and other IT-related changes. By conducting such a pragmatic analysis we are in a position to develop a common vision, thus maintaining the integrity of the whole edifice. Furthermore, this brings security of information systems to center stage and engenders a subculture for security.
2. **A secure organization lays emphasis on the quality of its operations.** A secure state cannot be achieved by considering threats and relevant countermeasures alone. Equally important is maintaining the quality and efficacy of the business operations. There is no quantitative measure for an adequate level of quality, as it is an elusive phenomenon. The definition of quality is constructed, sustained, and changed by the context in which we operate. In most companies, the attitude for maintaining the quality of business operations is extremely rationalist in nature. The management have made an implicit assumption that by adopting structured service quality assurance practices, it is possible for them to maintain the quality of the business operations. In most cases the top management assumes that their desired strategy can be passed down to the functional divisions for implementation. However, this is a very "tidy" vision of quality, whereas in reality the process is more diffuse and less structured. In fact, the "rationalist approaches" adopted by the management of many corporations causes discontentment, rancor, and alienation among different organizational groups. This is a serious security concern. A secure organization therefore has to lay emphasis on the quality of its business operations.
3. **A security policy denotes specific responses to specific recurring situations and hence cannot be considered as a top-level document.** To maintain the security of an enterprise, we are told that a security policy should be formulated. Furthermore, top managements are urged to provide support to such a document. However, the very notion of having such a document is problematic. Within the business management literature a policy has always been considered as a tactical device aimed at dealing with specific repeated situations. It may be unwise to elevate the position of a security policy to the level of a corporate strategy. Instead, corporate planning should recognize secure information systems as an enabler of businesses (refer to Figure 5.1). Based on this belief a security strategy should be integrated into the corporate planning process, particularly with the information

systems strategy formulation. Depending on risk analysis and SWOT (strengths, weaknesses, opportunities, and threats) analysis-specific security policies should be developed. Responsibilities for such a task should be delegated to the lowest appropriate level.
4. **Information systems security planning is of significance if there is a concurrent security evaluation procedure.** In recent years emphasis has been placed on security audits. These serve a purpose insofar as the intention is to check deviance of specific responses for particular actions. In most cases, the whole concept of quality, performance, and security is defined in terms of conformity to auditable processes. The emphasis should be on expanding the role of security evaluation which should complement the security planning process.

## Summary

The aim of this section has been to clarify misconceptions about security policies. The origins of the term are identified and a systematic position of policies with respect to strategies and corporate plans is established. Accordingly various concepts are classified into three levels: corporate, business, and functional. This categorization prevents us from giving undue importance to security policies, and allows us to stress the usefulness of corporate planning and development of a security vision. Finally, a framework for information system security planning process is introduced. Underlying the framework are a set of four principles which help in developing secure organizations. The framework considers security aspects to be as important as corporate planning and critical to the survival of an organization. An adequate consideration of security during the planning process helps analysts to maintain the quality, coherence, and integrity of the business operations. It prevents security from being considered as an afterthought.

## In Brief

- Identification and development of structures of responsibility are a key aspect of formal information system security.
- Structures of responsibility define the pattern of authority, which is so essential in ensuring management of access.
- Organizational buy-in at all levels is key to the success of the information system security program in any organization.
- Security policies are an important ingredient of the overall security program.
- Proper security policy formulation and implementation is essential for the success of overall security.
- Planning for IS security entails developing a vision and a strategy for security.
- Security of IS should be thought of as an enabler to the smooth running of business.
- There are three classes of IS security decisions—strategic, administrative, and operational.
- Strategic IS security decisions deal with selecting an environment that ensures the smooth running of business.
- Administrative IS security decisions deal with creating adequate structures and processes to enable information handling.
- Operational IS security decisions relate to optimizing work patterns for efficiency gains.
- There are four core IS planning principles:
  - A well-conceived corporate plan establishes a basis for developing a security vision.
  - A secure organization lays emphasis on the quality of its operations.
  - A security policy denotes specific responses to specific recurring situations and hence cannot be considered as a top-level document.
  - IS security planning is of significance if there is a concurrent evaluation procedure.

# Questions and Exercises

**Discussion questions.** These are based on a few topics from the chapter and are intentionally designed for a difference of opinion. These questions can best be used in a classroom or a seminar setting.

1. What kind of executive-level support is essential for ensuring uptake of information system security? How should such support be generated? What strategies can be put in place to ensure that executive-level support is sustained over a period of time?

2. Security policies have always been considered the *sine qua non* of well-managed companies. Discuss.

3. Development of security policies and their implementation is the responsibility of different roles in organizations. Discuss the differences in opinion with respect to development and implementation of security policies.

4. How can engaging in double loop learning help in ensuring proper IS security?

5. How can the Orion strategy process help in addressing some of the fundamental problems that exist in IS security management?

**Exercise 1.** Procure a corporate policy of a company of your choice. Compare the corporate policy with the security policy. Comment on the discrepancies, if any. Suggest how alignment of corporate and security policies can be brought about.

**Exercise 2.** Identify examples in the popular press where a security breach has occurred because the security policy had not been followed. Undertake research to find reasons why the policy was not carried through and followed. Relate your findings to IS security planning principles discussed in this chapter.

## Short Questions

1. Usually security problems are a consequence of _____ breakdowns and lack of understanding of behaviors of various stakeholders.

2. The security management structure looks from the top down. Substantive actions required of members of the organization in the course of using the computer systems in place should take a _____ approach.

3. The effectiveness of the security policy is a function of the level of support it has from an organization's _____ _____.

4. A strategy of "locks and keys" becomes inadequate if people _____ the organization open those locks (i.e. subvert the controls).

5. The security policies determine the processes and techniques required to provide the security but not the _____.

6. Following the implementation there is a constant need to _____ the security processes and techniques.

7. In business management practices, the term _____ was in use long before _____ but the two are often used interchangeably, despite having very different meanings.

8. In practice, implementing a _____ can be delegated, while for implementing a _____, executive judgment is required.

9. At a _____ level the security strategy determines key decisions regarding investment, divestment, diversification, and integration of computing resources in line with other business objectives.

10. At a _____ level, the security strategy looks into the threats and weaknesses of the IT infrastructure.

11. The emphasis should be to develop a _____ security vision that brings the issue of security to center stage and binds it to the organizational objectives, but this does not mean that organizations should not have any security policies sketching out _____ procedures.

12. Relegating IS security decisions to the operational levels of the firm could result in lack of _____ by the top management.

13. One of the fundamental problems with respect to security is for a firm to choose the right kind of an _____ to function in.

14. Allocation of _____ among competing needs can become a critical problem in terms of strategizing about security.

15. While many organizations have engaged in identifying security issues and created relevant security policies, there is a clear mismatch between what the _____ mandates and what is done in practice.

16. To a large extent high _____ processes are a consequence of adequate planning and policy implementation.

17. Careful _____ and establishing proper checks and balances are perhaps the cheapest of the operational level security practices.

18. Maintaining integrity of business processes is a function of adequate _____ and _____ structures.

## Case Study: The Hack at UC Berkley

Hackers broke into a computer at the University of California at Berkley recently and gained access to 1.4 million names, social security numbers, addresses, and dates of birth that were being used as part of a research project. The FBI, the California Highway Patrol, and California Department of Social Services were investigating the incident. Security personnel were performing a routine test of intrusion detection when they noticed that an unauthorized user was attempting to gain access to the computer. A database with a known security flaw was exploited, and a patch was available that would have prevented the attack. The negligence in attending to the known security flaw appears to be a common mistake among institutes of higher learning in the state. Banks, government agencies, and schools are known to be the top targets for hackers. Hackers may attack financial institutions in an effort to profit from the crime, and government agencies to gain notoriety. Private companies generally have made at least some effort to ensure that data is secure, but hackers attack institutes of higher learning often because there are frequent lapses in security. This not only presents a problem for the university, but also is a danger to other entities, since denial of service attacks may be generated from the compromised university computers. One of the problems at universities may be the lack of accountability or of an overarching department that has authority to oversee all systems, and limit modifications. In the name of learning, many less qualified individuals, sometimes students, are given authority to make modifications to operating systems and applications. This presents a continuing problem for administrators and represents a threat to all who access the Internet.

1. Name policies and procedures that would enable universities to limit vulnerabilities while still allowing students access to systems.
2. Ultimately, who should be held accountable for ensuring a sound security policy is in place?
3. Who at your school is responsible for maintaining a security policy and how often is it updated?

Source: Based on "Hack at UC Berkeley Potentially Nets 1.4 Million SSNs," *eWeek.com*, October 20, 2004.

# References

1. Ansoff, H.I., *Corporate strategy*. 1987, Harmondsworth, UK: Penguin Books.
2. Armstrong, H., *A soft approach to management of information security*, in *School of Public Health*. 1999, Curtin University: Perth, Australia. p. 343.
3. Backhouse, J. and G. Dhillon, *Structures of responsibility and security of information systems*. European Journal of Information Systems, 1996. **5**(1): p. 2-9.
4. Checkland, P.B. and J. Scholes, *Soft systems methodology in action*. 1990, Chichester: John Wiley.
5. Dhillon, G., *Managing information system security*. 1997, London: Macmillan.
6. Dhillon, G., *The Challenge of Managing Information Security*. International Journal of Information Management, 2004. **24**(1): p. 3-4.
7. Dhillon, G. and J. Backhouse, *Managing for secure organizations: a review of information systems security research approaches*, in *Key issues in information systems*, D. Avison, Editor. 1997, McGraw Hill.
8. Dhillon, G. and S. Moores, *Computer Crimes: theorizing about the enemy within*. Computers & Security, 2001. **20**(8): p. 715-723.
9. Dhillon, G. and G. Torkzadeh. *Value-focused assessment of information system security in organizations*. in *International Conference on Information Systems*. 2001. New Orleans, LA.
10. Fiol, C.M. and M.A. Lyles, *Organizational learning*. Academy of Management Review, 1985. **10**: p. 803-813.
11. Greenwald, J., *Jack in the box*, in *Time*. 1994.
12. Greenwald, J., *A blown billion*, in *Time*. 1995. p. 60-61.
13. Mattia, A. and G. Dhillon. *Applying Double Loop Learning to Interpret Implications for Information Systems Security Design*. in *IEEE Systems, Man & Cybernetics Conference*. 2003. Washington DC.
14. Mintzberg, H., *Crafting strategy*. Harvard Business Review, 1987(July-August).
15. Perry, W.E., *Developing a computer security and control strategy*. Computers & Security, 1982. **1**(1): p. 17-26

16. Ramachandran, S. and G.B. White. *Methodology To Determine Security ROI.* in *Proceedings of the Tenth Americas Conference on Information Systems.* 2004. New York, New York, August: AIS.
17. Rawnsley, J, *Going for broke: Nick Leeson and the collapse of Barings Bank.* 1995, London: Harper Collins.
18. Solms, B.v and R.v Solms, *The 10 deadly sins of information security management.* Computers & Security, 2004. **23**: p. 371-376.
19. Wrapp, H.E., *Good managers don't make policy decisions*, in *The strategy process*, H. Mintzberg and J.B. Quinn, Editors. 1991, Prentice-Hall: Englewood Cliffs. p. 32-38.

# CHAPTER 6

# Risk Management for Information System Security

> *When you can measure what you are speaking about, and express it in numbers, you know something about it; but when you cannot measure it, when you cannot express it in numbers, your knowledge is of a meager and unsatisfactory kind; it may be the beginning of knowledge, but you have scarcely, in your thoughts, advanced to the stage of science.*
>
> —William Thompson, Lord Kelvin (1821–1907),
> *Popular Lectures and Addresses 1861–1894*

> *If the mind treats a paradox as if it were a real problem, then since the paradox has no "solution," the mind is caught in the paradox forever. Each apparent solution is found to be inadequate, and only leads on to new questions of a yet more muddled nature.*
>
> —David Bohm

Over the past few weeks Joe Dawson had made a lot of progress. Not only had he been able to conceptualize the needs and wants of SureSteel with respect to maintaining security, but he also knew how various plans could be laid. He had also become fairly comfortable in identifying the security objectives of various stakeholders and integrating them somewhat into a security policy.

One aspect of security that really worried Joe was his limited ability to forecast and think through the range of risks that might exist. Should he simply hire a risk management consultant to help him out and design security into the systems? Or, should he buy an off-the-shelf package to identify business impacts? Joe was aware of numerous such packages. He had seen a number of advertisements in the popular press on business impact analysis as well. Clearly there was a need for SureSteel to identify major information assets and perhaps attribute a dollar value to them. This would help Joe to better undertake resource allocation.

Although Joe knew what he had to do, he was unsure of the way in which he should proceed. As he thought about what needed to be done, Joe started browsing through the *CIO Magazine*. An article on "The Importance of Mitigating IT Risks" caught his eye. As he scanned the content it became clear that indeed he had to learn more. The *CIO Magazine* clearly stated that one of the main problems in risk management was the lack of awareness of the concerned executives. The article read:

> Enterprises face ongoing exposures to risk in the forms of application, electronic records retention, event, platform, procedure, and security exposures. However, many IT executives lack sufficient knowledge and data about their vulnerabilities and potential losses from failure. Continuously evolving legislative and regulatory requirements, increasing business reliance on data, and regional and global uncertainty dictate corporations regularly appraise the solidity of business, operational, and technical capabilities to support these requirements and palliate risk. IT executives should work with executive, internal audit, LOB, and their own teams to assess where exposures exist, establish mitigation requirements and governance procedures, gauge the importance of critical infrastructure components, and judge potential outage costs. (*CIO Magazine*, October 27, 2003)

"This was so very true," Joe thought. He had recently seen survey results from a Canadian market research firm, Ipsos (www.ipsos.ca), where it was found that just 42% of Canadian CEOs said that protecting against cyberterrorism is of moderate concern; 19% had not even considered any sort of protection to be a priority. Just 30% of these CEOs said their security measures were effective. And most of these companies had had one kind of a security breach or another (45% had been inflicted by a computer virus; 22% were victims of computer theft, and 20% said their systems had been hacked). Awareness was something that was clearly lacking at SureSteel. Joe did not want to be included in the group of people who did not consider security risk management to be important.

Joe also had to convince the IT people at SureSteel that they had to take security risk management seriously. Although he himself was not entirely comfortable with the subject matter, he wanted to make a case in favor of risk management and begin discussions as to how the vision could be realized. One way of proceeding would be to make a presentation. Joe sat his computer and started making some slides. He scanned the Internet to collect information on security breaches. Some of the facts he included in his presentation were,

- 8%–12% of IT budgets will be devoted to IS security by 2006 (META Group).
- 47% of security breaches are caused by human error (CompTIA).
- $50 billion is the estimated direct cost of damage that a "bad" worm could cause.
- 43% of large companies have employees assigned to read outbound employee email (Forrester Research).
- $45 billion is the annual amount of losses incurred by US businesses because of identity theft (Federal Trade Commission).
- The cost of identity theft to individuals ranges from $500 to $1,200 (Federal Trade Commission).

Joe knew these figures were going to strike a chord with many employees.

Risks exist because of inherent uncertainty. Risk management with respect to security of IT systems is a process that helps in balancing operational necessities and economic costs associated with IT-based systems. The overall mission of risk management is to enable an organization to adequately handle information. There are three essential components of risk management: **risk assessment, risk mitigation,** and **risk evaluation**. The risk assessment process includes the identification and evaluation of risks so as to assess the potential impacts. This helps in recommending risk-reducing strategies. Risk mitigation involves prioritizing, implementing, and maintaining an acceptable level of risk. Risk evaluation deals with the continuous evaluation of the risk management process such that ultimately successful risk management is achieved.

Security risk management is not a stand-alone activity. It should be integrated with the systems development process. Any typical systems development is accomplished through the following six steps: initiation, requirements assessment, development or acquisition, implementation, operations/maintenance, and disposal. Failure to integrate risk management with systems development results in "patchy" security. Some of the risk management activities accomplished at each of the systems development stages are identified and presented below:

- **Initiation**. At this stage the need for an IT system is expressed and the purpose and scope established. Risks associated with the new system are explored. These feed into project plans for the new system.
- **Requirements assessment**. At this stage all user and stakeholder requirements are assessed. Security requirements and associated risks are identified alongside the system requirements. The risks identified at this stage feed into architectural and design trade-offs in systems development.
- **Development or acquisition**. Make or buy and other sourcing decisions are taken. The IT system is designed or acquired. Relevant programming or other development efforts are undertaken. Controls identified in requirements assessment are integrated into system designs. If systems are being acquired, necessary constraints are communicated to the developers or third parties.
- **Implementation**. The system is implemented in the given organizational situation. Risks specific to the context are reviewed and implementation challenges considered. Contingency approaches are also reviewed.
- **Operations or maintenance**. This phase relates to typical maintenance activities, where constant changes and modifications are made. Relevant upgrades are also instituted. Risk management activities are performed periodically. Re-accreditation and re-authorizations are considered, especially in light of legislative requirements (e.g. the Sarbanes-Oxley Act in the US).
- **Disposal**. In this stage legacy systems are phased out and data is moved to new systems. Risk management activities include safe disposal of hardware and software such that confidential data is not lost. Any system migration needs to take place in a secure and systematic manner.

# Risk Assessment

Risk assessment is the process by which potential threats throughout the system development process are determined. The outcome of risk assessment is the identification of appropriate controls for minimizing risks. Risk assessment considers **risk** to be a function of the **likelihood** of a given threat resulting in certain **vulnerabilities**. Such vulnerabilities may have an adverse impact on an organization.

In order to determine the likelihood of future adverse events, the threats, vulnerabilities, and controls are evaluated in conjunction with each other. The interplay between a threat, vulnerability, and control is the impact that an adverse event might have. The level of impact is a function of the outcome of a given activity and the relative value of IT assets and resources. The US National Institute of Standards and Technology (Publication 800-30) identifies the following nine steps to be integral to risk assessment:

1. System Characterization
2. Threat Identification
3. Vulnerability Identification
4. Control Analysis
5. Likelihood Determination
6. Impact Analysis
7. Risk Determination
8. Control Recommendations
9. Results Documentation

## System Characterization

A critical aspect of risk assessment is to determine the scope of the IT system. System characterization helps in identifying the boundaries of the system, i.e. what functions and processes of the organization the system might deal with and what resources a system might be consuming. System characterization also helps in scoping the risk assessment task. System characterization can be achieved by collecting system-related information. Such information relates to the kind of hardware and software present, the system interfaces that exist (both internal and external to the organization), and the kind of data that might reside in the system. Understanding the kind of data is very important since it helps in evaluating the true business impact there might be because of the losses.

Besides an understanding of technical aspects of the system, the related roles and responsibilities need to be understood. Roles and responsibilities that relate to critical business processes also need to be identified. It is also important to define the system's value to the organization. This would mean a clear definition of system and data criticality issues. A definition of the level of protection required in maintaining confidentiality, integrity, and availability of information is also important. All the information collected helps in defining the nature and scope of the system.

Various pieces of operational information are also essential. Such operational information relates to the functional requirements of the system, the stakeholders of the system, and security policies and architectures governing the IT system. Other necessary operational information includes network typologies, system interfaces, system inputs and output flows, technical and management controls in place (e.g. identification and authentication protocols, access controls, encryption methods). An assessment of the physical security environment is also essential. This is often overlooked and emphasis is placed on technical controls.

Information for system characterization can be gained by any number of methods. These include questionnaires, in-depth interviews, secondary data review, and automated scanning tools. At times more than one method may be necessary to develop a clear understanding of the IT systems in place. Sample interview questions appear in Box 6.1.

---

**Box 6.1. Sample Interview Questions**

Who are the valid users?
What is the organization's mission?
Where does the system fit in given the organization's mission?
What is the relative importance of the system to other IT-based systems?
What are the information requirements of the organization?
How critical is the information for the organization?
What are the data flows?
What are the sensitivity levels of the information?
Where is the information stored?
What types of storage mechanisms are in place?
What is the potential business impact if the information were disclosed?
What are the effects on the organization if the information is not reliable?
To what extent can system downtime be tolerated?

---

## Threat Identification

Threat is an indication of impending danger or harm. Threats get exercised through vulnerabilities. Vulnerability is a weakness that can be accidentally triggered or intentionally exploited. A threat is really not a risk unless there is a vulnerability. If any likelihood of a threat is to be determined, the potential source of the threat, vulnerability, and the existing controls need to be understood.

Identification of a threat source results in compilation of a list of threat sources that might be applicable to a given IT system. A threat source is any circumstance or event which has a potential to cause harm to the IT system. Threat sources may reside in an organization's environment, *viz*, earthquakes, fire, floods, etc. Or they can be of a malicious nature. In attempts to identify all sorts of threats, it is useful to consider them as being intentional or unintentional.

Intentional threats reside in the motivations of humans to undertake potentially harmful activities. These might result in systems being compromised. Prior research has shown opportunities, personal factors, and work situations to have an impact on internal organizational employees attempting to subvert controls for their advantage. Deliberate attacks can occur because of a malicious attempt to gain unauthorized entry to a system, which might result in compromising the confidentiality, integrity, and availability of information. Attacks on systems can also be benign instances that attempt to circumvent system security.

Various threat sources can be classified into five categories: Hackers/crackers; computer criminals; terrorists; industrial espionage; insiders. Motivations and threat actions for each of these classes are presented in Table 6.1.

**Table 6.1.** Threat classes and resultant actions (Based on NIST Special publication 800-30)

| Threat type | Intention | Resultant action |
| --- | --- | --- |
| Hackers | Challenge, rebellion | Hacking, system intrusion, computer break-ins, unauthorized access |
| Computer criminals | Destruction of information, Illegal disclosure of information, Monetary gain, Unauthorized modification of data | Computer crimes, Computer frauds, Spoofing, System intrusion |
| Terrorism | Blackmail, Extortion, Destruction, Revenge | Information warfare, Denial of service, System tampering |
| Industrial espionage | Competitive advantage | Economic exploitation, Information theft, Intrusion of personal privacy |
| Insiders | Work situation, Personal factors, Opportunity | Computer abuse, Fraud and theft, Falsification, Planting of malicious code, Sale of personal information |

The threat sources and the intentions, as presented in Table 6.1, are a rough guide only. Organizations can tailor these for their individual needs. For instance, in certain geographic areas there is greater danger of earthquakes than floods. In other environments, such as the military, there is a greater probability of hacking attacks. The nature and significance of certain kinds of attacks keeps on changing with time. It is therefore prudent to tune into the latest developments. Good sources of information include the Computer Security Institute website (www.gocsi.org) and the CERT Coordination Centre (www.cert.org). In addition, there are a number of security portals that share the latest happenings in the security world.

## Vulnerability Identification

Vulnerability assessment deals with identifying flaws and weaknesses that could possibly be exploited because of the threats. Generally speaking, there are four classes of vulnerabilities that might exist. Figure 6.1 identifies the four classes and paragraphs below discuss these in detail.

The first class of vulnerabilities are **behavioral and attitudinal vulnerabilities**. Such vulnerabilities are generally a consequence of people-based issues. In many cases individuals tend to subvert organizational controls because of their past experiences with the firm. Perhaps there may be too many strict controls imposed by their bosses, or the work situation created was too constraining. In some cases if a promotion of a certain employee was denied, s/he might end up being disgruntled. In other cases sheer greed could be the reason for subverting controls. It is therefore important that a range of behavioral and

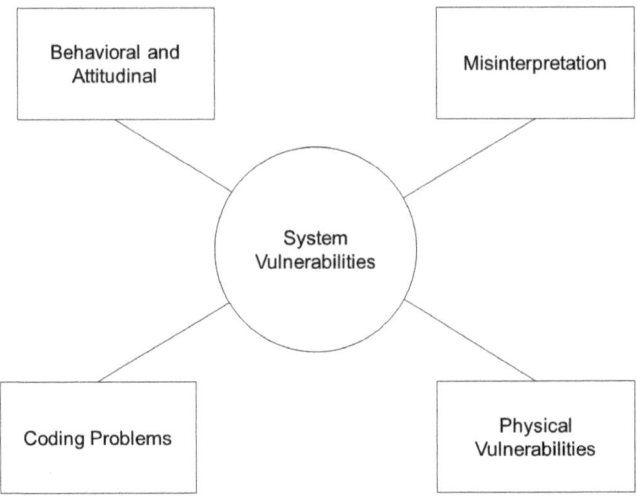

**Figure 6.1.** Vulnerability classes

attitudinal factors are considered. Such factors help in understanding the vulnerabilities that might exist.

Behavioral and attitudinal vulnerabilities have been well studied in the literature. Based on the theory of reasoned action [1], two major factors affecting a person's behavioral intentions have been identified: attitudinal (personal) and social (normative). That is, a person's intention to perform or not perform a particular behavior is determined by his/her attitude towards the behavior and the subjective norms. Further, a person's attitude towards the behavior is determined by the set of salient beliefs s/he holds about performing the particular behavior. Although subjective norms are also a function of beliefs, these beliefs are of a different nature and deal with perceived prescriptions. That is, subjective norms deal with the person's perception of the social pressures placed on individuals to perform or not perform the behavior in question. It has been argued that attention to variables such as personality traits alone is misplaced and instead the focus should be on behavioral intentions and the beliefs that shape those intentions. Clearly the key to predict behavior lies in the intentions and beliefs.

The second class of vulnerabilities relates to **misinterpretations**. Misinterpretations are a consequence of ill-defined organizational rules. The rules may relate to flawed organizational structures and reporting patterns or to the manner in which system access is managed. In many cases the manner in which access to systems is prescribed by the IT system is different from the way in which organizational members report or are part of the organizational hierarchy. This leads to a lot of confusion. Misinterpretation also occurs at the time of systems development where user requirements may be interpreted differently or wrongly by the analysts. This results in a flawed or an inadequate system design, which eventually gets reflected in the implementation.

The third type of vulnerabilities is a consequence of **coding problems**. Such vulnerabilities have their origin in misinterpretations of requirements by analysts, which subsequently get reflected in the code. Coding problems also arise because of flaws and programmer errors. In certain cases, even when the system has been implemented, coding problems emerge because of lack of software updates and other legacy system problems. In 1999 the Y2K problem was a major cause of concern to companies. This can be attributed to vulnerabilities arising because of coding problems.

The fourth category of vulnerabilities are **physical**. These usually result from inadequate supervision, negligent persons, and natural disasters such as fires, floods, etc. For most of the vulnerabilities there are standard methodologies for ensuring safety. There are also system certification tests and evaluations that could be undertaken. It is important to be aware of such tests and certifications specific to a given industry and be involved with them.

A large number of vulnerabilities have been known for a while and various groups have put together lists and checklists. Prominent among these is the NIST I-CAT vulnerability database (http://icat.nist.gov). There are also numerous other security testing evaluation guidelines (e.g., the Network Security Testing guidelines developed by NIST—NIST SP 800-42).[1]

## Control Analysis

The purpose of this step is to analyze and implement controls that would minimize the likelihood of threats exploiting any vulnerability. Controls are usually implemented either for compliance purposes or simply because an organization feels that it is prudent to do so. Compliance-oriented controls are externally driven. These may be mandated because of legislation (e.g. HIPAA and Sarbanes-Oxley in the US) or required by trading partners. Self-control is the other kind of control, which members of the organization choose to institute because they feel it's important to do so. Both compliance and self-controls can be implemented for information utilization and information creation.

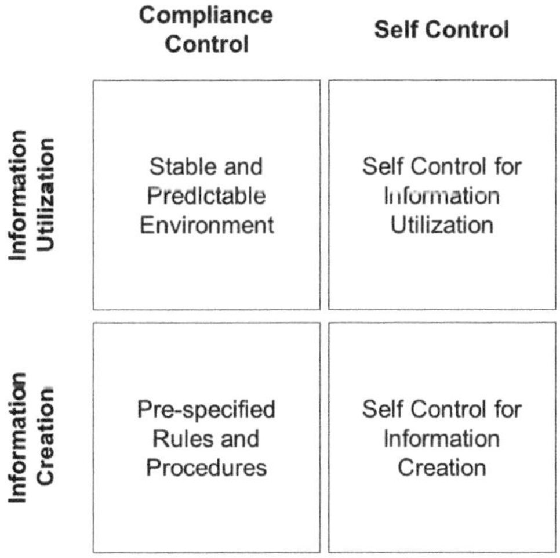

|  | Compliance Control | Self Control |
|---|---|---|
| Information Utilization | Stable and Predictable Environment | Self Control for Information Utilization |
| Information Creation | Pre-specified Rules and Procedures | Self Control for Information Creation |

**Figure 6.2.** Classes of controls

---

[1] The document can be downloaded from http://csrc.nist.gov/publications/nistpubs/800-42/NIST-SP800-42.pdf.

Figure 6.2 presents a summary of control types as they relate to information processing. In a situation where new information is being created and the organization operates in a compliance control environment, there are pre-specified rules and procedures and hence little is left to uncertainty. Even where new information is not created, systems and procedures are put in place thereby reducing uncertainty. Such situations are fairly stable and the environment is predictable.

If there is no compliance control, the organization really needs to be well disciplined in order to manage its information. Such situations pose the most danger as there are no regulatory constraints. If managed properly, such organizational environments are also very productive and innovative.

## Likelihood Determination and Impact Analysis

There are three elements in calculating the likelihood that any vulnerability will be exercised. These include:

- Source of the threat, motivation, and capability
- Nature of the vulnerability
- Effectiveness of current controls

Simply stating definitions of likelihood and assessing the appropriate level would help in determining the level of likelihood. Table 6.2 defines the likelihood levels.

**Table 6.2.** Likelihood determination

| Likelihood level | Definition |
|---|---|
| High | The source of threat is highly motivated and capable of realizing the vulnerability. The prevalent controls are ineffective in dealing with the threat. |
| Medium | The source of threat is highly motivated and capable of realizing the vulnerability. However, controls exist, which may prevent the vulnerability being exercised. |
| Low | The source of threat lacks motivation and capability. Sufficient controls are in place to prevent any serious damage. |

Once the likelihood has been determined, it is important to assess the business impact. A business impact analysis is usually conducted by identifying all information assets in the organization. In many enterprises, however, information assets may not have been identified and documented. In such cases sensitivity of data can be determined based on the level of protection required for confidentiality, integrity, and availability. The most appropriate approach for assessing business impact is by interviewing those responsible for the information assets or the relevant processes. Some interview questions may include:

- What is the effect of unintentional errors? Things to consider include typing wrong commands, entering wrong data, discarding the wrong listing, etc.

- What would be the scale of loss because of willful malicious insiders? Things to consider include disgruntled employees, bribery, etc.
- What would happen because of outsider attacks? Things to consider include unauthorized network access, classes of dial-up access, hackers, individuals sifting through trash, etc.
- What are the effects of natural disasters? Things to consider include earthquakes, fire, storms, power outages, etc.

Based on these questions the magnitude of impact could be assessed. A generic classification and definitions of impacts is presented in Table 6.3.

**Table 6.3.** Magnitude of impact

| Magnitude of impact | Definition |
| --- | --- |
| High | A vulnerability may result in (1) loss of major tangible assets and resources; (2) significantly violate, harm or impede an organization's mission and reputation; (3) human death or serious injury |
| Medium | A vulnerability may result in (1) loss of some tangible assets and resources; (2) some violation, harm, and impediment of an organization's mission and reputation; (3) human injury |
| Low | A vulnerability may result in (1) limited loss of tangible assets and resources; (2) limited violation, harm, and impediment of an organization's mission and reputation; (3) human injury |

An assessment of magnitude of impact results in estimating the frequency of the threat and the vulnerability over a specified period of time. It is also possible to assess costs for each occurrence of the threat. A weighted factor based on a subjective analysis of the relative impact of a threat can also be developed.

## Risk Determination

Risk determination helps in assessing the level of risk to the IT system. The level of risk for a particular threat or vulnerability can be expressed as a function of:

- The likelihood of a given threat exercising the vulnerability
- The magnitude of the impact of the threat
- The adequacy of planned or existing security controls

Risk determination is realized by developing a *Level of Risk Matrix*. Risk for a given setting is calculated by multiplying ratings for threat likelihood and threat impact. NIST prescribes risk determination to be done based on Table 6.4. Further classes of likelihood and impact can also be developed, i.e. the scale may be 4- or 5-point rather than the 3-point scale (High, Medium, Low) used in the illustration.

**Table 6.4.** Level of Risk Matrix (based on NIST SP 800-30)

| Threat Likelihood | Impact | | |
|---|---|---|---|
| | Low (10) | Medium (50) | High (100) |
| High (1.0) | Low<br>10 × 1.0 = 10 | Medium<br>50 × 1.0 = 50 | High<br>100 × 1.0 = 100 |
| Medium (0.5) | Low<br>10 × 0.5 = 5 | Medium<br>50 × 0.5 = 25 | Medium<br>100 × 0.5 = 50 |
| Low (.1) | Low<br>10 × 0.1 = 1 | Medium<br>50 × 0.1 = 5 | Low<br>100 × 0.1 = 10 |

Risk Scale: High (>50 to 100); Medium (>10 to 50); Low (1 to 10)

The risk scale with ratings of High, Medium, and Low represents the degree of risk to which an IT system might be exposed. If a high level of risk is found then there is a strong need for corrective measures to be initiated. A medium level of risk means that corrective mechanisms should be in place within a reasonable period of time. A low level of risk suggests that it is up to the relevant authority to decide if the risk is acceptable.

## Control Recommendations and Results Documentation

Control recommendation deals with suggesting appropriate controls given the level of risk identified. Any recommendation of a control is guided by five interrelated factors:

- The effectiveness of recommended controls. This is generally considered in light of system compatibility.
- Existing legislative and regulatory issues.
- The current organizational policy.
- The operational impact the controls might have.
- The general safety and reliability of the proposed controls.

Clearly all kinds of controls can be implemented to decrease chances of a loss. However, a cost-benefit analysis is usually required in order to define the requirements. While economics offers a number of principles for undertaking a cost-benefit analysis, there are also commercial methodologies available. Prominent among these is COBRA, which is also discussed in a later section of this chapter.

> **Box 6.2. Sample Risk Assessment Report (based on NIST SP 800-30)**
>
> **Executive Summary**
> 1. Introduction
>    - Purpose and scope of risk assessment
> 2. Risk Assessment Approach
>    - Description of approaches used for risk assessment
>    - Number of participants
>    - Techniques used (questionnaires, etc.)
>    - Development of the risk scale
> 3. System Characterization
>    - Characters of the system including hardware, software, system interfaces, data, users, input and output flowcharts, etc.
> 4. Threat Statement
>    - Compile the list of potential threat sources and associated threat actions applicable to the situation
> 5. Risk Assessment Results
>    - List of observations, i.e. vulnerability/threat pairs
>    - List and brief description of each observation. E.g., Observation: User passwords are easy to guess
>    - Discussion of threat sources and vulnerabilities
>    - Existing security controls
>    - Likelihood discussion
>    - Impact analysis
>    - Risk rating
>    - Recommended controls
> 6. Summary

Once risk assessment is completed, the results should be documented. Such documents are not set in stone, but should be treated as evolving frameworks. Documented results also facilitate ease of communication between different stakeholders — senior management, budget officers, operational and management staff. There are numerous ways in which the output of risk assessment could be presented. The NIST-suggested format appears in Box 6.2.

# Risk Mitigation

Risk mitigation involves the process of prioritizing, evaluating, and implementing appropriate controls. Risk mitigation and the related processes of sound internal risk control are essential for the prudent operation of any organization. Risk control is the entire process of policies, procedures, and systems that an institution needs to manage all risks resulting from its operations. An important consideration in risk mitigation is to avoid conflicts of interest. Risk control needs to be separated from and sufficiently independent of the business units.

In many organizations risk control is a separate function. Inability to recognize the importance of risk mitigation and appropriate control identification results in:

- Lack of adequate management oversight and accountability. This results in failure to develop a strong control culture within organizations.
- Inadequate assessment of the risk.
- The absence or failure of key control activities, such as segregation of duties, approvals, reviews of operating performance, etc.
- Inadequate communication of information between levels of management within an organization, especially in the upward communication of problems.
- Inadequate or ineffective audit programs and other monitoring services.

In dealing with risks and identifying controls, the following options may be considered:

- **Do nothing.** In this case the potential risks are considered acceptable and a decision is taken to do nothing.
- **Risk Avoidance.** In this case the risk is recognized, but strategies are put in place to avoid the risk by either abandoning the given function or through system shutdowns.
- **Risk Prevention.** In this case the effect of risk is limited by using some sort of control. Such controls may minimize the adverse effects.
- **Risk Planning.** In this case a risk plan is developed, which helps in risk mitigation by prioritizing, implementing, and maintaining the range of controls.
- **Risk Recognition.** In this case the organization acknowledges the existence of the vulnerability and attempts to undertake research to manage and take corrective actions.
- **Risk Insurance.** At times the organization may simply purchase insurance and transfer the risk to someone else. This is usually done in cases of physical disasters.

Implementation of controls usually proceeds in seven stages. First, the actions to be taken are prioritized. This is based on levels of risk identified in the risk assessment phase. Top priority is given to those risks that are clearly unacceptable and have a high risk ranking. Such risks require immediate corrective actions. Second, the recommended control options are evaluated. Feasibility and effectiveness of recommended controls is considered. Third, a cost-benefit analysis is conducted. Fourth, controls are selected based on the cost-benefit analysis. A balance between technical, operational, and management controls is established. Fifth, responsibilities are allocated to appropriate people who have expertise and skills to select and implement the controls. Sixth, a safeguard implementation plan is developed. The plan lists the risks, recommended controls, prioritized actions, selected controls, resources required, responsible people, start and end dates of implementation, and maintenance requirements. Seventh, the identified controls are implemented and an assessment of any residual risk is made. Figure 6.3 presents a flow chart of risk mitigation activities as proposed by NIST.

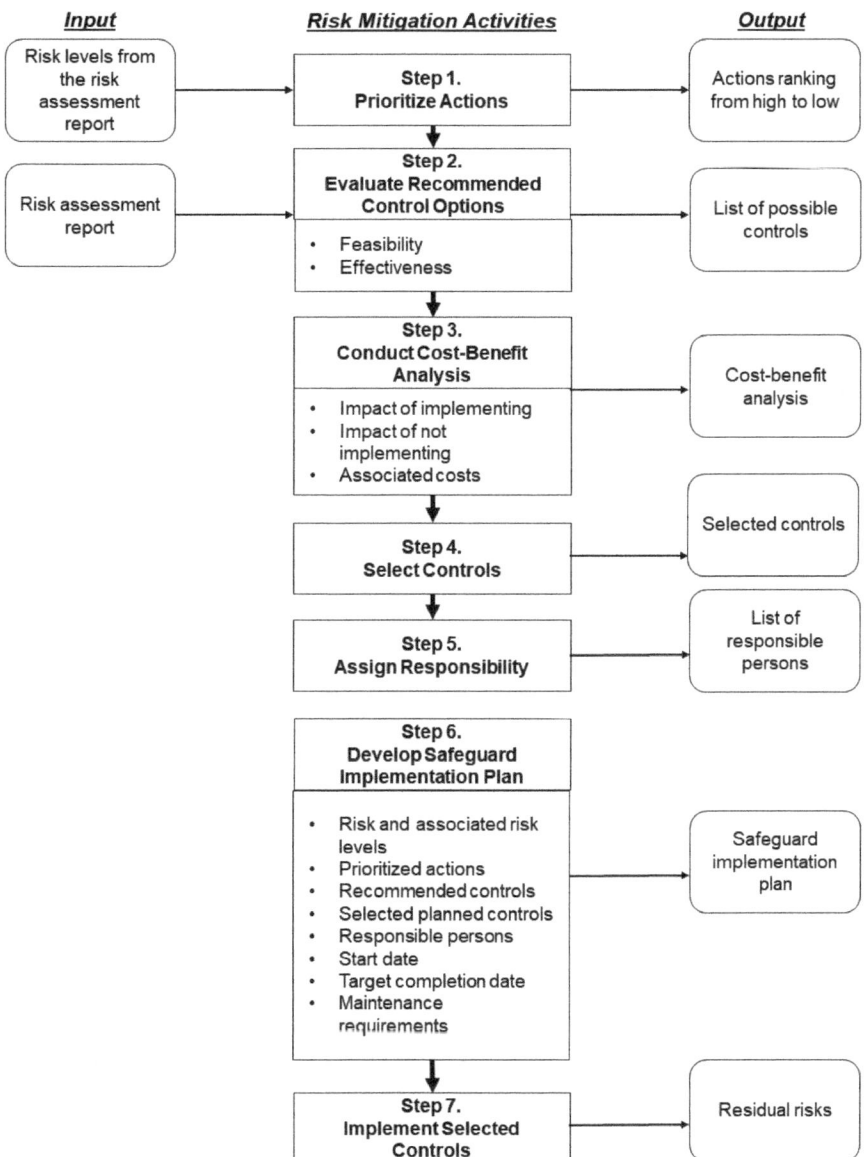

**Figure 6.3.** Risk mitigation flow of activities (NIST SP 800-30)

## Control Categories

Security controls when used properly help in preventing and deterring the threats that an organization might face. Control recommendation involves choosing among a combination of technical, formal, and informal interventions (Table 6.5).

There may be a lot of trade-offs that an organization might have to consider. Implementation of technical controls might involve add-on security software to be installed,

**Table 6.5.** Summary of technical, formal, and informal controls

| Technical Controls | Formal Controls | Informal Controls |
|---|---|---|
| **Supportive Controls**<br>• Identification—implemented through mandatory and discretionary access control<br>• Cryptographic key management<br>• Security administration<br>• System Protections—object reuse, least privilege, process separation, modularity, layering | **Preventive**<br>• Security responsibility allocation<br>• Security plans and policies<br>• Personnel security controls—separation of duties, least privileges<br>• Security awareness and training | **Preventive Controls**<br>• Security awareness program<br>• Security training in both technical and managerial issues<br>• Increasing staff competencies through ongoing educational program<br>• Developing a security subculture |
| **Preventive Controls**<br>• Authentication—tokens, smart cards, digital certificates, Kerberos<br>• Authorization<br>• Access control enforcement—sensitivity labels, file permissions, access control lists, roles, user profiles<br>• Non-repudiation—digital certificates<br>• Protected communication—virtual private network, Internet Protocol Security, cryptographic technologies, secure hash standard, packet sniffing, wiretapping<br>• Transaction privacy—Secure Socket Layer, secure shell | **Detection management controls**<br>• Personnel security controls—background checks, clearances, rotation of duties<br>• Periodic effectiveness reviews<br>• Periodic system audits<br>• Ongoing risk management<br>• Address residual risks | **Detection**<br>• Encourage informal feedback mechanisms<br>• Establish reward structures<br>• Ensure formal reporting structures match informal social groupings |
| **Detection and Recovery**<br>• Audit<br>• Intrusion detection and containment<br>• Proof of Wholeness<br>• Restore secure state<br>• Virus detection | **Recovery management controls**<br>• Contingency and disaster recovery plans<br>• Incident response capability | **Recovery**<br>• Provide ownership of activities<br>• Encourage stewardship |

while formal controls may be taken care of simply by issuing new rules through internal memorandums. Informal controls require culture change and development of new normative structures. A discussion of the three kinds of controls was also presented in Chapter 1.

The output of the control analysis phase is a list of current and planned controls that could be used for the IT system. These would mitigate the likelihood of realizing vulnerabilities and hence reduce chances of adverse events. Sample technical, formal, and information controls are summarized in Table 6.5.

## Risk Evaluation and Assessment

New threats and vulnerabilities emerge on a rather ongoing basis. At the same time people in organizations change. So do business processes and procedures. Such continual change suggests that the risk management task needs to reevaluate itself on a continuing basis. Not only do newer vulnerabilities need to be understood, so do the processes linked with its management.

Continuous support of senior management needs to be stressed, and also the support and participation of the IT team. The skill levels of the team and the general competence of the risk management organization need to be reassessed on a regular basis. In the US, a three-year review cycle for federal agencies has been suggested. Although the law requires risk assessment to be undertaken especially for the federal agencies, it is generally good practice for such an assessment to be done and integrated into the systems development life cycle.

Evaluation of the risk management process and a general assessment of the risks is a means to ensure that a feedback loop exists. Clearly establishing a communication channel that identifies and evaluates the range of risks is a good practice. Many of the risk management models and methods do integrate evaluation and assessment as part of their processes. Some of the models are discussed in the remainder of this chapter.

## COBRA: A Hybrid Model for Software Cost Estimation, Benchmarking, and Risk Assessment

Proper planning and accurate budgeting for software development projects is a crucial activity, impacting the lives of all software businesses. Accurate cost estimation is the first step towards accomplishing the aforementioned activities. Equally importantly, the software business needs to plan for risks in order to effectively deal with the "uncertainty" factor associated with all typical software projects, besides benchmarking their project to gauge its productivity against their competitors in the market place.

The two major types of cost estimation techniques available today are, on the one hand, developing algorithmic models, and on the other, informal approaches based on the judgment of an experienced estimator. However, each of these is plagued by some inherent problems. The former makes use of extensive past project data. Statistical surveys show that 50% of organizations do not collect data on their projects, rendering the construction of an algorithmic model impossible. The latter approach has more often than not led to over- or underestimation, each of which translates into a negative impact on the success

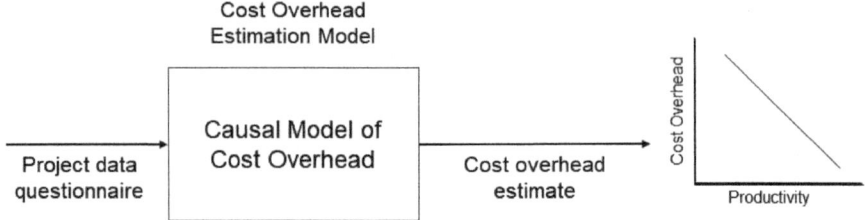

**Figure 6.4.** Overview of productivity estimation model

of the project. Also, it is not always possible to find an experienced estimator available for the software project.

At the Fraunhofer Institute for Experimental Software Engineering in Germany, Lionel Briand, Khaled Emam, and Frank Bomarius [2] developed an innovative technology to deal with the abovementioned issues confronting software businesses. They devised a tool called COBRA (Cost estimation, Benchmarking and Risk assessment), which utilizes both expert knowledge (experienced estimators) and quantitative project data (in a limited amount) to perform cost modeling (Figure 6.4).

At the heart of COBRA lies the productivity estimation model. It comprises two components: a causal model of estimating cost overhead and a productivity model, which calculates productivity using the output of the causal model as its input. The following figure depicts this model.

The hybrid nature of the COBRA approach is depicted in its productivity estimation model. While the cost overhead estimation model is based on the project manager's expert knowledge, the productivity estimation model is developed using past project data.

The relationship between productivity (P) and Cost Overhead (CO) is defined as:

$$P = \beta_0 - (\beta_1 \times C_o)$$

Where, $\beta_0$ is the productivity of a nominal project[2]
And $\beta_1$ is the slope between $C_o$ and P

The $\beta$ parameters can be determined using only a small historical set of data (around 10). This is a significant advantage considering the fact that the absence of large amounts of historical data is not an issue, unlike traditional algorithmic approaches of cost modeling.

### Estimating the Cost of the Project

The researchers have assumed the relationship between effort and size is linear, considering the evidence provided by recent empirical analysis. This relationship can be expressed as:

### Effort = α × Size

The α value in turn is given by:

---

[2] A nominal project is a hypothetical ideal; it is a project run under optimal conditions. A real project always deviates from the nominal to a significant extent.

$$\alpha = \frac{1}{\beta_0 - (\beta_1 \times C_O)}$$

Using $\alpha$ (calculated using $C_O$) and Size, the effort for any project can be estimated.

## Project Cost Risk Assessment

The assessment of risk associated with a given project is the probability that the project will overrun its budget. The authors have used $C_O/\alpha$ values to compute the probability. The model defines the maximum tolerable probability that can be defined, so also categorizes projects into high-risk entities or otherwise.

## Project Cost Benchmarking

During this process the $C_O$ value of a given project is compared to a historical data set of similar projects. This helps the business analyze if the given project will be more difficult to run than the typical project, and if it would entail more $C_O$ than the typical project.

While the cost of software projects may be affected by a number of factors, not all of them have the same impact. This study uses a total of 12 cost drivers/factors to develop the cost risk model. (The authors zeroed in on these 12 after careful assessment by project managers of all cost drivers, separating them into the ones that would have the largest impact and the ones that would have relatively less impact.)

Software reliability, software usability, and software efficiency emerged as the factors that would cause the greatest impact on cost, while database size and computational operations were categorized as the ones that would cause the least impact.

In developing the causal model the authors took into consideration the effect caused by interaction of cost drivers. Subsequently, a project data questionnaire was developed that helps the project managers characterize their projects in order to effectively use the cost estimation model. As a next step, the qualitative causal model was quantified, using the advice of experiences managers, followed by operationalizing the cost overhead estimation model.

Using size and effort data for six recently completed projects, the cost overhead estimate was obtained and the relationship between the cost overhead and productivity was determined for each of these projects. Statistical analysis showed that on an average the model will over/underestimate by just 9% of the actual. Results show that COBRA is a convenient method when needing to develop local cost estimation and risk models. It is however not suitable when there is a large data set of project data.

# A Risk Management Process Model

Originally developed by Alexander Korzyk [4] in his doctoral dissertation, but later published in the *Journal of Information System Security* [5], the model integrates risk analysis into IS development and specification of security requirements at the initial stage of system development.

## Three Levels of I2S2 Model

The I2S2 model is based on the strategic event response model, which is a combination of two military organizational processes (Deliberate Planning Process and Crisis Action Planning Processes) with Michael Porter's [6] five competitive forces. I2S2 is a conceptually designed meta-model with three levels that integrate six primary components. At level zero, the I2S2 model combines the submodels of six primary components, namely, threat definition, information acquisition requirements, scripting of defensive options, threat recognition and assessment, countermeasure selection, and post-implementation activities. The level zero components have many complex relationships that will be further simplified as the model expands by level. The integration of the six subcomponents is achieved using the three channels developed by Galbraith [3] that facilitate the integration—structural, procedural, and instrumental—with the three detailed I2S2 model levels: level one, level two, and level three, respectively.

Level one of the I2S2 model shows high-order inter-relationships between the six components of I2S2 (see Figure 6.5). However, it should be noted that the relationship among

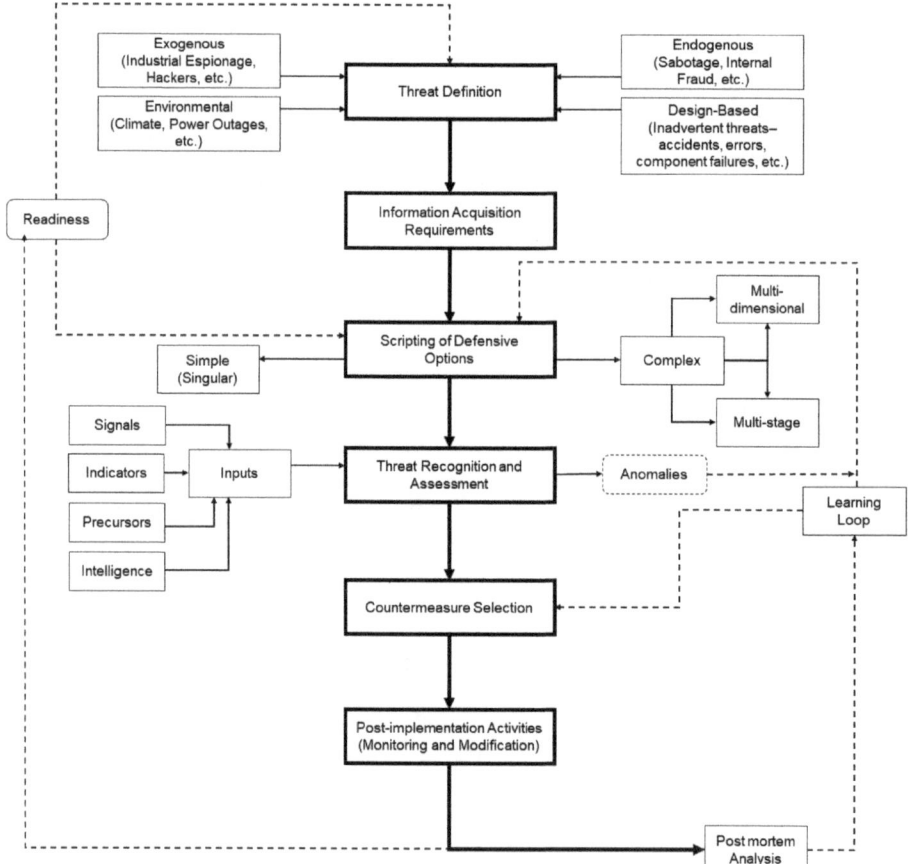

**Figure 6.5.** I2S2 Model at level one

the six components is not completely linear. The upper half of the I2S2 model (the first three components) deals with the notional threats and vulnerabilities, while the lower half (the last three components) deals with the real security incidents related to those notional threats and vulnerabilities. The learning loop that is part of level one enhances the selection of defensive options and countermeasure selection. The same learning loop even improvises, if need be, the threat definition so that the future security incidents consider the lessons learned from the past. Moreover, at this level the structural integration is achieved since the components have inter- and intra-relationships among themselves.

Levels two and three are discussed in more detail, and in relation with the components, in the following paragraphs. It is important to note that level two considers the performance of the components to achieve the procedural integration. Level three is finer and specifies the technical integrative facilities and mechanisms. The I2S2 model is clearly useful since it integrates IS security design issues with systems development. In most system development efforts, such integration does not take place, thus resulting in development duality problems.

## Six Components of I2S2 Model

As stated previously, the I2S2 model consists of six primary components:

Component-1: Threat definition
Component-2: Information acquisition requirements
Component-3: Scripting of defensive options
Component-4: Threat recognition and assessment
Component-5: Countermeasure selection
Component-6: Post-implementation activities

### Component-1: Threat Definition

The threat definition component provides the foundation for the successive submodels. Though it would be difficult to define the word "threat" in definitive terms, a clear understanding of the submodel cannot be formed without a definition. Threat can be defined as an indicator or pointer that something might happen in future to a particular asset. Though the severity varies greatly, all threats can be categorized into four types, namely, environmental, exogenous, endogenous, and design-based.

*Environmental threats* primarily relate to climatic threats, power outages, and various other natural disasters such as earthquakes, hurricanes, floods, tornadoes, and landslides. Most of these environmental threats happen without any warning. The environmental threats relate to the physical equipment, such as computers, laptops, etc., that store information. The consequences of any environmental threat might range anywhere between the temporary malfunction of the physical equipment and the total annihilation of the equipment. The obvious solutions to environmental threats include environmental scanning, weather advisories, and various other types of alerts that can be used to prevent the environmental threats.

*Exogenous threats* include cybercrimes committed by various perpetrators for the purpose of industrial espionage and to thwart the competition of other companies. Since these threats expose business secrets of the companies, the affected companies very rarely report these types of abuses as any reporting of such crimes adversely affects the companies' reputation and future profitability.

*Endogenous threats* include sabotage, fraud, or other types of financial or non-financial embezzlements caused by disgruntled employees who had been either denied promotion or a pay raise. If the disgruntled employee is incapable or unwilling to cause the threat directly, he/she would pass the critical information to outsiders or competitors who would take advantage of the tip-off. Periodic audits and inspection, forced vacation of the employees who occupy crucial positions in the company, are some of the remedial measures.

*Design-based threats* can be either hardware related or software related. Software-related threats are for the most part inadvertent threats that happen at the time of designing information systems or during the critical updates of the systems. These threats are often due to lack of adequate training for the system developers. Maintaining a help desk or CERT alert mechanisms can counteract any software-related threats. The hardware-related threats are either due to initial design or due to insufficient maintenance of the equipment. Safety reports and regular maintenance are some of the solutions to overcome the hardware-related threats.

A particular set of threats come together to form threat scenarios. Essentially, any threat scenario consists of six types of elements that detail who caused the threat, what capabilities and resources are necessary, why the adversaries are motivated to induce the threat, when, where, and how the threats get manifested. Each threat scenario results in one of four outcomes: unauthorized disclosure, unauthorized modification, denial or disruption of service, and fraud or embezzlement. While the threat scenarios form the basis for monitoring the new threat patterns in the coming future, the threat scenarios also take a feedback loop from the third component of the I2S2 model, the scripting of defensive options.

While level two of the threat definition component expands to include threat scenario facilities and instruments, level three of threat definition details the building of a threat scenario using the relevant modules, such as threat probability module and threat impact module. With regard to the probability of a threat and its impact on the organization, it will be decided whether a particular threat is active or inactive. If it is an inactive threat, nothing will happen, but if the threat is an active threat, it would be added to the ordered threat array and will be taken to the next level in the I2S2 model, the development of information acquisition requirements.

## *Component-2: Information Acquisition Requirements*

For each known and active threat included in the ordered threat array, information acquisition can be done with one or more classes of information: signals, precursors, indicators, and intelligence. While a signal is an irrefutable direct data typically received during the emergent threat time period to the information system, an indicator is an indirect data source that is also identified during the threat period. A precursor is a set of events or conditions suggestive of a future threat. Intelligence is very closely related to indicators except that intelligence is softer or non-definitive in character. The above classification of information forms the basis of the information acquisition requirements component. Each threat may have more than one class of information and each class of information may relate to more than one threat. Attaching value to these pieces of information is an important step in the information acquisition requirements.

Quantifying the value of organizational data is very complex and depends on the size and nature of the organization. This is required for various purposes including developing a data security strategy. There are many methods to quantify the organizational information, such as insurance policy quantification approach, cost of attack–based approach, sensitive data approach, etc. However, any attempt to quantify information involves a trade-off between the cost and benefit of this quantification. The value of information depends on the specific application and versatility of the information.

Depending on the nature and value of the information, an organization can decide on the kind of protection mechanisms it needs to institute. Linking value to information also helps in considering information as an asset, thus allowing for adequate data security policies and procedures to be instituted.

With respect to the identification of the four classes of information that relate to a threat scenario, an organization can make use of many methods that are prevalent in the industry. Pattern recognition and/or matching is a common method, where organizations accumulate all types of data and use various technologies to observe and unearth the hidden patterns. Signatures such as IP addresses attached to the computer or network, virus definitions, etc., can also help the identification of information.

Classification, quantification, and identification of signals, indicators, precursors, and intelligence information lead to the next level (level three) in the information acquisition requirements component in the I2S2 model. At this level each class of information is subdivided into four categories of threat identification. For example, a signal can lead to either a single threat or multiple threats and a set of signals can lead to either a single threat of multiple threats. This categorization is necessary for the next component of scripting defensive options.

There are two important relationships that exist among these four classes of information which are made explicit in level three of the information acquisition requirement component. Intelligence information might lead to the identification of any of the other three classes of information. Also, each of these four classes of information might influence the other types of information which might eventually lead to a threat scenario. Eventually, all the classes of information that are relevant to a threat form the aggregated information acquisition requirement array which will be used in scripting defensive options.

## *Component-3: Scripting of defensive options*

Developed from the previous component, the information acquisition requirements contain information that is very closely related to the time at which the threat scenario is initiating the task of defensive script generation. This guides the goals of defensive script options. For example, prior to the threat, the goal would be to prevent the threat, and during and after the threat, the focus would be damage control and containment. Whatever may be the goal of the defensive script generation, there are primarily two stages in the scripting of defensive options: the initial and final scripting of defensive options. These two stages match with level one and level two of the scripting of defensive options component.

The initial scripting of defensive options depends on the nature of the threat scenario. If the threat scenario is simple, the final scripting of defensive options is achieved and the threat is dealt with accordingly. However, if the threat scenario is complex, then the script should follow one of these three options: multi-dimensional, or multi-stage, or multi-dimensional and multi-stage. The bottom line is that each threat scenario is linked with more than one security incidents and hence influences the allocation of information agents to deal with the threats.

When it comes to final scripting of defensive options, each threat scenario is matched with the specific countermeasures to see if there is a full match or partial match or no match at all. Based on the matching patterns and the options available, many things can happen at this level of scripting defensive options component. We either accept the defensive script and move on to the next component or reject the script and update the information acquisition requirements components.

A cybernetics approach is used to generate the defensive scripts. Cybernetics is the science of organization, planning, control, and communication between man, machines, and organizations. Defensive scripts based on this approach can cater to an organization's internal security needs and can handle the incompleteness and uncertainty of the situations.

### Component-4: Threat Recognition and Assessment

The threat recognition and assessment component is organized in three modules. The information acquisition targets the inputs in the form of signals, indicators, precursors, and intelligence that directly feed into the threat recognition facilities module. At this module, there are three levels of threat recognition. If the threat is certain (i.e., the probability of the threat is 1), then the countermeasure selection component is performed for the threat. If the threat is completely unknown or unrecognizable (i.e., the probability of the threat is 0), then a new template is built for future reference and identification. However, if the likelihood of the threat is ambiguous (i.e., the probability of threat is between 1 and 0), then the next module—threat/security incident assessor—kicks in. At this stage, the time period associated with the occurrence of the threat is also considered and the threats are assessed as projected or emergent or occurring or just threats.

Level three of threat recognition and assessment analyzes the threats at a deeper level. With the feed from a signal, indicator, precursor, or intelligence, the threat/situation monitoring module moves the information to the security incident reporting and assessment module. Here occurrences are either entered with the appropriate scenario in the repository and associated signals, indicators, precursors, or intelligence information; or the occurrences may be discarded because they may be highly unlikely. If the scenarios are selected, then that information is sent to the countermeasure selection component.

### Component-5: Countermeasure Selection

A countermeasure is a safeguard added to the system design such that it mitigates or eliminates the vulnerability. A countermeasure reduces the probability that any damage would occur. Countermeasures are the resources that are at the disposal of the organization to counter threats or to minimize the damage from a security threat. Countermeasures can be active, passive, restorative, preventive, and pre-emptive. Each of the above five countermeasures are deployed in different situations depending on the severity of the threat and the quantified value of the information assets that are involved.

No matter which countermeasure is selected to thwart the potential threat, the concept behind the deployment of any countermeasures is Cooperative Engagement Capability. This concept involves the centralization of the countermeasure allocation in the hands of a command center which monitors the overall situation of all the facilities within an organization. For example, if there is a potential threat scenario that is emerging at a particular location, it is clear that the particular location would require more resources than they have at their disposal. In this situation, the central command center allocates the necessary resources from other locations so that the threat can be prevented or the damage can be minimized. Once the situation becomes normal, all the resources are ploughed back into the respective locations.

At level three of countermeasure selection, attention is paid to the cost aspect since the use of any countermeasures involves some cost to the organization. Thus, the main criteria to choose the resources to counter a threat would be efficiency rather than averting the threat or damage control. However, care must be taken to balance cost and benefits. Obviously, resources need to be diverted to places where there is most need. If the cost of a countermeasure is high and the importance of the information asset is marginal, then a managerial decision needs to be taken with regard to the suitability, nature, and scope of the countermeasure. There is clearly no need to spend an exorbitant amount of money to protect a relatively trivial asset.

The timing of allocation of resources to avert the threat plays a very important role as well. Countermeasures deployed too early may change the attacker's strategy for the worse. Hence, organizations must watch and let the attacker reach a point of no return so that the countermeasure will be effective in its implementation of remedial measures.

## *Component-6: Post-implementation Activities*

While most of the tools that help in analysis of risk and threats do not consider the post-implementation review, it is important that such an analysis is undertaken. There are three reasons why such an analysis is important. First, post-implementation review allows for an organization to reconsider the efficacy of the countermeasures. This allows for the controls to be fine-tuned and aligned with the purpose and intent. Second, the post implementation review allows for the feedback to inform any improvements that might be instituted. Lessons learned can be formalized and made available to the decision-makers. Such learning can also be integrated into the I2S2 model at various levels.

Third, a post-implementation review acts as a real-time crisis management mechanism. There may be some situations in which the attacker uses new techniques or technologies that are unfamiliar to the organization's countermeasures. In those situations, the post-implementation activities component of I2S2 module feeds back into the defensive scripting options. This helps with categorizing threat scenarios and improvising scripting options and countermeasure selection.

## Concluding Remarks

In this chapter the concepts of risk assessment, risk mitigation, and risk evaluation are introduced. Various threat classes and resultant actions are also presented. Descriptions and discussions are based on NIST Special publication 800-30, which in many ways is considered the standard for business and government with respect to risk management for IS security. The discussion sets the tone for a comprehensive management of IS risks.

Two models, which bring together a range of risk management principles, are also presented. The models form the basis for a methodology for undertaking risk management for IS security. The first model incorporates software cost estimation and benchmarking for risk assessment. Commonly referred to as COBRA, the model in many ways is an industry standard, especially for projects where cost estimation is a major consideration. The second model emerges from doctoral-level research and sets the stage for incorporating risk analysis in IS development and specification. Referred to as the I2S2 model, it incorporates threat definition, information acquisition requirements, scripting of defensive options, threat recognition and assessment, countermeasure selection, and post-implementation activities. All together the concepts suggest a range of principles that are extremely important for undertaking risk management.

## In Brief

- There are three essential components of risk management: risk assessment, risk mitigation, and risk evaluation
- Risk assessment considers risk to be a function of the likelihood of a given threat resulting in certain vulnerabilities
- The US National Institute of Standards and Technology (publication 800-30) identifies the following nine steps to be integral to risk assessment:
  - System Characterization
  - Threat Identification
  - Vulnerability Identification
  - Control Analysis
  - Likelihood Determination
  - Impact Analysis
  - Risk Determination
  - Control Recommendations
  - Results Documentation
- There are four classes of vulnerabilities:
  - Behavioral and attitudinal vulnerabilities
  - Misinterpretations
  - Coding problems
  - Physical
- There are three elements in calculating the likelihood that any vulnerability will be exercised. These include:
  - Source of the threat, motivation, and capability
  - Nature of the vulnerability
  - Effectiveness of current controls
- Risk mitigation involves the process of prioritizing, evaluating, and implementing appropriate controls
- Evaluation of the risk management process and a general assessment of the risks is a means to ensure that a feedback loop exists.

# Questions and Exercises

**Discussion questions.** These are based on a few topics from the chapter and are intentionally designed for a difference of opinion. These questions can best be used in a classroom or a seminar setting.

1. What is the systematic position of risk management in ensuring the overall security of an enterprise? Discuss giving examples.

2. If an organization were to make risk management central to its security strategy, what would be the positive and negative aspects associated with such a move?

3. "Risk management is really a means to communicate about the range of risks in an organization." Comment on the statement giving examples.

**Exercise.** Use one of the prescribed risk management approaches in this chapter to calculate the level of risks your department might face. Calculate the business impact and the extent of financial loss. After undertaking the exercise, comment on the usefulness of risk assessment in identifying potential risks and generating relevant management strategies.

See Appendix A (after the Index) for a set of Short Questions for this chapter.

## Case Study: Insiders Play a Role in Security Breaches

In April 2005, a high-tech scheme was uncovered to steal an estimated £220m from Mitsui Bank in London. It was claimed that members of the cleaning crew had placed hardware devices on the keyboard ports of several of the bank's computers to log keyboard entries. Investigators found several of the devices still attached at the back of the computers. The group also had erased or stopped the recording of the CCTV system recorder to cover their tracks. Police in Israel have detained a member of the cleaning crew for questioning, after he attempted to transfer £23m into his bank account. The suspect is believed to be a junior member of the gang, and the others remain at large. Attempts to trace the gang through forensics have failed thus far, but the investigation continues. According to a spokesperson for London's High Tech Crime Unit, the criminals have not been able to transfer any funds so no money has been lost thus far. The concentration of the forensic investigation is now focused on the key logger devices that were inserted into the USB keyboard ports. The devices can be purchased at several spy shops on the Internet, and can be used to download passwords or other data used to authenticate users on secure systems. Wireless keyboards are also a threat, and it is believed the bank has since banned their use. It is rumored that some banks are super-gluing keyboards and other devices into the ports, but this seems to be a short-sighted attempt to resolve the problem. Most computers have additional USB ports, and other devices such as flash drives or even iPods can exploit this weakness. According to figures from software auditors Centennial Software, a vast majority of IT managers surveyed took no action to prevent such devices coming into the workplace even though many of them recognized USB storage devices were a threat. "External security risks are well documented, but firms must now consider internal threats, which are potentially even more damaging," said Andy Burton, chief executive of Centennial Software.

Software is available which may mitigate this risk, but policies should also be in place which limit access to sensitive computers. Portable USB devices and key loggers are only some of the high-tech methods criminals are using to exploit computer weaknesses. Computer criminals are growing increasingly sophisticated as organized crime gangs have found cybercrime to offer greater rewards with less risk.

In 2014, a California school, Corona del Mar, expelled 11 students when it was discovered that a private tutor had directed the students to change their grades on the teacher's computer. The pupils had used a key logger as a means to gain access. The private tutor, Timothy Lai, was implicated and a huge swathe of grade tampering was discovered. Interestingly, hardware key loggers are legal and there have been several instances were cases of "domestic key logging," when family members use such devices to spy on each other, have been reported.

1. What might be the reason that many IT managers are ignoring the threat from portable USB devices?
2. What measures could be taken to limit insider threats such as the one at Mitsui Bank?
3. Should key logging devices be legal? What could be possible beneficial uses of key logging devices?

Source: various news items appearing in:

Warren, P., and Streeter, M. 2005. Mission impossible at the Sumitomo Bank. *The Register*, Apr. 13. http://goo.gl/Jiz0yR. Accessed Mar. 30, 2017.

Dunn, J. 2014. US school expels pupils for using hardware keyloggers to change grades. *Techworld*, Feb. 5. http://www.goo.gl/XUJeMw. Accessed Mar. 30, 2017.

# References

1. Ajzen, I. and M. Fishbein, *Attitude-behaviour relations: a theoretical analysis and review of empirical research.* Psychological Bulletin, 1977. **84**(5): p. 888-918.
2. Briand, L.C., K.E. Emam, and F. Bomarius. *COBRA: a hybrid method for software cost estimation, benchmarking, and risk assessment.* in *20th international conference on Software engineering.* 1998. Kyoto, Japan: IEEE Computer Society Washington, DC, USA.
3. Galbraith, J.K., *A journey through economic time : a firsthand view.* 1994, Boston: Houghton Mifflin.
4. Korzyk, A., *A Conceptual Design Model for Integrative Information System Security*, in *Information Systems Department.* 2002, Virginia Commonwealth University: Richmond, VA.
5. Korzyk, A., J. Sutherland, and H.R. Weistroffer, *A Conceptual Model for Integrative Information Systems Security.* Journal of Information System Security, 2006. **2**(1): p. 44-59.
6. Porter, M., *Competitive strategy.* 1980, New York: Free Press.

# CHAPTER 7

# Information Systems Security Standards and Guidelines

> *If one wants to pass through open doors easily, one must bear in mind that they have a solid frame: this principle, according to which the old professor had always lived is simply a requirement of the sense of reality. But if there is such a thing as a sense of reality—and no one will doubt that it has its raison d'être—then there must be something that one can call a sense of possibility. Anyone possessing it does not say, for instance: Here this or that happened, will happen, must happen. He uses his imagination and says: Here such and such might, should or ought to happen. And if he is told that something is the way it is, then he thinks: Well, it could probably just as easily be some other way. So the sense of possibility might be defined outright as the capacity to think how everything could "just as easily" be, and to attach no more importance to what is than to what is not.*
>
> —R Musil (1930–1942),
> *The Man without Qualities*

Every single time Joe Dawson's IT director came to see him, he mentioned the word *standard*. He always seemed to overrate the importance of security standards—stating that they had to adopt 27799 or have to be in compliance with the NIST guidelines. Joe had heard of 27799 so much that he became curious to know what it was about and what standards could do to help improve security. Joe had a real problem with the arguments presented. He always walked out of the meetings thinking that security standards were some sort of magic that, once adopted, all possible security problems would get solved. But adopting a standard alone is not going to help any organization improve security.

Joe remembered history lessons from his high school days. For him, standards had been in existence since the beginning of recorded history. Clearly some standards had been created for convenience derived from harmonizing activities with environmental demands. Joe knew that one of the earliest examples of a standard was the calendar, with the modern day calendar having its origins with the Sumerians in the Tigris and Euphrates valley. Security standards therefore would be no more than a means to achieve some common understanding and perhaps comply with some basic principles. *How could a standard improve security when it was just a means to establish a common language?* Joe thought. At least this is something that one of Joe's high school friends, Jennifer, had mentioned. Jennifer had joined Carnegie Mellon University and had worked on the earlier version of the "Computer Emergency Response Teams." CERT, as they were often referred to, had some role in defining the cyber security version of the very famous Capability Maturity Model.

Although Joe did not think much of the standards, he felt that there was something that he was missing. Obviously there was a lot of hype attached to standardizations and certifications. This was to the extent that a number of his employees had requested that SureSteel pay for them to take the CISSP exam. Joe felt that he had to know more. The person he could trust was Randy, his high school friend who worked at MITRE.

Joe called Randy and asked, "Hey, what is all this fuss about security standards? I keep on hearing numbers such as 27799, etc. Can you help, please?"

---

In the ever-increasing organizational complexity and the growing power of information technology systems, proper security controls have never been more important. As the modern organization relies on information technology–based solutions to support its value chain and to provide strategic advantages over its competitors, the systems that manage that data must be regarded as among its most valuable assets. Naturally, as an asset of an organization, information technology solutions must be safeguarded against unauthorized use and/or theft; therefore proper security measures must be taken. An important aspect related to security measures and safeguards is that of standardization. Standards in general have several advantages:

1. Standards provide a generic set of controls, which form the baseline for all systems and practices.
2. Standards help in cross-organizational comparison of the level to which a given control structure exists.
3. Standards help in formulating necessary regulations.
4. Standards help with compliance to existing practices.

In this chapter we discuss the systematic position of standards in defining the security of systems within an organization. We link the role of standards in the overall protection of the information assets of a firm. Finally, we discuss the most important information systems: security standards.

# The Role of Standards in Information Systems Security

Defining standards is akin to establishing controls. In the realm of information systems security, it's not easy to establish a generic set of controls that would cover all kinds of systems and organizations. Yet there are some generic principles that can be adhered to. Before we begin exploring each of the types of security standards, it's important to understand the nature of controls. For it is the controls that eventually get standardized and principles developed. There are four types of control structures that are essential to take note of in specifying security in most enterprises. These are auditing, application controls, modeling controls, and documentation controls.

## Auditing

The process of auditing a system is one of the most fundamental control structures that may be used to examine, verify, and correct the overall functionality of that system. In order to audit an event, that event must necessarily have occurred. Fundamentally, the audit process verifies that a system is performing the functions that it should. This verification is of critical importance to security processes. It is not sufficient merely to implement security procedures: the organization must have a way to verify that the security procedures work. No manager or IT professional should be satisfied to say "We have not had a security breach; therefore our procedures are adequate."

The audit process can be accomplished in two primary forms. Although both forms are actually variations of the same concept, most organizations distinguish the two for implementation purposes. The first form of the audit function is the record of changes of state in a system (i.e., events and/or changes in the system are recorded). The second form of the audit function is a systematic process that examines and verifies a system (i.e., the system is evaluated to determine if it is functioning correctly). In order for the audit control structure to be successful, both forms of the audit control structure need to be in place. It is difficult to perform a systematic evaluation of a system or event if they are not recorded. Furthermore, if changes of state are not recorded, there is no reason to record the changes. Therefore successful auditing control procedures should

- Record the state of a system
- Examine, verify, and correct the recorded states

All too often, the audit function within an organization is invoked only during the production stage of a process. In reality, the audit process should be invoked at all stages of the system life cycle. Generally speaking, the costs associated with the correction of an error in a system increase as the system life cycle progresses. While the word "audit" conjures images of a certified public accountant combing through files looking for wrongdoing within the organization, the scope of the audit process encompasses every aspect of a system. Clearly this could become prohibitively expensive if independent auditors were to verify every aspect of and every event within a system. Therefore, who should be performing the audit function? Everyone should be performing the audit function. It is the responsibility of everyone working on a system, whether during design or production, to think critically and ask, "Is the system doing what it is supposed to be doing?"

## Application Controls

In most general terms, application controls are the set of functions within a system that attempts to prevent a system failure from occurring. Application controls address three general system requirements:

- Accuracy
- Completeness
- Security

Accuracy and completeness controls both address the concept of correctness (i.e., that the functions performed within a system return the correct result). Specifically, accuracy controls address the need for a function to perform the correct process logic, while completeness controls address the need for a function to perform the process logic on all of the necessary data. Finally, security controls attempt to prevent the types of security breaches previously discussed.

Application controls are categorized based on their location within a system, and can be classified as

- Input controls
- Processing controls
- Output controls

Although many system failures are the result of input error, processing logic can be incorrect at any point within the system. Therefore, in order to meet the general system requirements stated previously, all three classes of system controls are necessary in order to minimize the occurrence of a system failure.

While security controls often concentrate on the prevention of intentional security breaches, most breaches are accidental. As authorized users interact with a system on a regular basis, the likelihood of accidental breaches is much higher than deliberate breaches; therefore the potential cost of accidental breaches are also much higher.

Application controls are typically considered to be the passwords and data integrity checks embedded within a production system. However, application controls should be incorporated at every stage of the system life cycle. In the highly automated development environments of today, application controls are just as necessary to protect the integrity of system development and integration—for example, the introduction of malicious code at the implementation stage could quite easily create costly system failures once a system were put into production. In order to best minimize the occurrence of a system failure, application controls and audit controls should be coordinated as part of a comprehensive strategy covering all stages of the system life cycle. Application controls address the prevention of a failure, and audit controls address the detection of a failure (i.e., audit controls attempt to determine if the application controls are adequate).

## Modeling Controls

Modeling controls are used at the analysis and design stages of the systems life cycle as a tool to understand and document other control points within a system. Modeling controls allow for the incorporation of audit and application controls as an integral part of the systems process, rather than relying on the incorporation of controls as add-on functionality. Just as with other modeling processes within a system, modeling controls will take the form of both logical and physical models: logical controls illustrate controls required as a result of business rules, and physical controls illustrate controls required as a result of the implementation strategy.

As an example, prudence dictates that an online banking system would necessarily require security. However, it is not sufficient to simply "understand" during the initial system development that security would be required. Modeling controls show how the control would interact with the overall functionality of the system, and locates the control points within the system. In an online banking system, for example, the logical control model might include a control point that "authenticates users." The implementation control model might include a control point "verify user id and password."

In addition to the inclusion of the control point, the model should demonstrate system functionality, should the tests of the control point fail. In the online banking example noted previously, if the authentication test fails, will the user simply be allowed to try an indefinite number of times? Will the user account be locked after a certain number of failed attempts? Will failed attempts be logged? Will the logs be audited to search for potential security threats? The answers to all of these questions should be included and modeled from the first stages of the system development process.

Given the industry and the kind of application in question, appropriate standards need to be adhered to. While security standards such as ISO 27002 prescribe principles for user authentication, they do not specifically go into the details for, say, the banking industry. In such cases, specific financial standards come into play to define the appropriate security controls. Such may also be the case for other industries.

## Documentation Controls

Documentation is one of the most critical controls that can be used to maintain integrity within a system; ironically, it is also one of the most neglected. Documentation should exist for all stages of the system life cycle, from initial analysis through maintenance. In theory, a system should be able to be understood from the documentation alone, without requiring study of the system itself. Unfortunately, many IT professionals consider documentation to be secondary to the actual building of the system itself—to be done after the particular development activity has been completed. After all, the most important result is the construction of the system!

In reality, documentation should be created in conjunction with the system itself. Although the documentation of results after a particular phase is important, document controls should be in place before, during, and after that phase. Furthermore, while it is true that documentation is a commitment of resources that could be used toward other development activities, proper and complete documentation can ultimately save resources by making a system easier to understand. These savings are particularly apparent during the production and maintenance stages. How many programmers have had to reverse engineer a section of program code in order to learn what the code is supposed to be doing? Good documentation dramatically increases the accuracy and reliability of other controls, such as auditing. In the previous example, by already knowing the purpose of a section of a code, the programmer could spend his time verifying that the code was performing its intended purpose.

Good documentation controls not only answer what the functions of a system are and how those functions are being accomplished; the controls address the question of why the system is performing those particular functions. Specifically, documentation should show the correlation between system functionality and business/implementation needs. For example, an application control may be placed within a system in order to meet a specific requirement. If that requirement were to change, the control may no longer be needed and would become a waste of resources. Worse yet, the control may actually conflict with new requirements.

# Process Improvement Software

As proper controls are a requirement in order to foster improvement in organizational systems, there are tools available to assist with the successful implementation of controls within the development strategy. In addition, these automated solutions can assist with a general environment of improvement within the software development life cycle. Software tools would include classes such as automated learning and discover tools, which assist with the analysis of data and program structures to help determine exactly what is happening within a system. A second class of automated improvement tools is the program enhancement environment. These software environments assist with the management of the software life cycle from beginning to end; rather than using different tools at each stage, a single development suite can integrate multiple stages into a comprehensive package.

As was discussed, proper documentation is one of the most critical controls that can be integrated into the system development cycle. Managing and tracking documentation within large systems can be a daunting task. Two classes of tracking software can provide invaluable assistance in large systems development processes: change tracking software and requirements tracking software. As the name implies, change tracking software tracks changes through the development process. Although its purpose is relatively straightforward, the benefits are gained by the sheer volume of information that can be managed.

All systems development projects are designed to meet certain business requirements. Through the analysis and design phases, these business processes are translated and modeled into logical and physical constructs that function to satisfy the requirement. During implementation, these models are transformed into data stores, classes, and procedures. Requirements tracking is the process of connecting business requirements, through analysis and design, to the program artifacts that support the requirement. Essentially, requirements tracking is a process that proves that a system meets the business requirements for which it was designed. Although not all system development projects require the rigors of requirements tracking, the process can prove valuable. Furthermore, requirements tracking can be a condition placed upon contractors performing development services: the contractor may not receive the bid if the tracking is not performed, or the contractor may not be compensated if satisfactory tracking is not demonstrated. As systems grow in size and complexity, it can become impossible to maintain these connections manually, and requirements tracking software may become the only feasible solution.

# The SSE-CMM

The Software Engineering Institute has done some pioneering work in developing the System Security Engineering Capability Maturity Model (SSE-CMM). The model guides improvement in the practice of security engineering through small incremental steps, thereby developing a culture of continuous process improvement. The model is also a means of providing a structured approach for identifying and designing a range of controls. One of the primary reasons for organizations to adopt SSE-CMM is that it renders confidence in the organizations practices. The confidence also reassures stakeholders, both internal and external, to the organization, since it's a means to assess what an organization can do relative to what it claims to do. An added benefit is the assurance of the developmental process. Essentially this follows the concept of something being on time and within budget. The SSE-CMM is a metric for determining the best candidate for a specified security activity.

The SSE-CMM is a model that is based upon the requirements for implementing security in systems or a series of such related systems. The SSE-CMM model could be defined in various ways. Rather than invent an additional definition, the SSE-CMM Project chose to adapt the definition of systems engineering from the Software Engineering Capability Maturity Model as follows:

> Systems security engineering is the selective application of scientific and engineering efforts to: transform a security policy statement into a description of a system that best satisfies the security policy according to accepted measures of effectiveness (e.g., functional capabilities) and need for assurance; integrate related security parameters and ensure compatibility of all environmental, administrative, and technical security disciplines in a manner which optimizes the total system security design; integrate the system security engineering efforts into the total system engineering effort. (System Security Engineering Capability Model Description Document, Version 3.0, 2003, Carnegie Mellon University)

It addresses a special area called system security, and SSE-CMM is designed using the generalized framework provided by the Systems Engineering CMM as a foundation (see Figure 7.1 for CMM levels). The model architecture separates the specialty domain from process capability. In the case of SSE-CMM, it is a specialty domain with system security engineering process areas separated from the generic characteristics of the capability side. Here the generic characteristics relate to increasing process capability.

A question that is often asked is why is security engineering important? Clearly information plays an important role in shaping up the way business is being conducted in this era of the Internet. Information is an asset that has to be properly deployed to get the maximum benefits out of it. In mundane day-to-day operational decisions, information can be

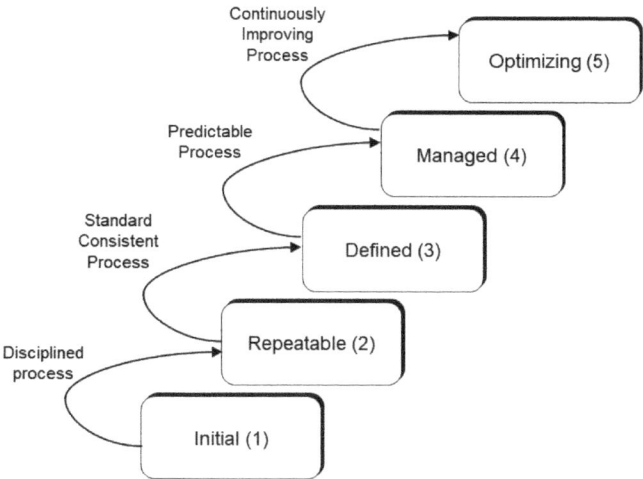

**Figure 7.1.** CMM levels

used to provide strategic directions to the corporation. Thus acquiring the relevant data is important, but the security of the vital data acquired is also an issue of paramount concern. Many systems, products, and services are needed to maintain and protect information. The focus of security engineering has expanded the horizons of the need for data protection—hence security—from classified government data to broader applications, including financial transactions, contractual agreements, personal information, and the Internet. These trends have increased the need for security engineering, and by all probabilities these trends seem to be there to stay.

Within the information technology security (ITS) domain, the SSE-CMM Model is focused on processes that can be used in achieving security and the maturity of these processes. It does not show any specific process or way of doing particular things—rather, it expects organizations to base its processes in compliance with any ITS guidance document. The scope of these processes should incorporate the following:

1. System security engineering activities used for a secure product or a trusted system. It should address the complete life cycle of the product, which includes:
   a. Conception of idea
   b. Requirement analysis for the project
   c. Designing of the phases
   d. Development and integration of the parts
   e. Proper installation
   f. Operation and maintenance
2. Requirements for the developers (product and secure system) and integrators, the organizations that provide computer security services and computer security engineering
3. It should be applicable to various companies that deal with security engineering, academia, and government.

SSE-CMM promotes the integration of various disciplines of engineering, since the issue of security has ramifications across various functions.

Why was SSE-CMM developed in the first place? Why was the need to have a reference model like SSE-CMM felt? When we venture into the context of the development of this type, we realize that there could be various reasons that called for this kind of an effort. Every business is interested in increasing efficiency, a practical way for which could be to have a process that provides a high quality product with minimal cost. Most statistical process controls suggest that higher quality products can be produced most cost-effectively by emphasizing on the quality of processes that produce them, and the maturity of the organizational practices inherent in these processes. More efficient processes are warranted, given the increasing cost and time required for the development of secure systems and reliable products. These factors again can be linked to people who manage the technologies.

As a response to the problems identified previously, the Software Engineering Institute (SEI) began developing a process maturity framework. This framework would help organizations improve their software processes and guide them in becoming mature organizations. A mature software organization possesses an organization-wide ability for managing software development process. The software process is accurately communicated to the staff, and work activities are carried out according to the planned process. A disciplined process is consistently followed and always ends up giving better quality controls, as all of the participants understand the value of doing so, and the necessary infrastructure exists to support the process.

Initially the SEI released a description of the framework along with a maturity questionnaire. The questionnaire provided the tool for identifying areas where an organization's software process needed improvement. The initial framework has evolved over a period of time because of ongoing feedback from the software community, into the current version of the SEI CMM for software. The SEI CMM describes a model of incremental process improvement. It provides organizations with a sequence of process improvement levels called *maturity levels*. Each maturity level is characterized by a set of software management practices. Each level provides a foundation to which the practices of the next level are added. Hence the sequence of levels defines a process of incremental maturity. The primary focus of the SEI CMM is the management and organizational aspects of software engineering. The idea is to develop an organizational culture of continuous process improvement. After years of assessment and capability evaluations using SEI CMM, its benefits are being realized today.

Results from implementation of the SEI CMM concepts indicate that improved product quality and predictable performance can be achieved by focusing on process improvement. Long-term software industry benefits have been as good as tenfold improvement in productivity and hundredfold improvement in quality. The return on investment (ROI) of process improvement efforts is also high. The architecture of SSE-CMM was adopted from CMM, since it supports the use of process capability criteria for specialty domain areas such as system security engineering.

The objective of the SSE-CMM Project has been to advance the security-engineering field. It helps the discipline to be viewed as mature, measurable, and defined. The SSE-CMM model and appraisal methods have been developed to help in:

1. Making investments in security engineering tools, training, process definition, and management practice worthwhile. It helps in improvements by engineering groups.
2. Providing capability-based assurance. Trustworthiness is increased based on confidence in the maturity of an engineering group's security and practices.
3. Selecting appropriately qualified providers of security engineering through differentiating bidders by capability levels and associated programmatic risks.

The SSE-CMM initiative began as a National Security Agency (NSA) sponsored effort in April 1993 with research into existing work on capability maturity models (CMMs) and investigation of the need for a specialized CMM to address security engineering. During this early phase, a "straw man" security engineering CMM was developed to match the requirement. The information security community was invited to participate in the effort at the First Public Security Engineering CMM Workshop in January 1995. Representatives from more than 60 organizations reaffirmed the need for such a model. As a result of the community's interest, project working groups were formed at the workshop, initiating the development phase of the effort. The first meeting of the working groups was held in March 1995. Development of the model and appraisal method was accomplished through the work of the SSE-CMM steering, author, and application working groups, with the first version of the model published in October 1996 and of the appraisal method in April 1997.

In July 1997, the Second Public Systems Security Engineering CMM Workshop was conducted to address issues relating to the application of the model, particularly in the areas of acquisition, process improvement, and product and system assurance. As a result of issues identified at the workshop, new project working groups were formed to directly address the issues. Subsequent to the completion of the project and the publication of version 2 of the model, the International Systems Security Engineering Association (ISSEA) was formed to continue the development and promotion of the SSE-CMM. In addition, ISSEA took on the development of additional supporting materials for the SSE-CMM and other related projects. ISSEA continues to maintain the model and its associated materials as well as other activities related to systems security engineering and security in general. ISSEA has become active in the International Organization for Standardization and sponsored the SSE-CMM as an international standard ISO/IEC 21827. Currently version 3.0 of the model is available. Further details can be found at http://www.sse-cmm.org.

## Key Constructs and Concepts in SSE-CMM

This section discusses various SSE-CMM constructs and concepts. SSEE-CMM considers process to be central to security development and also a determinant of cost and quality. Thus ways to improve processes is a major concern for the model. SSE-CMM is founded on the premise that the level of process capability is a function of organizational competence to a range of project management issues. Therefore process maturity emerges as a key construct. Maturity of a process is considered in terms of the ability to explicitly define, manage, and control organizational processes. Using the CMM framework, an engineering organization can turn from less organized into a highly structured and effective enterprise. The SSE-CMM model was developed with the anticipation that applying the concepts of statistical process control to security engineering will promote secure system development within the bounds of cost, schedule and quality.

Some of the key SSE-CMM constructs and concepts are discussed in the following sections.

## Organizations and Projects

It is important to understand as to what is meant by *organizations* and *projects* in terms of SSE-CMM. This is because the terms are usually interpreted in an ambiguous way.

## Organization

An organization is defined as a unit or subunit within a company, the whole company or any other entity like a government institution or service utility, responsible for the oversight of multiple projects. All projects within an organization typically share common policies at the top of the reporting structure. An organization may consist of geographically distributed projects and supporting infrastructures. The term *organization* is used to connote an infrastructure to support common strategic, business, and process-related functions. The infrastructure exists and must be utilized and improved for the business to be effective in producing, delivering, supporting, and marketing its products.

## Project

The project is defined as the aggregate of effort and other resources focused on developing and/or maintaining a specific product or providing a service. The product may include hardware, software, and other components. Typically a project has its own funding, cost accounting, and delivery schedule. A project may constitute an organizational entity completely on its own. It could also constitute a structured team, task force, or other group used by the organization to produce products or provide services. The categories of organization and project are distinguished based typically on ownership. In terms of SSE-CMM, one could differentiate between project and organization categories by defining the project as focused on a specific product, whereas the organization encompasses one or more projects.

## System

In the context of SSE-CMM, system refers to an integrated composite of people, products, services, and processes that provide a capability to satisfy a need or objective. It can also be viewed as an assembly of things or parts forming a complex or unitary whole (i.e., a collection of components organized to accomplish a specific function or set of functions).

A system may be a product that is exclusively hardware, a combination of hardware and software, or just software or a service. The term *system* is used throughout the model to indicate the sum of the products being delivered to the customer or user. In SSE-CMM, a product is denoted a system to emphasize the fact that we need to treat all the elements of the product and their interfaces in a disciplined and systematic way, so as to achieve the overall cost, schedule, and performance (including security) objectives of the business entity developing the product.

## Work Product

Anything generated in the course of performing a process of the organization could be termed as a "work product." These could be the documents, the reports generated during a process, the files created, the data gathered or used, and so on. Here, rather than listing the individual work products for each process area, SSE-CMM lists "Example Work Products" of a particular base practice, as it can elaborate further the intended scope of a base practice. These lists are illustrative only and reflect a range of organizational and product contexts.

## Customer

A customer, as defined in context of the model, is the entity (individual, group of individuals, organization) for whom a product is developed or service is made, or the entity (individual, group of individuals, organizations) that uses the product or service. The usage of customer in SSE-CMM context has an implication of understanding the importance of the users of the product, to target the right segment of consumers of the product.

In the context of the SSE-CMM, a customer may be either negotiated or nonnegotiated. A negotiated customer is entity who contracts with another entity to produce a specific product or set of products according to a set of specifications provided by the customer. A nonnegotiated, or market-driven, customer is one of many individuals or business entities who have a real or perceived need for a product.

In the SSE-CMM model, the individual or entity using the product or service is also included in the notion of customer. This is relevant in the case of negotiated customers, since the entity to which the product is delivered is not always the entity or individual that will actually use the product or service. It is the responsibility of the developers (at supply side) to attend to the entire concept of customer, including the users.

## Process

Several types of processes are mentioned in the SSE-CMM, some of which could be "defined" or "performed" processes. A defined process is formally described for or by an organization for use by its security engineers. The defined process is what is expected of the organization's security engineers to do. The performed process is what these people (security engineers) actually end up doing.

If a set of activities is performed to arrive at an expected set of results, then it can be defined as a "process." Activities may be performed iteratively, recursively, and/or concurrently. Some activities can transform input work products into output work products needed. The allowable sequence for performing activities is constrained by the availability of input work products and resources, and by management control. A well-defined process includes activities, input and output artifacts of each activity, and mechanisms to control performance of the activities.

## Process Area

A process area (PA) can be defined as a group of related security engineering process characteristics, which when performed in a collective manner can achieve a defined purpose. It is composed of base practices, which are mandatory characteristics that must exist within an implemented security engineering process before an organization can claim satisfaction in a given process area. SSE-CMM identifies 10 process areas. these are administer security controls, assess operational security risk, attack security, build assurance argument, coordinate security, determine security vulnerabilities, monitor system security posture, provide security input, specify security needs, and verify and validate security. Each process area has predefined goals. SSE-CMM process areas and goals appear in Table 7.1.

**Table 7.1.** SSE-CMM security engineering process areas (from [2])

| Process area | Goals |
| --- | --- |
| Administer security controls | • Security controls are properly configured and used. |
| Assess operational security risk | • An understanding of the security risk associated with operating the system within a defined environment is reached. |
| Attack security | • System vulnerabilities are identified and their potential for exploitation is determined. |
| Build assurance argument | • The work products and processes clearly provide the evidence that the customer's security needs have been met. |
| Coordinate security | • All members of the project team are aware of and involved with security engineering activities to the extent necessary to perform their functions.<br>• Decisions and recommendations related to security are communicated and coordinated. |
| Determine security vulnerabilities | • An understanding of system security vulnerabilities is reached. |
| Monitor system security posture | • Both internal and external security related events are detected and tracked.<br>• Incidents are responded to in accordance with policy.<br>• Changes to the operational security posture are identified and handled in accordance with security objectives. |
| Provide security input | • All system issues are reviewed for security implications and are resolved in accordance with security goals.<br>• All members of the project team have an understanding of security so they can perform their functions.<br>• The solution reflects the security input provided. |
| Specify security needs | • A common understanding of security needs is reached between all applicable parties, including the customer. |
| Verify and validate security | • Solutions meet security requirements.<br>• Solutions meet the customer's operational security needs. |

## Role Independence

When the process areas of the SSE-CMM are joined together as groups of practices and taken together, it achieves a common purpose. But the groupings are not meant to imply that all base practices of a process are necessarily performed by a single individual or role. This is one way in which the syntax of the model supports the use of it across a wide spectrum of organizational contexts.

## Process Capability

Process capability is defined as the range (which is quantifiable) of results that are expected or can be achieved by following a process. The SSE-CMM appraisal method (SSAM) is based on statistical process control concepts that define the use of process capability. The SSAM can be used to determine process capability levels for each process area within a project or organization. The capability side of the SSE-CMM reflects these concepts and provides guidance in improving the process capability of the security engineering practices that are referenced on the domain side of the SSE-CMM.

The capability of an organization's process is instrumental in predicting the ability of a project to meet goals. Projects in low capability organizations experience wide variations in achieving cost, schedule, functionality, and quality targets.

## Institutionalization

Institutionalization is the building of infrastructure and corporate culture that establishes methods, practices, and procedures. These established practices remain in place for a long time. The process capability side of the SSE-CMM supports institutionalization by providing a path and offering practices toward quantitative management and continuous improvement. In this way, the SSE-CMM asserts that organizations need to explicitly support process definition, management, and improvement. Institutionalization provides a means to gain maximum benefit from a process that exhibits sound security engineering characteristics.

## Process Management

Process management is the management of related set of activities and infrastructures that are used to predict, then evaluate, and finally control the performance of a process. Process management implies that a process is defined (since one cannot predict or control something that is undefined). The focus on process management implies that a project or organization takes into account all possible factors regarding both product- and process-related problems in the planning phase, at performance level, in evaluating and monitoring, and also corrective action.

## Capability Maturity Model

A capability maturity model (CMM) such as the SSE-CMM describes the stages through which processes show progress, as they are defined initially, implemented practically, and improved gradually. The model provides a way to select process improvement strategies by firstly determining the current capabilities of specific processes and then subsequently identifying the issues most critical to quality and process improvement within a particular domain. A CMM may take the form of a reference model to be used as a guide for developing and improving a mature and defined process.

A CMM may also be used for appraisal of the existence of a process and institutionalization of a defined process that implements referenced practices. A capability maturity model covers all the processes that are used to perform the tasks of the specified domain, (e.g., security engineering). A CMM can also cover processes used to ensure effective development and use of human resources, as well as the insertion of appropriate technology into products and tools used to produce them.

## SSE-CMM Architecture Description

The SSE-CMM architecture is designed to enable a determination of a security engineering organization's process maturity across the breadth of security engineering. The model evaluates each process area against common features. The goal of the architecture is to separate basic characteristics of the security engineering process from its management characteristics. In order to ensure this separation, the model has two dimensions, called "domain" and "capability" (described later). Of particular significance, the SSE-CMM does not imply that any particular group or role within an organization must undertake any of the processes described in the model. Nor does it require that the latest and greatest security engineering technique or methodology be used. The model does require, however, that an organization have a process in place that includes the basic security practices described in the model. The organization is free to create their own process and organizational structure in any way that meets its business objectives. The generic levels of SSE-CMM appear in Figure 7.2.

### The Basic Model

The SSE-CMM has two dimensions:

1. **Domain.** This consists of all the practices that together in a collective manner define security engineering in an organization. These practices could be called the "base practices."
2. **Capability.** This represents practices that indicate process management and institutionalization capability. These are also known as "generic practices," as they apply across a wide range of domains.

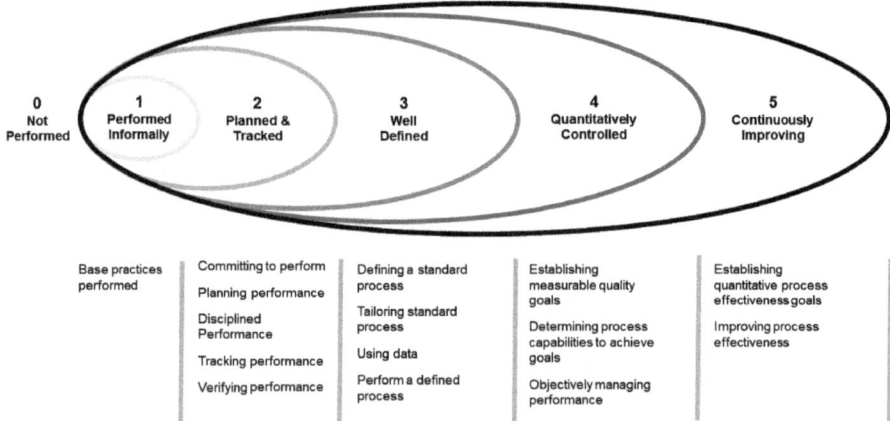

**Figure 7.2.** SSE-CMM levels (drawn from SSE-CMM documentation)

## Base Practices

The SSE-CMM contains 129 base practices, organized into 22 process areas. Of these, 61 base practices, organized in 11 process areas, cover all major aspects of security engineering. The remaining 68 base practices, organized in 11 process areas, address the project and organization domains. They have been drawn from the systems engineering and software CMM. They are required to provide a context and support for the systems security engineering process areas.

The base practices for security were gathered from a wide range of existing materials, practice, and expertise. The practices selected represent the best existing practice of the security engineering community.

It is a complicated task to identify security engineering base practices, as there are several names for the same activities. These activities could occur later in the life cycle, at a different level of abstraction, or individuals in different roles could perform them. However, an organization cannot be considered to have achieved a base practice if it is only performed during the design phase or at a single level of abstraction. Therefore the SSE-CMM has ignored these distinctions and tries to identify the basic set of practices that are essential to the practice of good security engineering.

Thus a base practice can have the following characteristics:

- Should be applied across the life cycle of the enterprise
- Should not overlap with other base practices
- Should represent a "best practice" of the security community
- Should be applicable using multiple methods in multiple business contexts
- Should not specify a particular method or tool

The base practices have been organized into process areas such that it meets a broad spectrum of security engineering requirements. There are many ways to divide the security-engineering domain into process areas. One might try to model the real world or create process areas that match security-engineering services. Other strategies attempt to identify conceptual areas that form fundamental security engineering building blocks.

## Generic Practices

Generic practices are activities by definition that should be applicable to all processes. They address all the aspects of the process: management, measurement and institutionalization. They are used for an initial appraisal, which helps in determining the capability of an organization to perform a particular process. Generic practices are grouped into logical areas called *common features* and are organized into five *capability levels*, which represent increasing organizational capability. Unlike the base practices of the domain dimension, the generic practices of the capability dimension are ordered according to maturity. Therefore generic practices that indicate higher levels of process capability are located at top of the capability dimension.

The common features here are designed in a way such that it helps in describing major shifts in an organization's manner of performing work processes (in this case, the security engineering domain). Each common feature has to have one or more generic practices.

Subsequent common features have generic practices, which helps in determining or assessing how well a project manages and improves each process area as a whole.

## The Capability Levels

The way in which the common features are ordered can be derived from the observation that implementation and institutionalization of some practices benefit from the presence of other practices. This is especially more applicable if practices are well-established. Before an organization can define, tailor, and use a process effectively, individual projects should have some experience managing the performance of that process. Before institutionalizing a specific estimation process across the entire organization, for example, an organization should first at least attempt to use the estimation process on a project. However, some aspects of process implementation and institutionalization should be considered together (not one ordered before the other), since they work together toward enhancing capability.

Common features and capability levels are important, both in performing and assessment of the current processes and improving an organization's process capability. In the case of an assessment where an organization has some but not all common features implemented at a particular capability level for a particular process, it usually operates at the lowest completed capability level for that process. An organization may not reap the full benefit of having implemented a common feature if it is in place, but not all common features at lower capability levels. An assessment team should take this into account in assessing an organization's individual processes. In the case of improvement, organizing the practices into capability levels provides an organization with an "improvement road map," should it desire to enhance its capability for a specific process. For these reasons, the practices in the SSE-CMM are grouped into common features, which are ordered by capability levels.

An assessment should be performed to determine the capability levels for each of the process areas. This indicates that different process areas can and probably will exist at different levels of capability. The organization will then be able to use this process-specific information as a means to focus improvements to its processes. The priority and sequence of the organization's activities to improve its processes should take into account its business goals.

It will be useful to think of capability maturity for security processes by using the informal, formal, technical model introduced in Chapter 1. The following discussion describes the extent of maturity at each level and positioning it in light of the informal, formal, technical model.

### *Level 1: Initial*

This level characterizes an organization that has ad hoc processes for managing security. Security design and development are ill-defined. Security considerations may not be incorporated in the systems design and development practices. Typically level 1 organizations would not have a contingency plan to avert any crisis, and at best security issues would be dealt in a reactive way. As a consequence, there is no standard practice for dealing with security, and procedures are reinvented for each project. Project scheduling is ad hoc, as are the budgets, functionality, and quality. There are no defined process areas for level 1 organizations.

## Level 2: Repeatable

At this level, an organization has a defined security policy and procedure. Such policies and procedures may be either for the day-to-day operations of the firm or specifically for secure systems development. The later applies more to software development shops and suggests that security considerations be integrated with regular systems development activities. Assurance can be provided since the processes and activities are repeatable. This essentially means that the same procedure is followed project after project, rather than reinventing the procedure every time. Process areas covered at level 2 include security planning, security risk analysis, assurance identification, security engineering, and security requirements. Figure 7.3, using the informal, formal technical model, illustrates the repeatability of security engineering practices.

## Level 3: Defined

As depicted in Figure 7.4, the defined level signifies standardized security engineering processes across the organization. Such standardization ensures integration across the firm and hence eliminates redundancy. A further benefit is in maintaining the general integrity of

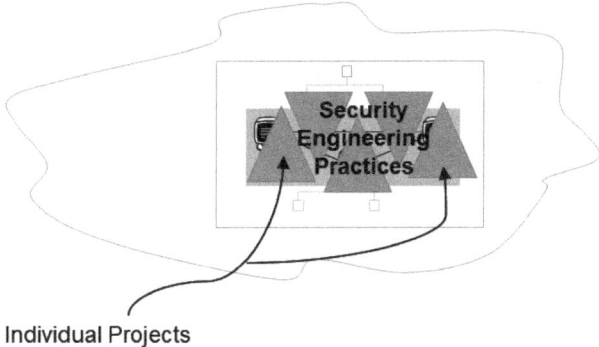

**Figure 7.3.** Level 2 with project-wide definition of practices

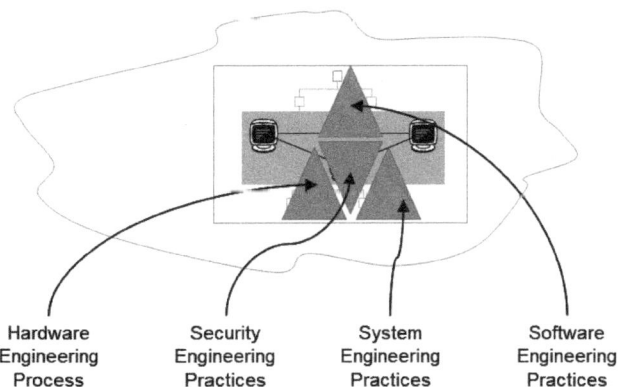

**Figure 7.4.** Level 3 with project-wide processes are integrated

the operations. Training of personnel usually ensures that the right kind of skill set is developed and necessary knowledge imparted. Since the security practices are clearly defined, it becomes possible to provide an adequate level of assurance across different projects. The process areas at level 3 include integrated security engineering, security organization, security coordination, and security process definitions.

### Level 4: Managed

Defining security practices is just one aspect of building capability. Unless there is competence to manage various aspects of security engineering, adequate benefits are hard to achieve. In case an organization has management insight into the security engineering process, it represents level 4 of SSE-CMM (Figure 7.5). At this level an organization should be able to establish measurable goals for security quality. A high level of quality is a precursor to good trust in the security engineering process. Examination of the security process measures helps in increasing awareness of the shortcomings, pitfalls, and positive attributes of the process.

### Level 5: Optimizing

This level represents the ideal state in security engineering practices. As identified in Figure 7.6, level 5 organizations constantly engage in continuous improvement. Such improvement emerges from the measures and identification of causes of problems. Feedback then helps in process modification and further improvement. Newer technologies and processes may be incorporated to ensure security.

This section has introduced two important concepts: The first deals with the issue of controls and how these need to be integrated in to the systems development processes. The importance of understanding and thinking through the process is presented as an important trait. The second concept deals with process maturity. The SSE-CMM is introduced as a means to think about processes and how maturity can be achieved in thinking about controls. The process areas, as identified in the SSE-CMM, are no more than the controls.

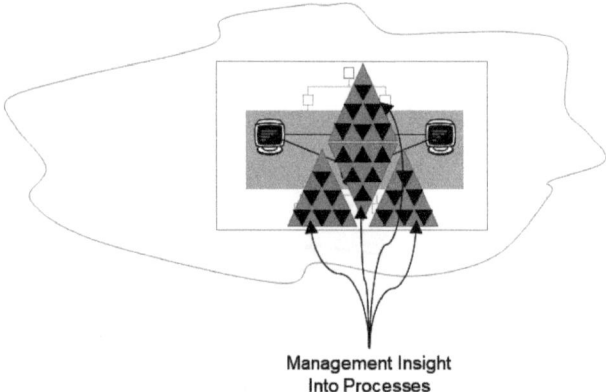

**Figure 7.5.** Level 4 with management insight into the processes

Usually implementation of adequate controls, consideration of security at the requirements analysis stage, and so on have been touted as useful means in security development and engineering. The SSE-CMM helps in conceptualizing and thinking through stages of maturity and capability in dealing with security issues. At a very basic level, it is useful to define a given enterprise in terms of its capability and then aspire for improving it, perhaps by moving up the levels in SSE-CMM.

## From the Rainbow Series to Common Criteria

In 1983 the US Department of Defense published the *Trusted Computer System Evaluation Criteria* (TCSEC), also known as the Orange Book (largely because of the color of the book). A second version of the criteria was published in 1995. The TCSEC formed the basis for National Security Agency (NSA) product evaluations. However, because of the mainframe and defense orientations, their use was restricted. Moreover, the TCSEC dealt primarily with ensuring confidentiality with issues such as integrity and availability being largely overlooked.

Skewed emphasis on confidentiality alone was in many ways a limiting factor, especially since the reality was of networked infrastructures. In later years TCSEC were interpreted for the network- and database-centric world. Thus in 1987 the National Computer Security Center (NCSC) published the *Trusted Network Interpretation*. This came to be known as the Red Book. Although the Red Book covered network specific security issues, there was limited coverage of database security aspects. This lead to another NCSC publication—*Trusted Database Management System Interpretation* (the Lavender Book).

The TCSEC suggests four basic classes that are ordered in a hierarchical manner (A, B, C, D). Class A is the highest level of security. Within each class, there are four sets of criteria: security policy, accountability, assurance, and documentation. The security aspects of the system are referred to as the *Trusted Computing Base*. Security requirements in each of the classes are discussed in the following sections.

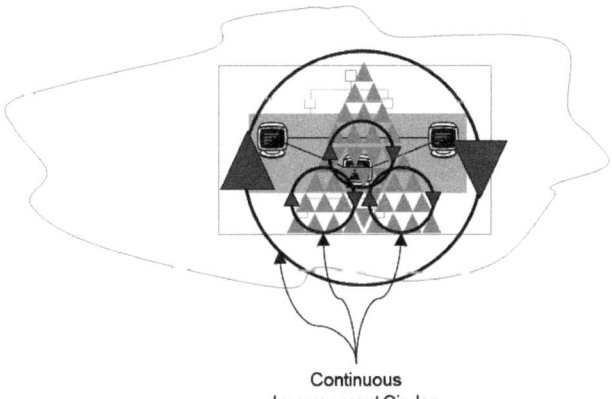

Continuous Improvement Circles

**Figure 7.6.** Level 5 with continuous improvement and change

## Minimal Protection (Class D)
Class D is the lowest level security evaluation. All it means that the systems have been evaluated but do not meet any of the higher level evaluation criteria. There is no security at this level.

## Discretionary Protection (Class C)
There are two subclasses in this category—C1 and C2. Class C1 specifies Discretionary Security Protection and requires identification and authentication mechanisms. C1 class systems make it mandatory to separate users and data. Discretionary access controls allow users to specify and control objects by named individuals and groups. Security testing is undertaken and ensures that protection mechanisms do not get defeated.

Class C2 specifies controlled access protection. This class enforces better articulated and finely grained discretionary access controls relative to C1. This gets manifested in making users accountable for login procedures, auditing of security relevant events, and allocation of resources. Class C2 also specifies rules for media reuse, ensuring that no residual data is left in devises that are to be reused.

## Mandatory Protection (Class B)
There are three subclasses within Class B—B1, B2, and B3. Class b1 is labeled security protection. B1 incorporates all security requirements of Class C. In addition, it requires an informal statement of security policy model, data labeling, and mandatory access control, especially for named subjects and objects. Class B1 mandatory access control policy is defined by the Bell La Padula model (discussed in Chapter 3).

Class B2 specifies structured protection. Class B2 specifies a clearly defined and documented security model requiring discretionary and mandatory access control that is enforced in all objects and subjects. B2 requires a detailed analysis of covert channels. Relative to the lower classes, authentication mechanisms are strengthened. The trusted computing base is classified into critical and noncritical elements. B2 class systems require the design to be such that a more thorough test is possible. A high level descriptive specification of the trusted computing base is required. It is required that there is consistency between the trusted computing base implementation and the top-level specification. At a B2 level, a system is considered as relative resistant to any kind of penetration.

Class B3 specifies the security domains. At this level, it is required that the reference monitor mediates access of all subjects to objects and is tamperproof. In addition, the trusted computing base is minimized by excluding noncritical modules. Class B3 systems are considered highly secure. The role of a system administrator is defined and recovery procedures are spelled out.

## Verified Protection (Class A)
There are two subclasses within this verified protection class. Verified design (Class A1) is functionally similar to Class B3. A1 class systems require the use of formal design specification and verification techniques. These raise the degree of assurance. Other Class A1 criteria include

- Formal security model
- Mathematical proof of consistency and adequacy
- Formal top-level specification
- Demonstration that formal top-level specification corresponds to the model
- Demonstration that the trusted computing base is consistent with formal top-level specification
- Formal analysis of covert channels

Beyond Class A1 category is futuristic with assurance criteria that are beyond the reach of current technology. A summary of key issues for each of the TCSEC classes is presented in Table 7.2.

**Table 7.2.** TCSEC classes

| Class | Subclass | Interpretation |
| --- | --- | --- |
| **Verified protection (A)** | Verified design (A1) | Formal design specification and verification is undertaken to ensure correctness in implementation. |
| **Mandatory protection (B)** | Security domains (B3) | All objects and subject access is monitored. Code not essential to enforcing security is removed. Complexity is reduced and full audits are undertaken. |
| | Structured protection (B2) | Formal security policy applies discretionary and mandatory access control. |
| | Labeled security protection (B1) | Informal security policy is applied. Data labeling and mandatory access control are applied for named objects. |
| **Discretionay protection (C)** | Controlled access protection (C2) | A lot of discretionary access controls are applied. Users are made accountable through login procedures, resource isolation, and so on. |
| | Discretionary security protection (C1) | Some discretionary access control. Represents an environment where users are cooperative in processing and protecting data. |
| **Minimal protection (D)** | Minimal protection (D) | Category assigned to systems that fail to meet higher levels. |

# ITSEC

The Information Technology Security Evaluation Criteria (ITSEC) are the European equivalent of the TCSEC. The purpose of the criteria is to demonstrate conformance of a product or a system (target of evaluation) against threats. The target of evaluation is considered with respect to the operational requirements and the threats it might encounter. ITSEC considers the evaluation factors as functionality and the assurance aspect of correctness and effectiveness. The functionality and assurance criteria are separated.

Functionality refers to enforcing functions of the security targets, which can be individually specified or enforced through predefined classes. The generic categories for enforcing functions of the security targets include

1. Identification and authentication
2. Access control
3. Accountability
4. Audit
5. Object reuse
6. Accuracy
7. Reliability of service
8. Data exchange

As per the ITSEC, evaluation of effectiveness is a measure as to whether the security enforcing functions and mechanisms of target of evaluation satisfy the security objectives. Assessment of effectiveness involves an assessment of suitability of target of evaluation functionality, binding of functionality (i.e., if individual security functions are mutually supportive), consequences of known vulnerabilities, and ease of use. The evaluation of effectiveness is also a test for the strength of mechanisms to withstand direct attacks.

Evaluation of correctness assesses the level at which security functions can or cannot be enforced. Seven evaluation levels have been predefined—E0 to E6. A summary of the various levels is presented in Table 7.3.

**Table 7.3.** ITSEC classes

| Evaluation level | Interpretation |
| --- | --- |
| E6 | Formal specification of security enforcing functions is ensured. |
| E5 | There is a close correspondence between detailed design and source code. |
| E4 | There is an underlying formal model of security policy. Detailed design specification is done in a semiformal manner. |
| E3 | Source code and hardware corresponds to security mechanisms. Evidence of testing the mechanisms is required. |
| E2 | There is an informal design description. Evidence of functional testing is provided. Approved distribution procedures are required. |
| E1 | Security target for a target of evaluation is defined. There is an informal description of the architectural design of the TOE. Functional testing is performed. |
| E0 | This level represents inadequate assurance. |

Relative to TCSEC, ITSEC offers the following significant changes and improvements:

- Separate functionality and assurance requirements
- New defined functionality requirements classes that also address availability and integrity issues
- Functionality that can be individually specified (i.e., ITSEC is independent of the specific security policy)
- Supports evaluation by independent commercial evaluation facilities

## International Harmonization

As stated previously, the original security evaluation standards were developed by the US Department of Defense in the early 1980s in the form of Trusted Computer Systems Evaluation Criteria (TCSEC), commonly referred to as the Orange Book. The original purpose of TCSEC was to evaluate the level of security in products procured by the Department of Defense. With time, the importance and usefulness of TCSEC caught the interest of many countries. This resulted in a number of independent evaluation criteria being developed for countries such as Canada, the United Kingdom, France, and Germany, among others. In 1990 the European Commission harmonized the security evaluation efforts of individual countries by establishing the European equivalent of TCSEC, the Information Technology Security Evaluation Criteria. The TCSEC evolved in their own capacity to eventually become the Federal Criteria. Eventually an international task force was created to undertake further harmonization of the various evaluation criteria. In particular, the Canadian Criteria, Federal Criteria, and ITSEC were worked upon to develop the Common Criteria (CC). ISO adopted these criteria to form an international standard—ISO 15408. Figure 7.7 depicts the evolution of the evaluation criteria/standards to ISO 15408 and the Common Criteria.

### Common Criteria

In many situations consumers lack an understanding of complex IT-related issues and hence do not have the expertise to judge or have confidence that their IT systems/products are

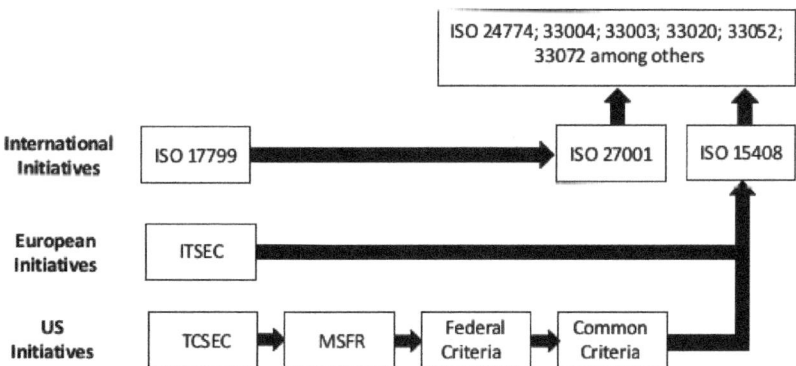

**Figure 7.7.** Evolution of security evaluation criteria

sufficiently secure. In yet other situations, consumers do not want to rely on the developer assertions, either. This necessitates the need for a mechanism to instill consumer confidence. As stated previously, a range of evaluation criteria in different countries helped instill this confidence. Today, the CC are a means to select security measure and evaluate the security requirements.

In many ways, the CC provide a taxonomy for evaluating functionality. The criteria include 11 functional classes of requirements:

1. Security audit
2. Communication
3. Cryptographic support
4. User data protection
5. Identification and authentication
6. Management of security functions
7. Privacy
8. Protection of security functions
9. Resource utilization
10. Component access
11. Trusted path or channel

The 11 functional classes are further divided into 66 families, each of which has component criteria. There is a formal process that allows for developers to add additional criteria. There are a large number of government agencies and industry groups that are involved in developing functional descriptions for security hardware and software. Commonly referred to as protection profiles (PP), these describe groupings of security functions that are appropriate for a given security component or technology. Protection profiles and evaluation assurance levels are important concepts in the Common Criteria. A protection profile has a set of explicitly stated security requirements. In many ways it is an implementation independent expression of security. A protection profile is reusable, since it defines product requirements both for functions and assurance. The development of PPs help vendors provide standardized functionality, thereby reducing the risk in IT procurement. Related to the PPs, manufacturers develop documentation explaining the functional requirements. In the industry, these have been termed *security targets*. Security products can be submitted to licensed testing facilities for evaluation and issuance of compliance certificates.

The Common Criteria, although an important step in establishing best practices for security, are also a subject for criticism. Clearly the CC do not define end-to-end security. This is largely because the functional requirements relate to individual products that may be used in providing a complex IT solution. PPs certainly help in defining the scope to some extent, but they fall short of a comprehensive solution. However, it is important to note that the CC are very specific to the target of evaluation (TOE). This means that for well-understood problems, the CC provide the best practice guideline. For new problem domains, it becomes a little difficult to postulate best practice guidelines.

Figure 7.8 depicts the evaluation process. The CC recommend that evaluation can be carried out in parallel with development. There are three inputs into the evaluation process:

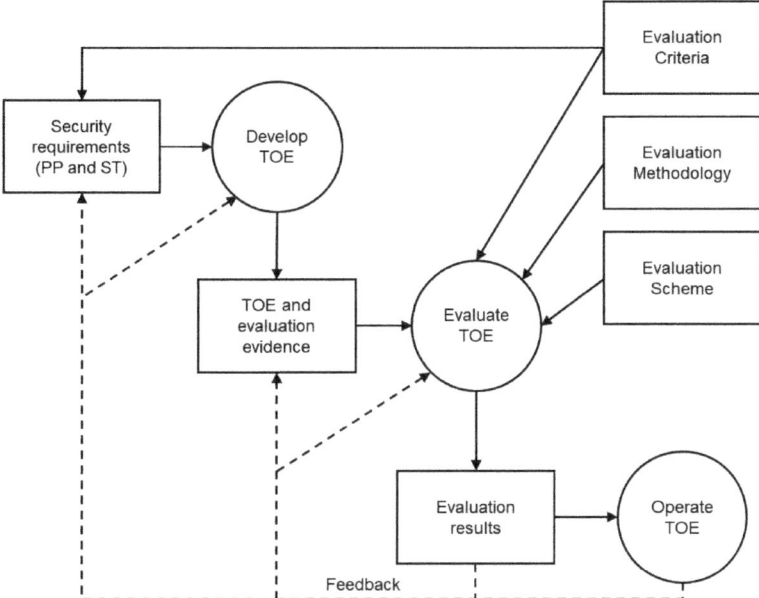

**Figure 7.8.** The evaluation process

- Evaluation evidence, as stated in the security targets
- The target of evaluation
- The criteria to be used for evaluation, methodology, and scheme

Typically the outcome of evaluation is a statement that the evaluation satisfies requirements set in the security targets. The evaluator reports are used as a feedback to further improve the security requirements, targets of evaluation, and the process in general. Any evaluation can lead to better IT security products in two ways. First, the evaluation identifies errors or vulnerabilities that the developer may correct. Second, the rigors of evaluation result in helping the developer better design and develop the target of evaluation.

Common Criteria are now widely available through the National Institute of Standards and Technology (http://csrc.nist.gov/cc/).

### Problems with the Common Criteria

The Common Criteria have gained significant importance in the industry, especially as a means to define the security needs of users. As suggested previously, this is achieved through the concept of protection profiles. Protection profiles inherit the assumptions that are typical of an operational environment. There are, however, some inherent deficiencies in the Common Criteria, which are important to understand if proper utilization is to be brought about. These are discussed in the following sections.

### Identification of the Product, TOE, and TSF

The Common Criteria lack clarity in defining what a product, target of evaluation, or target of evaluation security function might be. It can be rather overwhelming for the developers to differentiate between these and hence manage the nature and scope of their evaluations. Some of the reasons for this confusion are as follows:

- **Problem of TOEs spanning multiple products.** A rather practical problem with the concept of TOEs is that they span multiple products. More often than not, solutions are constituted of multiple products. There may also be situations when different components of various products (e.g., code libraries) may come together to provide a certain solution. In such cases, defining the target of evaluation becomes difficult. In the security domain, cryptographic code libraries are often reused across products. The resultant problem is more of the developer overlooking the TOE. So it becomes important for the developer to clearly understand the various components of the product comprising TOE and communicate this understanding through the development phases.
- **Problem of multiple TOEs in a product line.** For practical reasons, developers may want more than one product or variants of the same product evaluated at once. This happens mostly when there is one product, but for different operating systems. The difference between the variants makes TOE rather difficult to define. There may also be an issue where product variation may exist because different components of other products have been used. Consider the case of an online journal/magazine publisher selling online subscriptions. Although the base product is the same, there may be two or three variations in access control approaches. One kind of access control may be used for individual subscribers, there may be another kind for institutional subscribers, and yet another kind for libraries where traffic originating for a given IP address is given access. These are minor variations of access control for the same product.

In such situations, defining the product, TOE, and TSF is a difficult task. The developers need a clear understanding as to what constitutes a TOE and have an ability to define version-specific aspects of TOE.

- **Defining TOE within a product.** At most times, a small part of the product needs to be evaluated. It is, however, challenging to differentiate between parts comprising the TOE and the rest. At times there may be aspects of the system, not part of the TOE, that may end up compromising the security of the system. This is usually the case when there may be components outside the TOE that have a lower level of assurance. Ideally such parts of the product should be brought within the fold of the TOE in order to ensure overall safety and security of the product. If these are not brought within the scope of TOE, then the inherent assumptions need to be clarified. This will prevent misunderstandings.
- **Defining TSF.** Usually only a part of the TOE provides the target of evaluation security function. This means that those parts of TOE that are outside the TFS should also be evaluated to ensure that they do not compromise the security of the

entire product. If the assurance level is low, it is virtually impossible to differentiate between TOE and TSF.
- **Product design.** It is important to understand various product design attributes. This is especially true for products that have been designed at lower levels of assurance. Usually there is no documentation for such products, and hence it becomes difficult to define the TOE. This results in having security targets that may or may not have any relation to TOE. Although it is always advocated that developers need to prepare good documentation for all products, it is rarely accomplished in practice. There is no clear solution for this problem, apart from overstressing the importance of design documentation.

**Threat Characterization**

The Common Criteria lack a proper definition of threats and their characterization. In many ways the intent behind CC is to identify all information assets, classify them, and characterize assets in terms of threats. It is rather difficult to come up with an asset classification scheme, though. This is not a limitation of the CC per se, but an opportunity to undertake work in the area of asset identification, classification, and correlating these to the range of threats.

Some progress in this regard has been made, particularly in the risk management domain. The Software Engineering Institute at Carnegie Mellon University has been involved in building asset-based threat profiles. Development of the OCTAVE (Operationally Critical Threat, Asset, and Vulnerability Evaluation) method has been central to this work. In the OCTAVE method, threats are defined as having the following properties [1]:

- Asset—Something of value to the organization (could be electronic, physical, people, systems, knowledge)
- Actor—Someone or something that may violate the security requirements
- Motive—The actor's intentions (deliberate, accidental, etc.)
- Access—How the asset will be accessed
- Outcome—The immediate outcome of the violation

In terms of overcoming problem in the CC, it is useful to define the threats. Developers in particular need to be aware of the importance of threat identification and definition.

**Security Policies**

The Common Criteria make it optional whether to specify security policies or not. In situations where the product is being designed and evaluated for general consumption, a generic nature of the controls make sense—largely for wide distribution of the product. This suggests that not specifying any rule structures in CC is beneficial. However, with respect to organization specific products, a lack of clarity of security policies causes much confusion. More often than not, the developers incorporate their own assumptions for access and authentication rules into products. Many times these do not necessarily match organizational requirements. As a consequence, the rules enforced in the products have a mismatch with the rules specified in the organization. A straightforward solution is to make developers and evaluators aware of the nature and scope of the products, along with security policy requirements.

### Security Requirements for the IT Environment

The Common Criteria do not provide clearly details as to how the security requirements should be specified, although it is supposed to offer a requirements specification model (see Figure 7.9 for a requirement specification model of the CC). This poses problems in the evaluation stage of the product. The CC make a clear statement that all IT security requirements should be stated by reference to security requirements components drawn from parts 2 and 3 of the CC. If these parts explicitly state the requirement, then it is these that the evaluators look for. However, in case the requirement is not clearly stated, then these need to be clearly specified by the developers. This often does not happen. Moreover, such requirements may be included as assumptions of the TOE environment. In net effect, lack of clarity results in significant confusion.

There is also lack of clarity in auditing requirements. Although the CC identify the level of auditing, identify the types of events that are auditable by the TSF, and suggest the minimum audit-related information that is to be provided within various audit record types, there are aspects that can potentially confuse evaluators. The level of auditing is specified as minimal, basic, detailed, or not specified. However, the auditors need to be aware of a number of dependencies that can easily be overlooked. This is in spite of these having been clearly identified in part 2 of the CC. There are also issues because of the hierarchical nature of

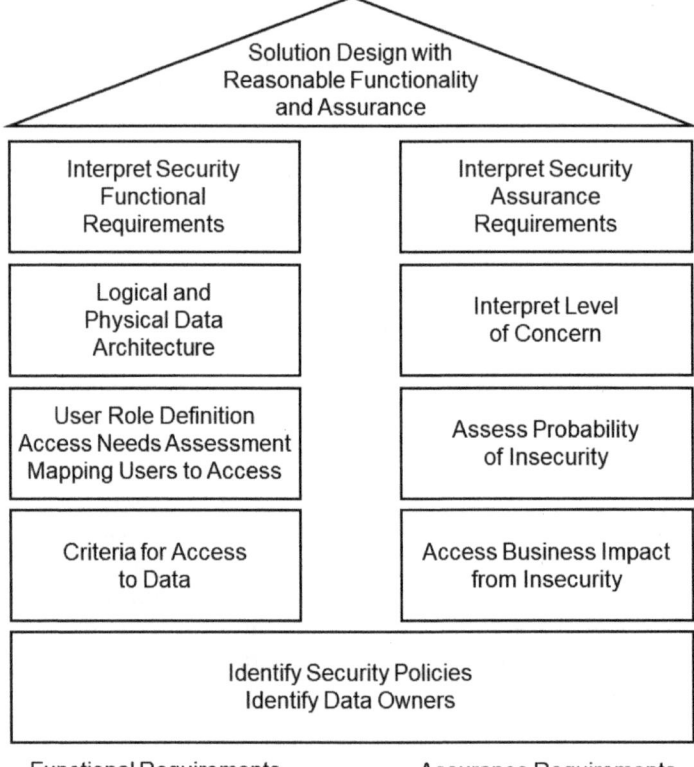

**Figure 7.9.** Requirements specification, as espoused by CC

the audit levels. Most of the problems stem from the lack of documentation within security targets, which causes the dependencies to be overlooked by developers.

## ISO/IEC 27002: 2013

ISO 27002 is perhaps one of the main internationally recognized IS security standards. The origins of ISO 27002 go back to predecessor standards—the ISO 17799 published in 2000 and the original British Standard BS 7799. ISO 27002 provides a comprehensive set of practices that all organizations should adopt. It is relevant to all types of organizations, which handle and rely on information such as commercial enterprises of all kinds, charities, government departments, and quasi-autonomous bodies.

The standard was extensively revised since its publication to include references to other security standards, in addition to incorporating information security best practices that came into existence since past publications. ISO 27002 is not a mandatory requirement or a certification standard, but merely a code of practice. The standard takes a very broad approach to the information security, meaning it is not just concerned with the IT systems security but also the computer data, intellectual property, knowledge, documentation, communication, messages, recordings, and photographs. Organizations are free to choose and implement controls other than ISO 27002. The organizations that follow ISO 27001 (which is a formal specification) also follow ISO 27002. The organizations using the ISO/IEC 27002 standard for guidance must elucidate their control objectives, evaluate their own information risks, and implement suitable controls.

As noted by the International Standards Organization, ISO 27002 is designed to be used by organizations that intend to[1]:

1. Select controls within the process of implementing an information security management system based on ISO/IEC 27001.
2. Implement commonly accepted information security controls.
3. Develop their own information security management guidelines.

While the original ISO 17799 focused on 10 control areas, ISO 27002 has fourteen control areas, defined as follows.

### *Information Security Policies*

The standard clearly identifies the importance of security policies, which are required to sketch out expectations and obligations of the management. A policy could also form a basis for ongoing evaluation and assessment. As has been identified in Chapter 5, there are certainly issues and concerns with how policies are created and what aspects need to be understood. The standard however does not go into such specific details.

### *Organization of Information Security*

Establishing a proper security organization is also considered central to managing security. Often there is lack of congruence between the control structures and the organizational structures. This control has two fundamental objectives: (1) to establish a framework to

---
1 https://www.iso.org/standard/54533.html

manage information security within your organization, and (2) to ensure the security of mobile devices and telework (work done away from the office at home or elsewhere). The control emphasizes the importance of creating a broad framework for security management, which includes education, training, and awareness.

## Human Resource Security

Controls related to human resource aspects focus on vetting and determining the suitability of employees and contractors to occupy certain roles. Since dual-reporting structures and lack of clarity of roles and responsibilities may facilitate a security breach, the control suggests that such aspects be adequately addressed. Hence the importance of organizational structures and processes being in synch is highlighted. Personnel security strives to meet three objectives: The first objective deals with security in job definition and resourcing. Its aim is to reduce the risks of human error, theft, fraud, or misuse of facilities, with the focus on adequate personnel screening for sensitive jobs. The second objective is to make all employees of the organization aware of information security threats and concerns through appropriate education and training. The third objective deals with minimizing the damage from security incidents and malfunctions. Here, the emphasis is on reporting the security incidents or weaknesses as quickly as possible through appropriate management channels. It also involves monitoring and learning from such incidents.

## Asset Management

The standard calls for organizational assets to be identified. The inherent argument is that unless the assets can be identified, they cannot be controlled. Asset identification also ensures an appropriate level of controls being implemented. Origins of asset management and its importance in information security can be traced to the findings of the Hawley Committee [3, 4]. (See Box 7.1 for a summary of the Hawley Committee findings.) Figure 7.10 identifies various attributes of information assets and presents a framework that could be used to account for and hence establish adequate controls.

---

**OBJECTIVE: ESTABLISH RESPONSIBILITY FOR CORPORATE ASSETS**

GOAL: To protect assets associated with information and processing facilities
MEMO: Define responsibilites for assets and ensure data owners are indentified

**Subgoal: Complete inventory of assets associated with information**
- Control: Identify all assets for each data owner.
- Control: Complete inventory of all information assets.
- Control: Maintain inventory of all information assets..

**Subgoal: Identify data owners and their responsibilities**
- Control: Identify span of control of each data owner.
- Control: Link data inventory to each data owner.
- Control: Define start and finish time for each data owner.

---

**Figure 7.10.** Asset management template

> **Box 7.1. Hawley Committee information assets**
>
> The Hawley Committee, in their extensive research, found that some information assets were consistently found across organizations. These were:
>
> - *Market and customer information.* Many utility companies, for instance, hold such information.
> - *Product information.* Usually such information includes registered and unregistered intellectual property rights kind of information.
> - *Specialist knowledge.* This is the tacit knowledge that might exist in an organization.
> - *Business process information.* This is information that ensures that a business process sustains itself. It is information that helps in linking various activities together.
> - *Management information.* This is information on which major company decisions are taken.
> - *Human resource information.* This could be the skills database or other specialist human resource information that may allow for configuring teams for specialist projects.
> - *Supplier information.* Trading agreements, contracts, service agreements, and so on
> - *Accountable information.* Legally required information, especially dealing with public issues (e.g., requirements mandated by HIPAA).

## *Access Control*

The aim of access control is to control and prevent unauthorized access to information. This is achieved by the successful implementation of the following objectives: control access to information; prevent unauthorized access to information systems; protect networked services; prevent unauthorized computer access; detect unauthorized activities; and ensure information security when using mobile computing and teleworking facilities. Access control policy should be established that lays out the rules and business requirements for access control. Access to information systems should be restricted and controlled through formal procedures. Similarly, policy on the use of network services should be formulated. Access to both internal and external network services should be controlled through appropriate interfaces and authentication mechanisms. Finally, the use and access to information processing facilities should be monitored and events be logged. Such system monitoring allows verification of effective controls.

## *Cryptography*

Cryptographic controls stress the importance of ensuring confidentiality, authenticity, and integrity of information. Chapter 3 of this book discusses in detail how various encryption and cryptographic approaches can be designed and implemented. In recent years, blockchain technologies have also gained prominence. It is of paramount importance for institutions to define a policy with respect to the use of cryptographic techniques in general, but blockchain in particular. PayPal, for instance, has publically acknowledged the relevance of blockchains and have accordingly shared their public policy (see Table 7.5).

**Table 7.5.** PayPal's public policy on blockchains (drawn in part)[2]

> Blockchains, like all new technologies, have the potential to disrupt multiple industries. In finance, for instance, consortium blockchains between banks could establish a distributed network controlled by participating financial institutions and disintermediate the SWIFT network for international transfers and settlements.
>
> PayPal believes that blockchains, while holding interesting potential, particularly in the world of finance, are still in their early days. We have not yet seen use cases in the financial space that are highly differentiated and particularly compelling, but we remain engaged with the broader ecosystem and are interested in how blockchain may result in demonstrable benefits for financial services.
>
> Governments and regulators should be careful to not rush into regulating blockchain. Rather, as other technological-based solutions (i.e., One Touch, tokenization), governments and regulators should look at how the technology is utilized in order to determine whether regulation is necessary. Where blockchains are used as a fully distributed platform, governments and regulators should also be aware that it will be challenging to regulate their use on a national or sub-national level, and we encourage standardization and consistency across the regulatory landscape.
>
> Blockchain technology originally was created to facilitate the cryptocurrency Bitcoin. Blockchains can be utilized for many things but Bitcoin could not exist without blockchain to facilitate, record and verify transactions.
>
> PayPal was one of the first payment companies to enable merchants to accept Bitcoin through Braintree, by way of partnerships with payment processors BitPay, GoCoin, and Coinbase. We also have integrated with Coinbase's virtual currency wallet and exchange so CoinBase users who sell Bitcoin can withdraw those proceeds to their PayPal accounts. These partnerships have provided us with valuable expertise and market insights that will shape our strategy and investments around Bitcoin and blockchain going forward.

### Physical and Environmental Security

These aspects deal with perimeter defenses, especially protecting boundaries, equipment, and general controls such as locks and keys.

### Operational Security

Both operational procedures and housekeeping involve establishing procedures to maintain the integrity and availability of information processing services as well as facilities. For housekeeping, routine procedures need to be established for carrying out the backup strategy, taking backup copies of essential data and software, logging events, and monitoring the equipment environment. On the other hand, advance system planning and preparation reduce the risk of system failures.

### Communication Security

Communication controls relate to network security aspects, especially dealing with confidentiality and integrity of data as it is transmitted. Issues related to system risks are particularly

---

[2] https://www.paypal.com/us/webapps/mpp/public-policy/issues/blockchain

considered important. Integrity of software and ensuring availability of data at the right time and place are identified as the cornerstones of communication and operations security. Business continuity through avoidance of disruption is another important aspect that is considered as part of this control. Besides, the general loss or modification or misuse of information exchanged between organizations needs to be prevented

### System Acquisition, Development, and Management
The overarching theme of the systems development and maintenance section is that the security requirements should be identified at the requirements phase of a project. Its main objective is to ensure security is built into application systems and into overall information systems. The analysis of security requirements and identification of appropriate controls should be determined on the basis of risk assessment and risk management. The use of cryptographic systems and techniques is advocated to further protect the information not adequately protected by such controls. In addition, IT projects and application system software should be maintained in a secure manner. As such, project and support environments need to be strictly controlled.

### Supplier Relationship
Given the interconnected nature of businesses, it is important to ensure that the security and integrity of supplier relationships is maintained. This particular control emphasizes the importance of supply chains and the inherent risks. Since security is as good as its weakest link, the control emphasizes that corporate information assets be ascribed proper responsibility for their protection. The controls also suggest that there should be an agreed-upon level to ensure service and security. There are other ISO standards that touch upon the supplier side of security management. For instance, ISO 30111 covers all vulnerability handling processes, irrespective of the threat being identified internal to the organization or from external sources. ISO 29147 covers vulnerability disclosures from sources that are external to the organization. These might include end users, security researchers, and hackers.

### Security Incident Management
ISO 27002 sketches out the importance of incident handling and reporting. The aspiration is to ensure that information security incidents are managed effectively and consistently. Security incident management can be accomplished by focusing on the following 10 aspects (additional details are available in Chapter 8):

1. Clarification of roles and responsibilities
2. Identification of people who are responsible, accountable, consulted, or informed before, during, and after the incident
3. Awareness of all responsibilities (Establishing items 1 and 2 is good, but the rest of the organization should be aware of the roles and responsibilities.)
4. Well-defined training programs
5. Checklists for operational maintenance responses (e.g., procedures for shutdown, startup, restoration, etc.)
6. Clearly defined dependencies and touch points for the security incidents and the related management policy

7. Well-defined evidence collection procedures
8. Definition of functional and forensics techniques (e.g., for quarantining, containment, real-time observation, investigation, analysis, and reporting)
9. A process to capture learning from an incident and subsequently update the vulnerability and risk repository
10. A method to measure the impact of the incident, and a well-defined communication and reporting structure to the management and other stakeholders

### Security Continuity Management

Business continuity management deals with dual objectives of counteracting interruptions to business activities and protecting critical business processes from the effects of major failures or disasters. It involves implementing the business continuity management process. Such a process would involve impact analysis, development, and maintenance of the continuity planning framework. These business continuity plans should be tested and reviewed regularly to ensure their effectiveness.

### Security Compliance Management

The final section broaches the issue of compliance. The first objective is to avoid breaches of any criminal and civil law, statutory, regulatory, or contractual security requirements. The second objective is concerned with ensuring compliance of systems with organizational security policies and standards. The final objective is to maximize the effectiveness of system audit processes. System audit tools should be safeguarded and access to them controlled to prevent apparent misuse. A summary of all ISO 27002 controls and their objectives appears in Table 7.6.

**Table 7.6.** Summary of ISO 27002 controls and their objectives

| Type of control | Objectives |
| --- | --- |
| **Security policy management**: Provide management direction and support | To provide management direction and support for information security activities |
| **Organization of information security**: Establish an internal information security organization; Protect your organization's mobile devices and telework | To establish a framework to manage information security within your organization<br>To ensure the security of mobile devices and telework (work done away from the office at home or elsewhere) |
| **Human resource security management**: Emphasize security prior to employment; Emphasize security during employment; Emphasize security at termination of employment | To ensure that prospective employees and contractors are suitable for their future roles<br>To ensure that employees and contractors meet their information security responsibilities<br>To protect your organization's interests whenever personnel terminations occur or responsibilities change |

**Table 7.6.** Summary of ISO 27002 controls and their objectives (*continued*)

| Type of control | Objectives |
|---|---|
| **Organizational asset control and management**: Establish responsibility for corporate assets; Develop an information classification scheme; Control how physical media are handled | To protect assets associated with information and information processing facilities<br>To provide an appropriate level of protection for your organization's information<br>To protect information by preventing unauthorized disclosure, modification, removal, or destruction of storage media |
| **Information access control**: Respect business requirements; Manage all user access rights; Protect user authentication; Control access to systems | To control access to your organization's information and information processing facilities<br>To ensure that only authorized users gain access to your organization's systems and services<br>To make your users accountable for safeguarding their own secret authentication information<br>To prevent unauthorized access to your organization's information, systems, and applications |
| **Cryptography**: Control the use of cryptographic controls and keys | To use cryptography to protect the confidentiality, authenticity, and integrity of information |
| **Physical security management**: Establish secure areas to protect assets; Protect your organization's equipment | To prevent unauthorized physical access to information and information processing facilities<br>To prevent the loss, theft, damage, or compromise of equipment and the operational interruptions that can occur |
| **Operational security management**: Establish procedures and responsibilities; Protect your organization from malware; Make backup copies on a regular basis; Use logs to record security events; Control your operational software; Address your technical vulnerabilities; Minimize the impact of audit activities | To ensure that information processing facilities are operated correctly and securely<br>To protect information and information processing facilities against malware<br>To prevent the loss of data, information, software, and systems<br>To record information security events and collect suitable evidence<br>To protect the integrity of your organization's operational systems<br>To prevent the exploitation of technical vulnerabilities<br>To minimize the impact that audit activities could have on systems and processes |
| **Network security management**: Protect networks and facilities; Protect information transfers | To protect information in networks and to safeguard the information processing facilities that support them<br>To protect information while it's being transferred both within and between the organization and external entities |

**Table 7.6.** Summary of ISO 27002 controls and their objectives (*continued*)

| Type of control | Objectives |
|---|---|
| **System security management**: Make security an inherent part of information systems; Protect and control system development activities; Safeguard data used for system testing purposes | To ensure that security is an integral part of information systems and is maintained throughout the entire life cycle<br>To ensure that security is designed into information systems and implemented throughout the development life cycle<br>To protect and control the selection and use of data and information when it is used for system testing purposes |
| **Supplier relationship management**: Establish security agreements with suppliers; Manage supplier security and service delivery | To protect corporate information and assets that are accessible by suppliers<br>To ensure that suppliers provide the agreed upon level of service and security |
| **Security incident management**: Identify and respond to information security incidents | To ensure that information security incidents are managed effectively and consistently |
| **Security continuity management**: Establish information security continuity controls; Build redundancies into information processing facilities | To make information security continuity an integral part of business continuity management<br>To ensure that information processing facilities will be available during a disaster or crisis |
| **Security compliance management**: Comply with legal security requirements; Carry out security compliance reviews | To comply with legal, statutory, regulatory, and contractual information security obligations and requirements<br>To ensure that information security is implemented and operated in accordance with policies and procedures |

# Other Miscellaneous Standards and Guidelines

## RFC 2196 Site Security Handbook

The Internet Engineering Task Force (IETF) Security Handbook is another guideline that deals with Internet Security Management specific issues. This site security handbook does not specify an Internet standard but rather provides guidance to develop a security plan for the site. It is a framework to develop computer security policies and procedures. The framework provides practical guidance to system and network administrators on security issues with lists of factors and issues that a site must consider in order to have effective security. Risk management is considered central to the process of effective security management, with identification of assets and corresponding threats as the primary tasks. Generally speaking, the handbook covers the issue of formulating effective security policies, security architecture, security services and procedures, and security incident response. Principles of security policy formulation, trade-offs, and mechanisms for regular updates are emphasized.

The section on architecture is classified into three major sections: The first section deals with objectives. It involves defining a comprehensive security plan, which is differentiated from a security policy as a framework of broad guidelines into which specific policies should fit. The necessity to isolate services into dedicated host computers is also stressed in this section. Importance of evaluating services and determining need is considered important. Finally, this section advises to evaluate all services and determine real need for them. The next section is network and service configuration. This section deals with the technical aspects of protecting the infrastructure, network, services, and security. The technical aspects are broached at the architecture level (or, at a higher level), rather than discussing the intrinsic technical details of implementing these security controls. The same is the case for firewalls discussed in the third section. This section provides a broad overview of firewalls, its working, composition, and importance to security. Firewalls are taken as just another tool for implementing system security providing certain level of protection.

Security services and procedures form the third section of the handbook. This section provides technical discussion on different security services or capability that may be required to protect the information and systems at a site. Again, the technical discussion is not concerned with intrinsic technical details on how to operationalize a control. For example, an overview of Kerberos, which is a distributed network security system, is provided while addressing the topic of authentication. But the details of its implementation or how its authentication mechanism works is not discussed. As such, this section provides a technical discussion or approach on how to achieve the security objectives. The topics addressed in this section include authentication, confidentiality, integrity, authorization, access, auditing, and securing backups.

The final major section of the handbook deals with security incident handling. It advocates the formulation of contingency plans in detail so that the security breaches could be approached in a planned fashion. The benefits for efficient incident handling involve economic, public relations, and legal issues. This section provides an outline of a policy to handle security incidents efficiently. The policy is composed of six major sections. Each section plays an important role in handling incidents and is addressed separately in detail. These critical sections include preparing and planning, notification, identifying an incident, handling, aftermath, and administrative response to incidents.

## ISO/IEC TR 13335 Guidelines for the Management of IT Security

Guidelines for the management of IT Security (GMITS) were developed by ISO/IEC JTC 1 SC 27 (standards committee). GMITS is only a technical report (TR), which means that this is actually a "suggestion" rather than a standard. The scope of GMITS is IT security and not information system security.

ISO/IEC TR 13335 contains guidance on the management of IT security. It comprises five parts:

- Part 1: Concepts and models for IT security
- Part 2: Managing and planning IT security
- Part 3: Techniques for the management of IT security

- Part 4: Selection of safeguards
- Part 5: Management guidance on network security

Part 1 presents the basic concepts and models for the management of IT security. Part 2 addresses subjects essential to the management of IT security and the relationship between these subjects. Part 3 provides techniques for the management of IT security. It also outlines the principles of risk assessment. Part 4 provides guidance on the selection of safeguards for the management of risk. Part 5 is concerned with the identification and analysis of communication factors that are critical in establishing network security requirements.

## Generally Accepted Information Security Principles (GAISP)

GAISP documents information security principles drawn from established information security guidance and standards that have been proven in practice and accepted by practitioners. It intends to develop a common international body of knowledge on security. This in turn would enable a self-regulated information security profession. GAISP has evolved from Generally Accepted System Security Principles (GASSP). The GASSP project was formed by the International Information Security Foundation (I²SF) in response to the first recommendation of the report "Computers at Risk" (CAR), "to promulgate comprehensive generally accepted security principles," published by the United States of America's National Research Council in 1991[3]. Generally Accepted System Security Principles version 1.0 was published in November 1995. The GASSP project was later adopted by the Information Systems Security Association (ISSA) and renamed Generally Accepted Information Security Principles (GAISP). The new name reflects the objective to secure information. GAISP version 3.0, which is an updated draft, was published in January 2004.

GAISP is organized into three major sections that form a hierarchy. The first section is the Pervasive Principles. This section targets governance and is based completely on the OECD Guidelines. It outlines the same nine principles advocated in the OECD Guidelines. The Broad Functional Principles form the second section, which targets management. It describes specific building blocks (what to do) that comprise the Pervasive Principles. These principles provide guidance for operational accomplishment of the Pervasive Principles. Fourteen Broad Functional Principles are outlined in the section, along with the rationale and an example. These fourteen principles are information security policy; education and awareness; accountability; information management; environmental management; personnel qualifications; system integrity; information systems life cycle; access control; operational continuity and contingency planning; information risk management; network and infrastructure security; legal, regulatory, and contractual requirements of information security; and ethical practices.

The third section is the Detailed Principles and targets information security professional. These principles provide specific (how-to) guidance for the implementation of optimal information security practices in compliance with the Broad Functional Principles.

---

[3] The report is available for a free download at http://nap.edu/1581.

## OECD Guidelines for the Security of Information Systems

The Organization for Economic Cooperation and Development (OECD) Guidelines were developed in 1992 by a group of experts brought together by the Information, Computer, and Communications Policy (ICCP) Committee of the OECD Directorate for Science, Technology, and Industry. The Guidelines for the Security of Information Systems form a foundation on which the framework for security of information systems could be developed. The framework would help in the development and implementation of coherent measures, practices, and procedures for the security of information systems. As such, it would include laws, codes of conduct, technical measures, management and user practices, and education and awareness activities. The guidelines strive to foster confidence and promote cooperation between the public and private sectors, as well as at the national and international level. It recognizes the commonality of security requirements across various organizations (public or private, national or international) and has developed an integrated approach. This integrated approach is outlined in the form of nine principles that are essential to the security of information systems and their implementation. These principles are accountability, awareness, ethics, multidisciplinary, proportionality, integration, timeliness, reassessment, and equity.

The accountability principle advocates that the responsibilities and accountability of stakeholders of information systems be explicit. The awareness principle is concerned with the ability of stakeholders to gain knowledge about the security of information systems without compromising security. The ethics principle deals with the development of social norms associated with security of information systems. The multidisciplinary principle stresses the importance of taking a holistic view to security that includes full spectrum of security needs and available security options. The proportionality principle is the common sense approach to information security. It states that the level and type of security should be weighed against the severity and probability of harm and its costs as well as the cost of the security measures. The integration principle emphasizes the importance of inculcating security at the design level of the information system. The timeliness principle stresses the need to establish mechanisms and procedures for rapid and effective cooperation in the wake of security breaches. These mechanisms should transcend both industry sectors and geographic boundaries. The reassessment principle suggests that security be reviewed and updated at regular intervals. The equity principle observes the principles of a democratic society. It recommends maintaining a balance between the optimal level of security and legitimate use and flow of data and information.

In terms of implementing the principles, the OECD Guideline call upon governments, the public sector, and the private sector to support and establish legal, administrative self-regulatory, and other measures, practices, procedures, and institutions for the security of information systems. This objective is further elaborated upon under the sections pertaining to policy development, education and training, enforcement and redress, exchange of information, and cooperation. The issues of worldwide harmonization of standards, promotion of expertise and best practices, allocation of risks and liability for security failures, and improving jurisdictional competence are discussed as part of policy development. This section also advocates adoption of appropriate policies, laws, decrees, rules, and international agreements.

**Table 7.7.** National Institute for Standards and Technology security documents

**Standard/guideline name**
- SP 800-12, Computer Security Handbook
- SP 800-14, Generally Accepted [Security] Principles and Practices
- SP 800-16, Information Technology Security Training Requirements: A Role- and Performance-Based Model
- SP 800-18, Guide for Developing Security Plans
- SP 800-23, Guideline to Federal Organizations on Security Assurance and Acquisition/Use of Tested/Evaluated Products
- SP 800-24, PBX Vulnerability Analysis: Finding Holes in Your PBX before Someone Else Does
- SP 800-26, Security Self-Assessment Guide for Information Technology Systems
- SP 800-27, Engineering Principles for Information Technology Security (A Baseline for Achieving Security)
- SP 800-30, Risk Management Guide for Information Technology Systems
- SP 800-34, Contingency Plan Guide for Information Technology Systems
- SP 800-37, Draft Guidelines for the Security Certification and Accreditation of Federal Information Technology Systems
- SP 800-40, Procedures for Handling Security Patches
- SP 800-41, Guidelines and Firewalls and Firewall Policy 4
- SP 800-46, Security for Telecommuting and Broadband Communications
- SP 800-47, Security Guide for Interconnecting Information Technology Systems
- SP 800-50, Building an Information Technology Security Awareness and Training Program (DRAFT)
- SP 800-42, Guideline on Network Security Testing (DRAFT)
- SP 800-48, Wireless Network Security: 802.11, Bluetooth, and Handheld Devices (DRAFT)
- SP 800-4A, Security Considerations in Federal Information Technology Procurements (REVISION)
- SP 800-35, Guide to IT Security Services (DRAFT)
- SP 800-36, Guide to Selecting IT Security Products (DRAFT)
- SP 800-55, Security Metrics Guide for Information Technology Systems (DRAFT)
- SP 800-37, Guidelines for the Security Certification and Accreditation (C&A) of Federal Information Technology Systems (DRAFT)

International Harmonization • 197

## Concluding Remarks

In this chapter we have reviewed and presented the various IS security standards. It is important to develop an understanding of all the standards, since they form the benchmark for designing IS security in organizations. While there are issues related to efficiency of having such a large number of standards, it is prudent nevertheless to develop a perspective as to where each of the standards fit in. Clearly some standards, such as ISO27002 have gained more importance in recent years, relative to some other standards such as ISO 13335. The point to note, however, is that all standards play a role in ensuring the overall security of the enterprise.

While ISO 27002 is essentially an IS Security management standard, the Rainbow Series and other evaluation criteria, including Common Criteria, seem to play a rather important role in evaluating system security. Similarly, the security development standards and SSE-CMM in particular indeed help in developing security practices that facilitate good, well-thought-out IS security development. Overall, security standards need to be considered in conjunction with each other, rather than competing standards.

Other guidelines and standards including the NIST 800 series publications (Table 7.7) and OECD guideline incorporate a wealth of knowledge as well. The problem, however, is the availability of a large number of standards, which leaves users confused as to the appropriateness of one standard over the other. As a user it is rather challenging to differentiate and align oneself with one set of guidelines over the other. This chapter logically classifies different standards—management, development, evaluation—and it is our hope that these will help users in identifying the right kind of standard for the task at hand.

### In Brief

- Security related to flawed systems development typically occurs because of
    - Failure to perform a function that should have been executed,
    - Performance of a function that should not have been executed, or
    - Performance of a function that produced an incorrect result.
- There are four categories of control structures: auditing, application controls, modeling controls, and documentation controls.
- Auditing controls record the state of a system and examine, verify, and correct the recorded states.
- Application controls look for accuracy, completeness, and general security.
- Modeling controls look for correctness in system specification.
- Documentation controls stress the importance of documentation alongside systems development rather than as an afterthought.
- SSE-CMM focuses on processes that can be used in achieving security and the maturity of these processes.

- The scope of the processes incorporate:
  - System security engineering activities used for a secure product or a trusted system. It should address the complete life cycle of the product, which includes the conception of an idea; requirement analysis for the project; designing of the phases; development; integration of the parts; proper installation; operation and maintenance.
  - Requirements for the developers (product and secure system) and integrators, the organizations that provide computer security services, and computer security engineering.
  - It should be applicable to various companies that deal with security engineering, academia, and government.
- SSE-CMM process areas include the following: administer security controls; assess operational security risk; attack security; build assurance argument; coordinate security; determine security vulnerabilities; monitor system security posture; provide security input; specify security needs; verify and validate security.
- SSE-CMM has two basic dimensions: base practices and generic practices.
- Security evaluation has a rich history of standardization. With origins in the US Department of Defense, the Rainbow Series of standards present assurance levels that need to be established for IS security.
- In the United States, the most prominent of the security evaluation standards has been the Trusted Computer System Evaluation Criteria (TCSEC).
- The TCSEC gave way to the European counterpart—the Information Technology Security Evaluation Criteria (ITSEC).
- While all individual evaluation criteria continue to be used, an International harmonization effort has resulted in the formulation of the Common Criteria (CC).
- Numerous other context-specific standards have been developed. Some of these include:
  - Internet Engineering Task Force (IETF) Security Handbook
  - Guidelines for the Management of IT Security (GMITS)
  - Generally Accepted System Security Principles (GASSP)
  - OECD Guidelines for the Security of Information Systems
  - 800 series documents developed by the National Institute for Standards and Technology

## Questions and Exercises

**Discussion questions.** These are based on a few topics from the chapter and are intentionally designed for a difference of opinion. These questions can best be used in a classroom or a seminar setting.

1. "SSE-CMM is the panacea of secure systems development." Discuss.

2. How does SSE-CMM ensure correctness of system specification leading to good system design? Is there a connection between good system design and security? If so, what is it? If not, give reasons for lack of such a relationship.

3. Establishing control structures in systems can best be achieved by focusing on requirement definitions and ensuring that controls get represented in basic data flows. Although such an assertion seems logical and commonsensical, identify and examine hurdles that usually prevent us from instituting such controls.

4. There are a number of independent security assurance and certification programs. Each claims itself to be the best in the industry and suggests that their certification allows companies and individuals to place a level of trust in the systems and practices. Can any security certification or assurance program guarantee a high level of success in ensuring security? Discuss the problem, if any, of multiple security schemes and certification bodies. You may also want to consider the issue of mandatory certification, especially for defense-related systems. Reference may be made to certifications such as TruSecure (http://www.trusecure.com/), SCP (http://www.securitycertified.net/), Defense Information Technology Systems Certification and Accreditation (http://iase.disa.mil/ditscap/), and the National Information Assurance Certification and Accreditation Process (http://www.dss.mil/infoas/).

**Exercise 1.** Think of a fictitious software house developing software for mission critical applications. Develop measures to assess the level of maturity for each of the SSE-CMM levels. Suggest reasons as to why your measures should be adopted.

**Exercise 2.** Make a list of all possible security standards that you can find. Try and cover at least standards in Europe and North America. Classify standards according to the systems development life cycle and comment on the usefulness of each of the standards.

## Short Questions

1. The four types of access that may be granted to a database are _____, _____, _____, _____ (CRUD), and also represent the four types of security breaches that may occur in systems.

2. Successful _____ control procedures should record the state of a system and then examine, verify, and correct the recorded states.

3. While security controls often concentrate on the prevention of intentional security breaches, most breaches are _____.

4. Application controls address the _____ of a failure.

5. Audit controls address the _____ of a failure (i.e., audit controls attempt to determine if the application controls are adequate).

6. Controls that are used at the analysis and design stages of the systems life cycle as a tool to understand and document other control points within a system are called _____ controls.

7. Good _____ controls not only answer what are the functions of a system and how those functions are being accomplished; the controls address the question of why the system is performing those particular functions.

8. Name a model used for assessing the security engineering aspects of the target organization.

9. The building of infrastructure and corporate culture that establishes methods, practices, and procedures is called _____.

10. A _____ practice should be applied across the life cycle of the enterprise, and should not specify a particular method or tool.

## Case Study: Remote Access Problems at the DHS

An audit of the Department of Homeland Security's system controls for remote access has found several deficiencies that put the DHS at risk of malicious hacker attacks.[4] The audit performed by the Office of the Inspector General is mandated by the new FISMA regulations that affect federal agencies.

The report indicates that "while DHS has established policy governing remote access, and has developed procedures for granting, monitoring, and removing user access, these guidelines have not been fully implemented." The report indicates that processes are being developed to implement the security policies, and they are awaiting software tools to assist in the implementation. Meanwhile, the DHS systems remain vulnerable to attack from outside sources. The report identified several specific deficiencies: (1) remote access hosts do not provide strong protection against unauthorized access; (2) systems were not appropriately patched; and (3) modems that may be unauthorized were detected on DHS networks. The report says that "due to these remote access exposures, there is an increased risk that unauthorized people could gain access to DHS networks and compromise the confidentiality, integrity, and availability of sensitive information systems and resources."

The report made three recommendations to assist DHS in remedying the deficiencies identified. Comment on the merits of each or make your own recommendations.

1. Update the DHS Sensitive Systems Handbook (DHS Handbook) to include implementation procedures and configuration settings for remote access to DHS systems.

2. Ensure that procedures for granting, monitoring, and removing user access are fully implemented.

3. Ensure that all necessary system and application patches are applied in a timely manner.

4. Who should be responsible for implementing the above recommendations?

---

[4] Based on "DHS Audit Unearths Security Weaknesses," eWeek.com, December 17, 2004. Accessed September 15, 2017.

# References

1. Alberts, C., and A. Dirifee. 2002. *Managing information security risks: The OCSTAVE (SM) approach.* Boston: Addison-Wesley Professional.
2. Ferraiolo, K., and V. Thompson. 1997. Let's just be mature about security! Using a cmm for security engineering. *CROSSTALK: The Journal of Defense Software Engineering,* Aug.
3. Hawley, R. 1995. Information as an asset: the board agenda. *Information Management and Technology,* 28(6): 237–239.
4. KPMG. 1994. *The Hawley Report: Information as an asset—the board agenda.* London: KPMG/IMPACT.

# CHAPTER 8

# Responding to an Information Security Breach

*The mantra of any good security engineer is: "Security is a not a product, but a process." It's more than designing strong cryptography into a system; it's designing the entire system such that all security measures, including cryptography, work together.*

—Bruce Schneier
Cryptographer and Computer Security Expert

Joe Dawson woke up in the morning only to read in the latest issue of the *Wall Street Journal* that the "Dark Web" sites had been hit in a cyber attack. *This is unbelievable*, Joe thought. We are now in an era where criminals are attacking criminals—something like the drug wars of the past. Joe was reminded of all the killings in Juarez, a border town in Mexico, in early 2010. Juarez was declared as one of the most dangerous cities in Mexico. The Mexican government had released figures that in the first nine months of 2011, nearly 13,000 people had been killed in drug-related violence. The only difference today was that the gang wars had gone high-tech.

Joe called his network administrator to understand what the Dark Web was and how his company could be affected. The network administrator sat down with Joe to explain things.

"Well, Joe, 'Dark Web,' 'Deep Wek,' 'Dark Net' are all spooky sounding phrases, but they all mean the same thing," explained Steve. He went on to draw a diagram and show how websites mask their IP address and how such sites can only be accessed using certain encryption-friendly tools, The Onion Router (TOR) being one of them. TOR scrambles the user's IP address through a distributed network, making it extremely difficult to figure out the exact location of the website. TOR project is an open-source community and also develops Tails, which was popularized by Edward Snowden. The program can run off a USB flash drive. Tails, in particular, provides additional layers of security such that browsing is not tied to a specific machine. With Tails, it is possible to store encrypted files, execute email programs, and launch and run the TOR browser.

"Really!" exclaimed Joe. "This means if we are attacked, it will be difficult to figure out where the traffic originated." "Correct," said Steve.

"So how can we protect SureSteel?" asked Joe.

Steve noted that it was not easy to stop the attacks, but a broader vulnerability management approach is necessary. Steve promised to write a memo to Joe explaining the dangers and what could be done. He took a couple of days to research the topic area and sent the following note to Joe.

FROM: Steve King, Senior Network Administrator, SureSteel
TO: Joe Dawson, CEO, SureSteel
SUBJECT: Dark Web, Vulnerability Management, and Managing a Breach

While the Dark Web sounds pretty grim, it is not very difficult to manage. Here is a description extracted from an article in *Dark Reading*:

> The Dark Web is veritably tiny in comparison to the more familiar public Web and minuscule when compared to the larger Deep Web that is not searchable by search engines. When most people think of the Dark Web, they immediately think of trade in drugs and pornography. While those are indeed the predominate commodities in a space built for illicit commerce and trade, the Dark Web offers other things too.
>
> If there is a silver lining in all of this, it's that most businesses already have all the tools on hand for starting a low-cost, high-return Dark Web intelligence operations within their own existing IT and cybersecurity teams. I have personally been a part of Dark Web data mining operations set-up, implementation and being productive in just a day's time.
>
> Setting up your Dark Web mining environment using TOR, private browsing on air gapped terminals via sequestered virtual machine clusters (VMs), is something that's well-understood among cybersecurity professionals already on your team. When you pair them with the security analysts and intelligence personnel you're hiring to staff up your cyber intelligence initiatives; it becomes something you can start almost in complete logistic (and fiscal) parallel with these other efforts.[1]

This was very informative, and Joe was thankful to Steve, who had clarified several issues. There was no doubt that SureSteel had to proactively think about protection and find ways and means of responding to a breach.

---

"We have been hacked!" These are the dreaded words no executive wants to hear. Yet this is exactly how the co-chairman of Sony Pictures Entertainment, Amy Pascal's, Monday morning started when the company discovered its entire computer system had been hacked

---

1 https://www.darkreading.com/analytics/the-dark-web-an-untapped-source-for-threat-intelligence-/a/d-id/1320983

by an organization called Guardians of Peace. This was one of the biggest attacks of 2014. Several others have followed in 2015 and 2016.

Over the past few years, the size and magnitude of cybersecurity breaches have been on an increase. The 2014 South Korean breach, where nearly 20 million (40% of the country's population) people were affected, epitomized the seriousness of the problem. More recently a cybersecurity breach was discovered in Ukrainian banks. Carbanak, a malware program, infected the bank's administrative computers. The breach resulted in banks of several countries, including the United States, Russia, and Japan getting infected. The seriousness of the problem can be judged from the 2016 Internet Security Threat Report published by Symantec. Nearly half a billion personal records were stolen or lost in 2015, and on an average one new zero-day vulnerability[2] was discovered each week. When a zero-day vulnerability is discovered, it gets added to the toolkit of cyber criminals.

An IBM study concluded that an average data breach costs about 3.52 to 3.79 million US dollars, and it keeps rising every year [5]. It is not just the dollar expense that matters in breach situations. It is very likely that the breach damages the company's reputation, and some smaller unprepared organizations might never recover from a major disaster. Cybersecurity breaches affect organizations in different ways. Reputational loss and decreased market value have often been cited as significant concerns. Loss of confidential data and compromising competitiveness of a firm can also cause havoc. There is no doubt that preventive mechanisms need to be put in place. However, when an IT security breach does occur, what should be the response strategy? How can the impact of a breach be minimized? What regulatory and compliance aspects should a company be cognizant of? What steps should be taken to avoid a potential attack?

Companies can defend themselves by conducting risk assessments, mitigating against risks that they cannot remove, preparing and implementing a breach response plan, and implementing best practices. Past events have shown that better prepared companies are able to survive an attack and continue their business operations. Experts recommend the board of director's involvement in data protection; active participation from senior decision makers can reduce the cost of data breach. There are several other ways managers can prevent, reduce, and mitigate against data breaches.

---

2 While there are several variants of zero-day vulnerabilities, in a more generic sense, this refers to a "hole" in a software that, though known to the software vendor, gets exploited by hackers. Vendors typically release patches to fill the holes.

## Box 8.1. Recent high-profile breaches

### Anthem: Another one bites the dust

On January 29, 2015, it was discovered that Anthem Inc, one of the nation's leading health insurers, was the victim of a cyberattack whereby cyberattackers attempted to gain access to personally identifiable information about current and former Anthem members. The hackers began accessing the information in early December 2014, and during a nearly 7-week window, perpetrators were able to gain access to nearly 80 million records [2]. Anthem has indicated that not only current members of Anthem were impacted. On its website, Anthem noted, "In addition, some members of other independent Blue Cross and Blue Shield plans who received health care services in any of the areas that Anthem serves may be impacted. In some instances, non-Anthem members and non-Blue Plan members may have been impacted if their employer offered Anthem and non-Anthem health plan options. Anthem is providing identity protection services to all individuals that are impacted."[3] Although Anthem maintains that no credit card or financial information was accessed, the threat to individuals' finances remains. The hackers were able to gain access to names of individuals, health care ID numbers, dates of birth, Social Security numbers, home addresses, email addresses, and employment information. With this data, it is easy to create identities and impersonate someone in a variety of settings.

### Home Depot: Sheer embarrassment

In the case of Home Depot, in September 2014 the company announced its payment systems were breached, which affected nearly 2,200 US and Canadian store locations in a cyberattack that may have started as far back as April 2014. Embarrassingly, Home Depot wasn't aware its payment systems were compromised until banks, and members of the law enforcement community, notified the company months after the initial data breach. The Home Depot security breach actually lasted longer than the Target breach spanning an estimated 4 months, resulting in thieves stealing tens of millions of customers' credit and debit card information. In the last six months of 2014, Home Depot processed approximately 750 million customer transactions that presented a treasure trove of information for hackers to focus on.

### Sony: Simple blame attribution

Sony faced a cyberattack prior to the expected release of the movie *The Interview* (2014), where hackers released usernames and passwords for staging and production servers located globally, in addition to the usernames/passwords and RSA SecurID tokens of Sony employees. Sony was forced to "turn off" its entire computer network infrastructure after it was also discovered the hackers posted information for all of Sony's routers, switches, and administrative usernames and passwords to log on to every server throughout the world. As a result of the Sony attack, an estimated 40% of large corporations will now have plans to deal with and address aggressive cybersecurity business disruption attacks. The Sony attack, in which hackers also posted embarrassing work emails of the Sony Pictures executives, has led to more buy-in from C-suite and executive boards across all corporations.

---

3 Following the breach, Anthem developed a dedicated website to share facts related to the breach (http://www.anthemfacts.com).

## Technicalities of a Breach

Now that the attack has happened and victims are reeling from the unsettling feeling that their personally identifiable information is out there somewhere, the real question is how did this all happen in the first place? To answer that question, we must first analyze the security policy that Anthem had in place at the time of their attack in early December 2014. At the time of the attack, there were several media reports[4] accusing Anthem of inadequate policies for accessing confidential information [3]. The insurer was also faulted for technical evaluation of software upgrades that verified authority of people or entities seeking access to confidential information. In addition to these accusations, the buzzword that surfaced after the attack seemed to be "encryption." Anthem was accused of storing nearly 80 million Social Security numbers without encrypting them. Some would argue that while encryption would make the data more secure, it may also render the data less useful.

The root of the issue is not a solitary smoking gun; there are a variety of technical factors that contributed to the inevitability of this security breach. First and foremost is the role of a security policy. As was mentioned previously, Anthem did a very poor job of formulating sound policies for granting access to various databases. Anthem also failed to implement adequate measures to ensure unauthorized users were denied access to client data. A related issue is undoubtedly about encryption. Anthem data were not encrypted. Had encryption been undertaken, the task of decrypting and making these data useful would have been a significantly more difficult task for the hackers. But let's assume for a moment that the benefit of using the data in their natural form outweighs the risk of leaving it unencrypted. But aren't there other ways of protecting the data? Many companies employ a variety of additional safeguards to protect their data, of which Anthem employed very few. Among these additional safeguards are random passcodes generated on a keyfob that change over a brief period of time, the use of IP-based access to remote servers, and the use of random IDs stored in a separate, unlinked database, to name a few. Anthem needs to take advantage of the veritable cornucopia of advanced security options to cover themselves from a technical vantage point or risk having disaster occur again.

Home Depot had similar issues and problems with their security policy. Once the attackers gained access to one of their vendor environments, they could use the login credentials of a third party vendor to then open the front door. Once on the network, it was easy for the hackers to exploit a known zero-day vulnerability in Windows. The vulnerability allowed the hackers to pivot from the vendor environment to the main Home Depot network. It was then possible to install memory scraping malware on the point of sales terminals. Eventually, in a total of 56 million records of credit and debit card data were stolen. The Home Depot vulnerability could have been prevented. While the network environment did have the Symantec Endpoint Protection, the Network Threat Protection feature had not been turned on. While this may not guarantee security, it would have certainly made life more difficult for the hackers. Moreover, the policy seemed to be deficient in terms of a proper vulnerability management program.

---

4 http://www.ktvu.com/business/4155658-story

## Policy Considerations

There is a variety of technical and human factors that contribute to the inevitability of a breach. In a majority of the cases, fingers have been pointed to the technical inadequacy of the enterprise. In the case of Anthem, it was the lack of encryption. For Home Depot, it was the lack of technical controls to prevent malware from collecting customer data. At Target, there was a basic networking segmentation error.

Occasionally we hear issues related to policy violations. In the case of Anthem, the US Department of Health and Human Services may impose a fine of some $1.5 million because of HIPAA violations. In many instances, efforts are made to ensure security policy compliance through rewards, punishment, or some behavioral change among employees. Rarely do we question the efficacy of the policy. Was the policy created properly? Was it implemented adequately? Were various stakeholders involved? Were there any change management aspects that were considered? These are some fundamental issues that need consideration.

Unfortunately these questions never get addressed. Security policies keep getting formulated and implemented in a top-down cookie-cutter manner. Organizational emphasis remains on punitive controls. And little attention is given to the content of the policy and how it is related. So how can organizations ensure that a coherent and secure strategic posture is developed?

- Security education, training, and awareness programs need to be established and monitored on an ongoing basis.
- All constituents are given access to cybersecurity strategic goals, which helps in inculcating ownership and hence compliance.
- Various stakeholders should be involved and encouraged to participate in cybersecurity decision-making, which helps with increased compliance.

## Reputation and Responsiveness

Reputational damage is significant following a data breach, particularly if a company fails to respond promptly. Following the Anthem, Sony, and Home Depot breaches, various social media outlets criticized the companies' delayed or inadequate responses regarding the breaches. In terms of crisis management, a three-day delay is considered significant. Post-crisis communication and a response strategy are essential to ensure that the right message gets through. Transparency in how the breach is being handled has its added importance.

Another well-publicized breach was that of JP Morgan, where hackers were able to steal confidential data for nearly 76 million US households. In a 2015 study, Syed and Dhillon collected Twitter data following the JP Morgan Chase breach in order to undertake a sentiment analysis [6]. The objective was to assess how individuals reacted to the breach. A total of 39,416 tweets were collected during the month of October 2014. Analysis of the results suggests that more than half of the tweets expressed negativity. Other significant findings included

- When a data breach responsibility is attributed to a company, it results in negative emotions, which in turn translates to negative word of mouth and even severing relationships with the enterprise.
- If the negativity related to the breach is high, it results in a quicker spread of the negative word of mouth sentiment (in our case, Twitter posting exhibited a shorter re-tweet time latency).
- The initial security breach responsibility shapes the reputation of the firm. Hence it is important to frame the message and security breach responsibility, since it has a direct reputational impact.

## Risk and Resilience

When a data breach occurs, postcrisis communication is perhaps the only opportunity that a company has to repair its reputation. Crisis situations can potentially have many negative consequences, ranging from losing customers, profitability, and market share to declining stock prices and job losses. A much less explored but very important factor is the impact of a crisis on organizational reputation. Corporate risk and resiliency planning are important for organizations to be able to bounce back from disruptions and thus retain stakeholder confidence. Understanding and identifying potential adverse events in computerized networks is important for planning and implementing resilient mechanisms to defend, detect, and remediate from such threats. The risk gets reduced when organizations implement both resilient technical and socioorganizational mechanisms. There is a need to integrate risk and resilience mechanisms into the organizational culture to prevent security breaches. There are four key characteristics of any risk and resilience approach:

1. The approach should provide a holistic framework, which assesses the systems and their interactions—from a system to the network, from the network to the organization and subsequently the societal impact.
2. The approach should emphasize the capacity to manage the range of hazards.
3. There need to be options for dealing with uncertainties, surprises, and any potential changes.
4. The focus should be on proactive management.

Thus a system that effectively reduces risks is going to be more resilient to the security breaches. Risk reduction means a deflection of risk and risk sharing. Also, an ability of an organization to prepare for the surprises and effectively responding to the breach incidents characterizes organizational resilience.

## Governance

Well-considered governance is at the core of any successful cybersecurity program. Many important aspects require consideration—policy, best practices, ethics, legality, personnel, technical, compliance, auditing, and awareness. Weak governance is often considered to be the cause of organizational crisis. Over the past several decades, we have observed that in institutions where governance was poor or the structures of accountability and responsibility were not clear, they have been susceptible to cybersecurity breaches—for instance, the multibillion-dollar loss experienced by Société Générale because of violation of internal controls by Jérôme Kerviel [7]. Similarly, the case of Barings Bank where Nick Leeson circumvented established controls [1]. Société Générale and Barings Bank showcase a lack of governance as the prime reason for the security breaches. Key principles for a sound and robust security governance include

- Senior leadership commitment to cybersecurity is essential for good security governance.
- Cybersecurity is considered strategic with due consideration of risk management, policy, compliance, and incident handling.
- Clear lines of communication are established between strategic thinkers and operational staff.

## Steps to Avoid a Potential Attack

Managers can take steps today to avoid potential breaches and mitigate damage when breaches occur. There is a vast amount of data from many sources that purports to answer exactly how to prepare for the inevitability of a cyber attack. Because the nature and purpose of every attack is different and the composition of every business is different, there is no single prescription for prevention. However, by reducing data from multiple sources, we can derive a list of high-level practices that all organizations should adopt.

- Executive buy-in
  - In order to create an optimal cybersecurity policy, support has to come from the top levels of the organization. Security must become a core part of the organizational culture.
- Fully understand your risk profile
  - By knowing your industry and its attack vectors, what is valuable to your organization and how to protect those assets, security personnel can effectively create, support, and promote cybersecurity initiatives.
  - Identify and classify different cyberattack scenarios.
- Take threats seriously
  - Many organizations understand the full extent of the damage that can be done during an attack as well as the aftermath. However, many companies choose to ignore the possibility of such an attack happening to them, or they are willing to accept the risk of not taking adequate precautions due to cost or complexity.
- Policy enforcement
  - Policies can be as simple as a strong password, but should ideally go well beyond passwords. Security policies should be documented and automated wherever possible to avoid human error or omission. Circling back to executive support, policies should be a part of the culture that everyone chooses to follow.
  - Keep things in simple terms that non-IT executives and users can understand.

- Training
  - Security awareness and policy enforcement is crucial in order to create a security culture within an organization. Awareness of policies and security should be of paramount concern to all organizations.
  - There should be specialized training for those that deal with the most sensitive data in the company.
- Employee screening
  - Not all possible employees possess the same moralities as the business owners and stakeholders. Employees should not only be screened to ensure that their skills meet the requirements of the positions but, more importantly, that their beliefs closely match those of the organization.
  - Remember that people are often the weakest link in a security chain.
- Offline backup of critical data
  - Data is the lifeblood of an organization. Data loss following a breach is often as damaging as monetary loss or brand reputation compromise. Many organizations never fully recover from data loss events; some go out of business entirely. A copy of critical data in a secure offsite location is one small step that should not be overlooked.
- Invest intelligently in security
  - Information overload prevents many organizations from making intelligent security decisions. There are a thousand vendors pitching a thousand variants of "best practice" security models. Create a plan based on the needs of the organization and implement policies and tools that augment the plan. Avoid tying your security policy to any vendor's software or hardware. There is no "one-size-fits-all" solution.
  - One of the more direct methods for avoiding a security breach is to implement application whitelisting. Application whitelisting[5] can prevent many forms of a breach where the spoofing of an application allows a virus or malware to traverse firewalls and scanners without detection.
- Keep systems updated
  - Another direct method for avoiding a breach is simply to apply security patches to software and hardware systems on a prompt and routine schedule. This may appear to most as a "no-brainer" but is often overlooked.

The detailed list provided describes concepts that every organization should consider to improve their cybersecurity preparedness. These concepts can be tailored to fit the individual organization culture and data protection requirements. Regardless of the specifics, every organization should understand the company's security chain. The CEO must enable the chief compliance officer (CCO), the chief privacy officer (CPO), the chief information officer (CIO), and so on, to ensure each understands their role before, during, and after an attack. Working together, these individuals must create and own an enterprise-wide incident

---

5 Whitelist is an index of approved entities. In cybersecurity, whitelisting works best in centrally managed environments, since systems typically have a consistent workload.

(or risk) management plan, a data management program, an incident response plan, and communication/reporting plans.

Once these initiatives are in place, more detailed workflows, such as the Continuous Diagnostics and Mitigation (CDM) program from the Department of Homeland Security (DHS), can be adopted. This program utilizes commercial off-the-shelf (COTS) software and hardware to continually monitor for security related events as well as continuously improve upon processes and risk prioritization. A CDM-style framework (see Figure 8.1) also provides a practical model that any organization can adopt and tailor to meet its specific cybersecurity requirements.

In this day and age managers have to be proactive in preventing an attack. No longer is the question asked *if* companies will be hacked but rather *when* they are hacked what will be the protocol. Being vigilant about even the smallest and seemingly insignificant changes can be extremely useful. To protect customers and employees from having their financial or private information stolen, both industry and governments have implemented regulations that help security against common cyber-attacks. To combat credit card fraud, the payment card industry created the Data Security Standard that requires merchants who process credit cards to take specific measures that help protect against hacking attacks. The European Union, United Kingdom, United States, and Canada are among the governments that have also instituted privacy acts meant to regulate how businesses protect their customer and employee data from malicious hackers.

In addition to the fees and legal ramifications that can come as a result of failing to comply with the different regulations, hacking attacks can also damage a company's reputation or brand to the point that they lose customers and revenue. A company who is in the news

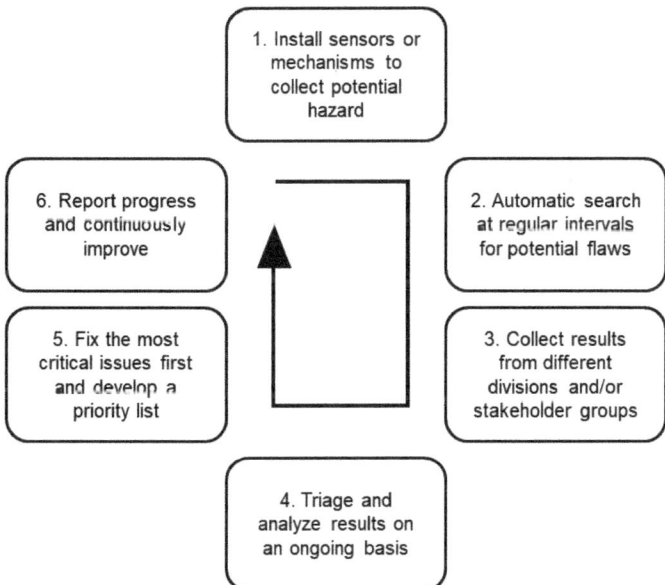

**Figure 8.1.** Continuous diagnosis and mitigation framework

because they have been hacked is sure to lose the trust of even their most loyal customers. The same happens with websites that are identified as containing spam or malicious scripts. Once this is known, most visitors will stay away. A company's brand, once damaged, may never be restored to its former status. Despite the restoration of services at Sony Pictures after the breach earlier this year, the Sony brand continues to fall under scrutiny. While monetary losses from the breach were significant, losses because of a damaged brand will continue to plague Sony for years.

# How to Respond When a Breach Occurs

As discussed previously, managers and organizations should take preventative steps to avoid the risk of a breach occurring. If after spending time planning, spending money, and training employees, someone still manages to break through the organization's security measures, what do you do then? Once a breach has been discovered, the organization should take the following immediate steps to limit the breach.

## Step 1: Survey the Damage

Following the discovery of the breach, the designated information security team members need to perform an internal investigation to determine the impact on critical business functions. This deep investigation will allow the company to identify the attacker, discover unknown security vulnerabilities, and determine what improvements need to be made to the company's computer systems.

## Step 2: Attempt to Limit Additional Damage

The organization should take steps to keep an attack from spreading. Some preventative strategies include

- Rerouting network traffic
- Filtering or blocking traffic
- Monitor all or parts of the compromised network

## Step 3: Record the Details

The information security team should keep a written log of what actions were taken to respond to the breach. The information that should be collected includes

- Affected systems
- Compromised accounts
- Disrupted services
- Data and network affected by the incident
- Amount and type of damage done to the systems

## Step 4: Engage Law Enforcement

A major breach should always be reported to law enforcement. The law enforcement agencies that should be contacted are

- Federal Bureau of Investigation (FBI)
- US Secret Service (USSS)
- US Immigration and Customs Enforcement (ICE)
- District attorney
- State and local law enforcement

Many companies wait until after a security breach before contacting law enforcement, but ideally the response team should meet with law enforcement before an incident occurs. The preliminary discussions would help an organization know when to report an incident, how to report an incident, what evidence to collect, and how to collect it. Once the incident is reported, the law enforcement agency may contact the media and ensure that sensitive information is not disclosed.

## Step 5: Notify Those Affected

If a breach puts an individual's information at risk, they need to be notified. This quick response can help them take immediate steps to protect themselves. However, if law enforcement is involved, they should direct the company as to whether or not the notification should be delayed to make sure that the investigation is not compromised. The individuals are usually notified via letter, phone, email, or in person. To avoid further unauthorized disclosure, the notification should not include unnecessary personal information.

## Step 6: Learn from the Breach

Since cybersecurity breaches are becoming a way of life, it is important to develop organizational processes to learn from breaches. This enables better incident handling, should a company be affected by a breach in the future. Some learning issues include

- Document all mistakes.
- Assess how the mistakes could have been avoided.
- Ensure training programs incorporate lessons learned.

These responses to an event in progress should not come as a surprise. These reactions should be rehearsed components of an organization's Cyber Incident Response Plan. Keep the plan up to date. Without a plan, confusion ensues and costly mistakes will likely be made. A working plan will show to law enforcement and the public that your intentions are good and will likely reduce fallout. Ensure the plan calls out key assets. Have those resources readily available. Identify tools and processes that will be used and followed. Knowing your industry and what is valuable to your organization (or what is valuable to someone looking to exploit your resources) will allow you to understand the attacker's intent and allow for proper assessment of the threat and proper plan execution. Have a postattack plan to ensure effective triage after the event. Use this plan to prioritize the efforts required to recover from a cyberattack, understand the extent of the damage, and minimize further damage. Close gaps in the environment and work the plan in such a way that it prevents causing more harm. Once again, document everything. Thorough documentation fosters credibility

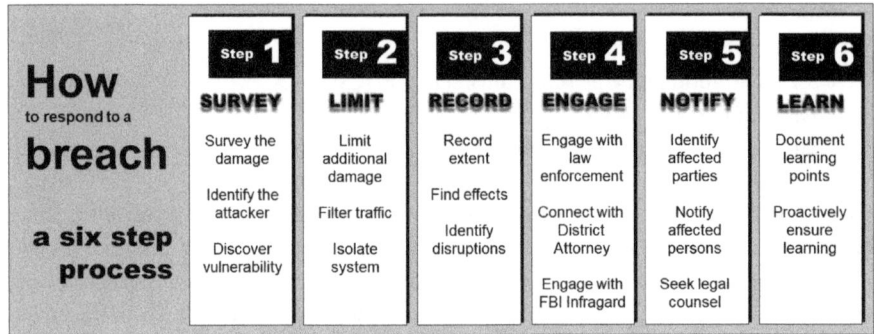

**Figure 8.2.** How to respond to a breach

for your organization, prevents repeats of mistakes, and produces confidence throughout the organization. Figure 8.2 summarizes the response process for a cybersecurity breach.

# Best Practices: How to Be Prepared for an Intrusion

Companies should use good judgment in avoiding, preparing for, and managing security events (even presumed events). History shows a wide variation in responses by major organizations. In Target's case, during the 2013 attack where payment card readers were infected, much evidence points to negligence by the retail giant. The retailer's FireEye malware detection product was found to be perfectly functional. In fact, the software found the same malware infection on consecutive attacks. There is an assumption based on these facts that Target either did not know how to read the data that the monitoring tools were reporting or intentionally neglected to report the breach. The net effect of either ignorance or negligence was huge brand damage to the retailer, and sales numbers dropped for some time. Having an adequate response plan along with notifying law enforcement and victims in a timely manner could have reduced the fallout for Target. In contrast, Anthem's response to the January 2015 breach was swift and thorough. Anthem found the breach themselves, reported it immediately to federal law enforcement, and set up a website and hotline for impacted customers. Anthem further offered its customers identity protection services. In comparison to Target, Anthem appeared to be poorly prepared for a breach, as the stolen data should have been encrypted, yet their response was almost textbook. Conversely, Target had the appropriate monitoring in place yet they either did not understand the reports or neglected to act on them appropriately. Despite Anthem's quick and forthright response, brand damage was still done due to their perceived lack of focus on cybersecurity. The 2014 Sony Pictures hack, however, was a different beast. The data breach affected mostly current and previous employees, not customers. It was a rapid and focused attack on Sony Pictures, bent on inflicting brand damage to the company. Sony has eluded scrutiny in their preparation and response to the attack due to the assumption that the hack was conducted by North Korean operatives and was sensationalized as an act of cyberterrorism. Sony did offer its employees identity protection services in response to the personal data loss.

**Table 8.1.** Must do's in managing a cybersecurity breach

| Must do's |
|---|
| Organizations must put the proper resources in place to ensure that any form of cybersecurity breach is dealt with swiftly and efficiently. |
| There should be an effective incident response plan. |
| Thoroughly check all monitoring systems for accuracy to ensure a comprehensive understanding of the threat. |
| Engage in continuous monitoring of their networks after a breach for any abnormal activity, and make sure intruders have been inhibited thoroughly. |
| It is important to perform a postincident review to identify planning shortfalls as well as the success in execution of the incident response plan. |
| Be sure to engage with law enforcement, and any other remediation support entity, soon after the threat assessment is made, to allow for containment of the breach and to inform any future victims. |
| Documentation is paramount. Thorough documentation from the onset of the breach through the clean-up must be a priority to ensure continual improvement of the incident response plan. |
| It is critical to the success of a business to integrate cybersecurity into its strategic objectives and to ensure that cybersecurity roles are defined in its organizational structure. |

Some best practices for being prepared for a cybersecurity breach are discussed in the following section.

## The Chief Information Security Officer Role

The chief information security officer's role is essential for large organizations. This position must be integrated into the organizational structure in such a way that it is included in all strategic planning discussions at the executive level. The inclusion ensures that information security is included in high-level strategic planning and that C-level executives are considering risk assessment along with other strategic planning objectives, including investments and business development. The CISO role will be able to assist in identifying an information security structure and building policies and procedures that protect an organization's most important assets.

## Business Continuity Plan with Cybersecurity Integration

As global cyberattacks increase, organizations must plan around this imminent danger. When we look at the broad impact of the cyberattack on Sony, it is apparent that organizations need to invest in business continuity plans with cybersecurity as a focal point. Many organizations are already doing this. However, an astonishing 4 in 10 companies have no formal security strategy [4]. In another survey, it was found that 83% of small- and medium-sized businesses have no formal cybersecurity plan. It goes without saying therefore that cybersecurity plans are important and need to be instituted. Plans can include a broad spectrum of responses, ranging from network auditing and segmentation to coordinating the release of information to media. Organizations must evaluate existing business continuity plans and ensure that information security strategy is included.

## Shift from Remediation to Prevention

Many large organizations have sophisticated network security mechanisms. These include security-oriented network designs, network intrusion prevention systems, and the most traditional systems like enterprise antivirus, firewall, group policy, and patch deployment. These systems have been effective for preventing most attacks and are somewhat effective for helping to identify where a breach may have occurred, but are not sophisticated enough for more advanced attacks, which are often the most damaging.

For an organization to move ahead of the threat of cyberattacks, it must go beyond traditional security systems and shift focus to more preventative solutions. Organizations must invest in tools that bring the organization to the front of cybersecurity, with a focus on prevention. Below are some examples of preventative tools and techniques that organizations can invest in:

- **Threat detection.** Organizations should focus on investigating and learning about breach attempts. An effective detection and response system should be implemented.
- **Network traffic inspection.** Network traffic inspection is essential for anticipating cyberattacks. A good network engineer should be asked to perform network traffic analysis as a daily routine.
- **Network segmentation.** Many organizations are segmenting business units from the network level, using VLAN technology. This type of segmentation ensures that in the event of a cyberattack, problem areas are isolated as they are investigated.
- **Penetration testing.** Penetration testing should be performed on a continual basis, to ensure that network security is maintained at the highest level. In addition to network penetration testing, social penetration testing should occur, to ensure that employees are trained on safe business communications practices.

## Auditing and Monitoring

As cybersecurity risks increase, it is important to ensure the organization's workforce is using information systems safely and securely. All too often, business units find themselves creating their solutions when IT is not involved, which leads to a significant security risk. An example would be the use of cloud storage services like Dropbox, which end users can very easily install and setup for sharing and storing business information. With the appropriate computer usage policies in place and the right governance structure, an effective CIO should be able to ensure that end users are complying with the right policies and procedures.

Even with perfectly functioning IT governance, it is important to check in with business units to ensure they are following policies and procedures. The best approach to ensuring security best practices is to perform continual IT assessments. Assessments, when supported by the organization, allow IT to review how individual business units are using technology to perform business functions. This assessment produces a report that allows the CISO to see overall compliance with security policy and identifies risks that can be mitigated. Logins should be tracked and reviewed for any activity outside of what is expected. Systems that don't automatically log activity should have such logs created. Programs that employees download and websites they visit should be reviewed for potential risks. Let employees know tracking mechanisms are in place to ensure cybersecurity—this will discourage them from engaging in non-work-related internet activities that can be risky. Informing employees informs the workforce that not only is monitoring most essential, but employee awareness of the practice is very important as well.

## Focus on Employees

While an organization can put in place state-of-the-art security infrastructure, with a security-focused organizational structure, these root-level improvements cannot prevent employees from causing harm. Research over the years suggests that employees are at the root of most cyberbreaches. Employees are most capable of an error—sending a confidential email to the wrong email address, forgetting to protect a sensitive document, or having their business-connected mobile device stolen. While IT policies can be implemented to prevent most of these occurrences, employees may not always follow the policy, and will inadvertently put the business at risk.

The best way to mitigate this risk is to put in place a security training program for the workforce. Many organizations already do this for various compliance requirements, like HIPAA. The objective is to provide the workforce with best practices for various processes and systems. Employees need to know how to recognize phishing attempts via phone, email, and other methods. Require strong passwords and enforce penalties up to and including termination for sharing them. Educate employees on how to recognize suspicious websites. Share stories of current security attacks in the news to explain how those companies were compromised and how the incident is affecting the business. Most employees are loyal to their company. They will gladly work to ensure its success if they are informed, understand how important their role is in cybersecurity, and feel as if they are part of the solution.

In some cases, a disgruntled employee may be at the root of a cyberattack. Disgruntled employees are capable of significant long-term damage to an organization. The following are a few solutions to mitigate this risk:

- Implement clear access rules to ensure employees have access to only the information they require.
- Put in place auditing process for access granted to business resources, including a reporting/review process.
- Ensure termination processes include functions for disabling access to business systems.

## Pay Attention to Personal Devices

In today's age, personal devices are a given risk. They have allowed companies to augment performance and throughput by giving access to systems outside of standard business hours and locations. But allowing access to secure systems via these devices can be extremely risky. Personal devices can range anywhere from cell phone devices to mini thumb drives. Within today's society, cell phones are considered dependencies with the mass majority of the corporate workforce. Phone calls constitute a small percentage of cell phone usage. Email, social media, and other applications are used more. Therefore, connecting these devices to Wi-Fi-based networks is an automatic "necessity" for most users. The logic for this ranges from the financial impact of cutting costs on the monthly utilization bill to speeding up the use of cell phone applications. Most organizations have "guest" Wi-Fi connections within their infrastructure, and these are not always secured to prevent potential spam infestations, major or minor. Controlling these connections with protocols that require users to register for usage is a good way to track these devices down when they cause risks.

## Concluding Remarks

Data breaches can happen to a wide range of organizations. Most likely, the attacker aims at bigger corporations that have a lot at stake. However, instigators may also target smaller organizations with malicious attacks. Statistics show us that the cost of an attack can be high and is increasing yearly. It is up to the company's management to adopt a cybersecurity policy and data breach response plan. Managers should evaluate their system and sensitivity to a potential data breach. They also need to keep in mind that the attacks do not just come from outside intruders. They can come from inside as well, and employees can either knowingly or unknowingly contribute to an attack. Sony's data breach constitutes a great example that an employee-generated data breach can go unnoticed for months and the outcome to the company may be grave. If a breach does occur, a good security response strategy should help mitigate the impact. A good plan should have response actions listed and responsibilities assigned to team members. The plan also details the contingency plan and prepares for business continuity. Every minute a company is not functioning, the revenue stream is impacted and the overall financial health is in jeopardy.

Managers have access to the best industry accepted practices that allow them to reduce infrastructure weaknesses and defend the company against potential attacks. Following best practices can also reduce the impact if an attack does occur, to aid in normalizing company operations. Managers cannot prevent cyber attacks, and due to the expanding technology, they are increasing in occurrence and cost every year. The best practice for any size company is to develop security measures and a response plan if a breach occurs.

## In Brief

- When a cybersecurity breach occurs, survey the damage; attempt to limit additional damage; record the details; engage law enforcement; notify those affected; develop a mechanism to learn from the breach.
- Organizations must put proper resources in place such that any cybersecurity breach is dealt with promptly.
- All monitoring systems should be thoroughly checked and reviewed. And there should be continuous monitoring of networks for abnormal activities.
- Engaging law enforcement is important. It ensures that threats are contained and future victims are adequately informed.
- Document all aspect of the breach from the onset. This ensures that a proper incident response plan is initiated at the right time.

# Questions and Exercises

**Discussion questions.** Discuss the following based on news reports and materials in this book.

1. If you are a CSIO and receive a request from an employee, who is a high powered individual within the organizations, to bring his personal router to set up a personal office, what would you do? The employee makes an argument that this enables him to sync all his devices seamlessly, which is important because of the nature of his work. Create a statement that needs to be distributed to various corporate employees as to what they should do when they receive an email attachment.

2. On September 22, 2016, Yahoo announced that hackers had stolen nearly 500 million records. The stolen data included email addresses, telephone numbers, dates of birth, encrypted passwords, and security questions. However, the stock market "yawned" at the hack. Discuss if boards of directors will take cyber risks seriously if the stock market merely shrugs off these breaches.

3. Since cyber insurance covers much of the cost associated with a cybersecurity breach, companies are complacent in instituting the right kind of controls to protect their resources. Discuss.

**Exercise.** Cybersecurity breaches are here to stay. And it is becoming important for companies to establish some kind of a rating system. Such a system could work more or less like the Moody's and Standard & Poor's creditworthiness rating. Conceptualize such a cybersecurity rating system for firms. Suggest how such a system could be implemented.

## Short Questions

1. A good network engineer should be asked to perform network _____ as a daily routine.
2. _____ ensures that in the event of a cyberattack, problem areas are isolated as they are investigated.
3. _____ ensures that employees are trained on safe business communications practices.
4. With the _____, an effective CIO should be able to ensure that end users are complying with the right policies and procedures.
5. Even with perfectly functioning IT governance, it is important to check in with business units to ensure they are following _____.
6. _____ should be tracked and reviewed for any activity outside of what is expected.
7. Systems that don't automatically _____ should have such logs created.
8. In some cases, a _____ employee may be at the root of a cyberattack.
9. What solutions can be implemented to mitigate long-term damage resulting from disgruntled employees?
    a. Implement clear access rules to ensure employees have access to only the information they require.
    b. Put in place an auditing process for access granted to business resources, including a reporting/review process.
    c. Ensure termination processes include functions for disabling access to business systems.
10. Why would you consider the role of a CSIO important for an organization? Can't a network administrator have the same job?
11. For an organization to move ahead of the threat of cyberattacks, it must go beyond traditional security systems and shift focus to more on _____.
12. A good plan should have response actions listed and _____ assigned to team members.
13. A good security plan should detail _____ and prepare for business continuity.

14. Identify at least five law enforcement agencies that work in the area of cybersecurity.

## Case Study: Equifax Breach

On September 7, 2017, Equifax announced a massive security breach. While the breach was originally discovered on July 29, the announcement was delayed by several months. An estimated 145 million US consumers were affected. The breach resulted in the loss of the following details:

- Names
- Social Security numbers
- Birth dates
- Addresses
- Driver license numbers (at least in some cases)

Equifax attributes the breach to a website application vulnerability that was exploited by criminals. The Apache Software Foundation believes that the vulnerability was possibly caused by the March Struts bug. Experts allege that once a vulnerability is exploited, it allows attackers to gain a foothold. Generally, following the exploit, the attacker becomes a system user and hence owns the web server process.

There are mounting concerns that Equifax could have prevented the breach if simple procedures and best practices were followed. Equifax has been accused of incompetence in regard to the protection of individual data and irresponsible behavior in responding to the breach. A patch for the website application vulnerability that was exploited was available several months before the attack, in March 2017. Even though Equifax had more than two months to take remedial actions and apply the patch, no action was taken.

There are several questions that emerge. Is Equifax competent enough to be the data steward for the public? Why did Equifax take so long to notify the public? Interestingly, the website set up by Equifax to address questions about the breach and offer free credit monitoring was itself vulnerable. Why was Equifax so negligent in handling and responding to the breach?

1. Develop an ideal response strategy for Equifax.
2. Suggest how:
    a. A technical security strategy could have helped Equifax
    b. A formally defined process could have helped Equifax
    c. A normatively developed approach could have helped Equifax
3. Following the breach, what could Equifax have done to protect their reputation?

# References

1. Dhillon, G., and S. Moores 2001. Computer crimes: Theorizing about the enemy within. *Computers & Security* 20(8): 715–723.
2. Hummer, C. 2015. Anthem says at least 8.8 million non-customers could be victims in data hack. *Reuters*, February 24.
3. Murphy, T. 2015. Hackers infiltrate insurer Anthem, access customer details. *Richmond Times Dispatch*, February 5.
4. O'Dwyer, P. 2016. Firms "lack cybersecurity plan." *Irish Examiner*, May 4.
5. Ponemon Institute. 2015. *Cost of data breach study: global analysis*. Traverse City, MI: Ponemon Institute.
6. Syed, R., and G. Dhillon. 2015. Dynamics of data breaches in online social networks: Understanding threats to organizational information security reputation. In *Proceedings of the International Conference on Information Systems*, Fort Worth, TX, December 13–16.
7. Udeh, I., and G. Dhillon. 2008. An analysis of information security governance structures: The case of Société Générale Bank. Paper presented at the Annual Symposium on Information Assurance (ASIA), Albany, NY.

# PART III

# INFORMAL ASPECTS OF INFORMATION SYSTEMS SECURITY

# CHAPTER 9

# Behavioral Aspects of Information System Security

*Ability is what you're capable of doing. Motivation determines what you do. Attitude determines how well you do it.*

—Lou Holtz

Joe Dawson sat in his SureSteel office and sipped his first cup of coffee of the day. Being an ardent reader, each morning he would scan through popular and specialist press. One of the magazines that came across his desk was *Computer Weekly*. Though everything was going online, Joe still enjoyed flipping through the tabloid-style magazine. As he flipped through the pages, the article "Are ethical questions holding back AI in France?" caught his attention. What amused Joe was the fact that EU parliament was considering granting AI systems the legal status of "electronic persons." If that were to happen, it would be interesting. The article had a quote from Frank Buytendijk, a Gartner research fellow Buytendijk argued about making robots behave like hums and have similar feelings. He noted: "This would help us by driving more civilized and caring behavior towards computer systems. If we spend all day talking coldly to robots, how are we going to talk to our children in the evening?"

Joe began thinking of various ramifications. In particular, he thought about research in the area of user behavior analytics. Behavior analytics was emerging to be an interesting area. Most companies generated vast amount of data from logs, user actions, server activity, network devices, and so on. But the challenge has always been to provide a context to this data and draw meaningful insights. If it were possible to do so, it would be a big help to SureSteel. *Well*, Joe thought. *That is another interesting endeavor.* He put the *Computer Weekly* down and continued with his morning coffee. Just as he did so, Patricia, his new executive assistant, knocked on the door.

"What's worrying you, boss?" she asked.

"Just thinking about behavioral analytics for security," said Joe.

"I know something about that," exclaimed Patricia. My boyfriend is a big Charlotte Hornets fan. He keeps talking about behavior analytics. Apparently the Hornets employ this technique to understand what their fans want. It's kind of a customer relationship management system. It combines millions of records to define a profile for each fan. This helps the Hornets have a better relationship with their fans, which eventually generates more revenue through improved relationship marketing.

Joe listened intently to Patricia. And then said, "Now imagine we could use the same technique to understand possible security threats. Because the majority of the threats are internal to the organization," Joe continued, "and I want to learn more about this."

## Employee Threats

Year over year, time and again, a business's employees have been found to represent the single greatest vulnerability to enterprise security. Organizations are putting their reputation, customer trust, and competitive advantage at risk by failing to adequately acknowledge and provide their staff with effective cybersecurity awareness training and the ability to defend against cyberattacks, both internal and external. According to a report by Axelos, a UK government / Capita joint venture, it found that 75% of large organizations suffered staff-related security breaches in 2015, with nearly 50% of the worst breaches being caused by human error. Worse yet, the research detailed that only a minority of executives responsible for information security training in organizations, with more than 500 employees believes their cybersecurity training is "very effective." Worse yet, numerous reports indicate that many organizations have no way of tracking sensitive internal information, enhancing their level of exposure to insider threats. One prominent example of this type of threat is that of Alan Patmore, a former employee of Zynga, a popular software/game developer for smartphones. When Patmore moved over to a small San Francisco startup, just before leaving Zynga, he created a Dropbox folder and used it to transfer approximately 760 files to the cloud. The data included a description of Zynga's methods for measuring the success of game features, an initial assessment of a popular game, and design documents for nearly a dozen unreleased games. All of this information was transferred without the knowledge or consent of his employer. This is only one of many examples, but serves as a clear demonstration of the threat insiders can present to an organization's information security. Hence it is important for organizations to be aware of the numerous types of insider threats, which their employees may present in order to be adequately prepared.

### Sabotage

Merriam-Webster defines sabotage as "an act or process tending to hurt or to hamper." While first thought when examining this definition might lead one to think of elaborate plots by rogue nations, many may actually be surprised to learn that sabotage is a fairly common occurrence in today's workplace. With this in mind, it is important to note that sabotage typically results in two forms; active and passive. In order to distinguish these two terms, one may consider active sabotage as actively engaging in an intentional act you shouldn't be doing that causes harm to the organization. Passive sabotage, on the other hand, can be thought of as intentionally not doing something you should be doing, which through this inaction results in harm to the organization. So what kind of employee commits acts of sabotage, either active or passive? Generally, if an employee is engaged and actively contributing to the success of an organization, then it is highly unlikely they would seek to cause it harm. However, an employee who is disengaged and unhappy in the workplace increases the chances of them committing an act of sabotage. Even so, acts of active sabotage among employees are rare, with many studies finding very low rates of occurrence. However, acts of passive sabotage are much more common, as it is far easier for a disengaged employee to simply ignore protocol, like changing their password regularly or leaving sensitive documents in the open. These forms of passive sabotage can result in data breaches, damage an organization's reputation, and result in a loss of customer confidence.

## Hacking

Hacking is the second most common employee threat, with nearly 40% of information security threats to an organization resulting from insider attacks. According to research conducted by the U.S. Computer Emergency Response Team (Cert), this is a serious problem and stems from disengaged employees having access to sensitive information. These employees are often technically proficient and possess authorized system access by which they can open backdoors into the system or deliver malicious software through the network. Additional studies have also found this assertion to be consistent with smaller businesses being uniquely vulnerable to IT security breaches due to their lack of the more sophisticated intrusion detection and monitoring systems used by large enterprises. For example, in 2002 an employee named Timothy Lloyd was sentenced to three-and-a-half years in prison for planting a software time bomb after he became unhappy with his employer, Omega. The result of the software sabotage was the loss of millions of dollars to the company and the loss of 80 employees' jobs. Hence hacking is a dangerous threat to any organization with respect to their own employees and can lead to tremendous devastation due to an employee's familiarity with their information system and authorized access.

## Theft

Unauthorized copying of files to portable storage devices, downloading unauthorized software, use of unauthorized P2P file-sharing programs, unauthorized remote access programs, rogue wireless access points, unauthorized modems, and downloading of unauthorized media all have one thing in common: They all pose a threat primarily in terms of loss of information, security breaches, and legal liability. All are commonly used to commit theft within an organization as, for example, unauthorized copying of files can lead to loss of confidential information, which would directly damage the business. This is well-demonstrated by the case of William Sullivan, a database administrator who in 2007 stole 3.2 million customer records, which included credit card, banking, and personal information from Fidelity National Information Services. Sullivan had authorized access to the system via his role as a database administrator, but had become disengaged at work. While Sullivan had authorized access, he did not have permission to take the records for any purpose and therefore directly engaged in the theft of Fidelity's secure information. This led to a great deal of public turmoil for Fidelity, with customers concerned about their stolen information being misused and a loss of confidence in Fidelity's ability to protect that information in the future.

## Extortion

Extortion happens all the time and places employers in a very difficult position—a current or ex-employee threatens to "blow the whistle" on some perceived employer misconduct to leverage the employer into providing a beneficial change at work or a hefty severance package. In some cases, the claim is bogus, yet the information possessed by the employee is still damaging to the organization if it is released. For instance, an employee may have access to medical records at a health care organization and take them without authorization. They might claim to the employer they contain evidence of medical malpractice and threaten to expose such criminal wrongdoing to the public, unless provided with some benefit. This is the very essence of extortion, and while the person in this scenario is clearly wrong, the release of this "evidence" may clear the organization of claims of medical malpractice yet result in other damaging consequences. The release of such private medical information about a patient could result in legal fines and penalties or damage the organization's reputation among its customer, who may no longer feel confident in their ability to keep medical records safe. Hence extortion, even if it has no real evidence of wrongdoing, can still be harmful to an organization.

## Human Error

Within the context of employee threats, unintentional acts are those with no malicious intent and consist of human errors. Human errors are by far the most serious threats to information security, as according to Gartner, 84% of high-cost security incidents occur when insiders send confidential data outside the company. It's easy to see why insiders pose the greater threat, as all the people inside a company can have ready access to customer, employee, product, and financial data. With confidential customer data and intellectual property just the slip of a keystroke away from the Internet, every organization should be considered at risk. While most human errors are unintentional and not malicious in nature, most enterprise systems are designed to prevent unauthorized access or accidental exposure of information from outside sources, not against internal threats. Therefore a careless or untrained employee may simply think they are doing their job or speeding up the process by sending secure information through an unsecured email attachment, for example. However, in reality they are making an error that could expose their organization to undue harm.

# Social Engineering Attacks

In the context of information security, the psychological manipulation of people into performing actions or divulging confidential information is known as Social Engineering. It is a type of confidence trick that is used for the purpose of information gathering, fraud, or gaining system access. Social engineering differs from a traditional "con" that one may generally think of, in that it is far more often one of several steps in a more complex scheme to commit fraud. Social engineering techniques are particularly effective because they are based on what can be considered a "bug" or flaw in the decision-making process of humans.

Someone using these types of attacks is attempting to exploit these inherent decision-making flaws to trick or influence an employee to do something or provide information they otherwise would not, failing to recognize a potential threat. There are various types of social engineering attacks, some of which target specific individuals at an organization, while others are blanket attacks that attempt to cast as wide a net as possible to increase the odds of successfully tricking someone.

## Phishing

A prominent social engineering technique, known as phishing, is often used as a technique for fraudulently obtaining private information. When an attacker attempts to use phishing as a technique for obtaining information from a target, typically they send an email that appears to come from a legitimate business—for example, a bank or credit card company. The hook is that the phishing attack requests some kind of "verification" of information and makes an explicit warning of dire consequence if it is not provided as soon as possible. In most cases the email uses official looking links, logos, and forms that are typical for requesting the desired information. As the request, especially at first glance, appears to be official, it will frequently elicit the desired reaction, as the recipient will do little to verify authenticity of the sender. While most organizations have spam filters intended to prevent this type of attack, it is impossible to stop all attempts, and it only takes one mistake to expose an entire organization. Further, as it is a relatively simple attack to conduct by spamming large groups of people, the "phisher" greatly increases the odds of a successful attack.

## Spear Phishing

Spear phishing is similar to phishing, as it is a technique that fraudulently obtains private information by sending emails to users. The main difference between these two types of attacks is in the way the attack is conducted. Phishing campaigns focus on sending out high volumes of generalized emails with the expectation that only a few people will respond, whereas spear phishing emails require the attacker to perform additional research on their targets in order to "trick" end users into performing requested activities. As spear phishing attacks are much more targeted and contain additional information specific to the target, they are much more successful. Users are much more likely to respond to these types of attacks, as the message is more relevant, but it requires more time and effort on the part of the attacker. As this type of attack is much more effective than typical spam phishing attacks, senior executives and other high-profile targets within businesses have become prime targets for spear phishing attacks, termed *whaling*. In the case of whaling, the masquerading web page / email will take a more serious executive-level form and is crafted to target upper management and the specific person's role in the organization. With the relative ease of conducting this type of attack, a high rate of success and the potential to snare lucrative targets with near limitless system access, spear phishing is a common threat organizations must deal with by training employees to recognize these dangers and think more critically about clicking links and divulging their passwords.

## Hoaxes

A hoax is similar to phishing in that it is an email message warning intended to deceive; however, it is usually sensational in nature and asks to be forwarded onward to new recipients.

In addition, it is often embedded with a virus and does not attempt to solicit information from any specific target. Since hoaxes are sensational in nature, they are easily identified by the fact that they indicate that the virus will do nearly impossible things, like cause catastrophic hardware failure or, less sensationally, delete everything on the user's computer. Often included are fake announcements claimed to originate from reputable organizations with claims of support from mainstream news media—usually the sources provide quotes or testimonials to give the hoax more credibility. The warnings use emotional language and stress the urgent nature of the threat in order to encourage readers to forward the message to other people as soon as possible. Generally hoaxes tend to be harmless overall, accomplishing little more than annoying people who identify it as a hoax and wasting the time of people who forward the message. However, for example, a hoax warning users that vital system files are viruses and encourage the user to delete the file could lead them to damage important files contained by the information system.

# Individual Motivation to Prevent Cyberattacks

Over the past few years a lot of researchers have focused on how individuals can prevent cybersecurity attacks. This research assumes that an organization has a well-defined and a well-crafted security policy. It also assumes that compliance with such a policy will ensure prevention of a cyberattack or a compromise. The research falls into two broad categories—extrinsic and intrinsic motivation.

## Extrinsic or Intrinsic Motivation?

Herath and Rao examined the influence of variables under both extrinsic and intrinsic motivation models of IS security policy compliance intentions [3], [4]. Their study reported that employees' intrinsic motivation, measured by perceived effectiveness, was positively associated with IS security compliance intentions. The extrinsic motivation model, measured by severity of penalty, certainty of detection, peer behavior, and normative beliefs, was partially supported. Overall, the findings of the research have suggested that both intrinsic and extrinsic motivators may influence the IS security behaviors of employees. However, the study did not predict the magnitude of contribution for each model.

Recent work by Son [7] examined the impact of perceived certainty and severity of sanctions (i.e., extrinsic motivation model), and perceived legitimacy and perceived value congruence (i.e., intrinsic motivation model) of IS security policy compliance among employees in the United States. Both extrinsic and intrinsic models were tested for their significance. The results of the study show that factors rooted in the intrinsic motivation model were significantly related to IS security policy compliance. However, contrary to expectations, both extrinsic factors were insignificant. More interestingly, the study predicted that both the extrinsic and intrinsic motivation models will explain significantly more employees' IS security policy compliance than variables from either the extrinsic or intrinsic motivation model. The results show that the extrinsic model explained 16% and the intrinsic model explained 41% of variance for IS security policies' compliance behavior. By simultaneously testing the relationships between the extrinsic factors, intrinsic factors, and IS security policy compliance intention, the model was significant in explaining the variance of IS security policies' compliance behavior. The results indicated that the contribution of the intrinsic motivation model exceeded that of the extrinsic motivation. The study proposed that intrinsic motivation may generate alternative explanations and solutions for compliance with organizational IS security policy. Thus organizations should increase their emphasis on intrinsic motivation-based approaches, and rely less on extrinsic-based approaches [7].

## Management Considerations

In sum, IS security policy compliance research has witnessed the rise of the extrinsic motivation argument. Scholars considering extrinsic factors to be important in ensuring compliance have largely examined the four dimensions of sanctions, rewards, monitoring, and social pressures. It goes without saying that individual compliance with IS security policy can be explained with respect to the extent of sanctions or rewards, how well they are being monitored, and what their social pressures might be.

Although most past and current research in IS security policy compliance has focused mainly on employees' value of extrinsic rewards, employees tend to value both intrinsic and extrinsic rewards. Intrinsic motivational factors such as self-efficacy, psychological ownership, perceived effectiveness, and perceived value congruence influence employees' decisions to comply with IS security policy. Interestingly, a recent study provides empirical evidence that the intrinsic factors could explain more of the variance in IS security policy compliance than the extrinsic factors [7]. This means that the factors constituting intrinsic motivation to comply with IS security policy are certainly promising. Unfortunately, relatively little research has been conducted within the intrinsic motivation paradigm; hence the call has be made to investigate other intrinsic factors [4], [8]. Further, researchers have acknowledged a few strategies or drivers, such as IS security training and IS security climate, to enhance employees' intrinsic motivation. However, no empirical work was found to investigate the drivers to enhance employees' intrinsic motivation. As a result, there is pressing need to investigate the impact of psychological empowerment, a factor rooted in intrinsic motivation model, on IS security policies' compliance intentions, and exploring the antecedents of psychological empowerment. Table 9.1 summarizes the theoretical basis for IS security policy compliance research.

# Cybercrime: Criminals and Mediums

Cybercrime is a form of computer crime, as it involves the use of a computer as an instrument to further illegal ends. This can include activities such as committing fraud, trafficking child pornography and intellectual property, stealing identities, or violating others' rights to privacy. Therefore cybercrime, which has grown in importance as the computer has become central to commerce, entertainment, and government, has become an integral part of the criminal's arsenal for committing crime. Technology provides an unprecedented ease of access to media, financial markets, and global communication to both society at large as well as criminals, which underpins and facilitates international criminal gangs and black markets. For example, the Russians, Nigerians, Ghanaians, and Chinese are some of the best-known cybercriminals, and while other groups may use similar tactics, their motivations, organizational structures, and cultures differ. The use of technology to the commit

**Table 9.1.** Summary of IS security policy compliance research

| Motivation | Factors | Description | Theory Used |
|---|---|---|---|
| EXTRINSIC | Sanctions | When people comply with security policies to avoid penalties | General deterrence theory; agency theory |
| | Monitoring | When people comply with security policies because they know their activities are being monitored | Control theory |
| | Rewards | People comply with security policies to attain rewards | Rational choice theory; theory of planned behavior |
| | Normative beliefs | People comply with security policies because others (superiors, IT management and peers) expect compliance | Protection motivation theory |
| | Social climate / Observation | People comply with security policies because the management, supervisors, and colleagues emphasize prescribed security procedures | Protection motivation theory |
| INTRINSIC | Perceived effectiveness | People comply with security policies because they perceive that their security actions will help in the betterment of the organization | Protection motivation theory |
| | Perceived self-efficacy | People comply with security policies because they perceive that they have the skills or competency to perform the security activities | Self-efficacy theory |
| | Perceived value congruence | People comply with security policies because they perceive that the security values/goals are congruent with personal values | General deterrence theory |
| | Perceived ownership | People comply with security polices because they perceive that the own the information assets | Protection motivation theory |

crime creates additional difficulties as well in combating these activities versus traditional avenues of activities such as fraud, including attribution, lack of international cooperation, and limited resources for law enforcement. An important reason why these gangs and the black markets in which they operate are so effective is that the barriers to entry and cost for cybercriminals is low, and hence cybercrime will continue to grow as a threat to international security, economic growth, and technological innovation as long as this remains the case.

## International Gangs

Cybercriminal gangs are as organized, well-resourced, and successful as many legitimate organizations. Criminal organizations are driven by profit, rather than personal ambition or sheer boredom, and therefore employ many of the same practices as legitimate businesses, which draws many people of tremendous skill and talent into such criminal enterprises. A prominent example of these types of groups are the Russians, who are some of the most successful and well-resourced organized cybercriminal groups. This talent is due to the employ of ex-KGB spies who are now using their skills and expertise for financial gain. They established what is known as the Russian Business Network (RBN) after the Iron Curtain lifted in the 1990s and have the patience and resources to allow members to hack information from high-ranking executives and government personnel, typically in the form of credit card and identity theft. With the increased proliferation of technology in business, the Internet of Things being an example of huge amounts of data capture, information systems are becoming a prime target for international gangs to exploit for monetary gain. Businesses keep both customer and employee personal information, which can sell on the black market for a large amount of money, especially when stolen in large quantities. This provides all the incentive international gangs require to engage in cybercrimes, which exploit any weaknesses within a business' information security practices.

## Black Markets

While cybercrime and international cybergangs are a threat to global information security, they could not exist without a market to facilitate the transactions and trade of stolen information. This has given rise to the use of black markets, defined as an underground economy or shadow economy in which some aspect of illegality or noncompliant behavior with an institutional set of rules occurs. For example, if a rule exists that defines the set of goods and services whose production and distribution is prohibited by law, noncompliance with the rule constitutes a black market trade since the transaction itself is illegal. Parties engaging in the production or distribution of prohibited goods and services are therefore members of the illegal economy. Hence these markets are the medium by which cybercriminals such as international gangs trade goods and services obtained through illegal means, irrespective of the laws governing such behavior. Examples often include drug trade, prostitution, illegal currency transactions, identity theft, and human trafficking. As these activities are illegal, cryptocurrency such as bitcoin have become, while intended to provide freedom from government controlled currency for legal purposes, tools for criminal enterprises to evade detection and engage in profiteering from crime in the digital world. One prominent black market that was shut down is that of the Silk Road. It was likely the first modern Dark Web market and was best known as a platform for selling illegal drugs. Being part of the Dark

Web and operated through Tor, a secure browser for searching the Dark Web, it was a hidden service that enabled users to browse it anonymously and securely without potential traffic monitoring. The Silk Road was launched in February 2011, and initially there was only a limited number of seller accounts available, which required prospective sellers to purchase an account via an online auction. In October 2013, the Federal Bureau of Investigation (FBI) shut down the website and arrested Ross William Ulbricht under charges of being the site's alleged pseudonymous founder "Dread Pirate Roberts." While attempts to revive the Silk Road were made, they were unsuccessful; however numerous other digital black markets, many unknown, still exist today.

# Cyberespionage

There is a fine line between what can be considered intelligence gathering and what would be termed espionage. Crane [1] suggests three criteria that could determine if indeed there was an ethical problem regarding the manner in which information was gathered:

1. The tactic relates to the manner in which information was collected. It might just be that the process was not deemed acceptable.
2. The nature of information sought is also an important consideration. Some basic questions need to be asked: Was the information private and confidential? Was it publicly available?
3. The purpose for which the information was collected. The following questions need consideration: How is the information going to be used? Is someone going to monetize it? Would it be used against public interest?

Tactics for gathering intelligent information take several forms. Whatever the kind of a tactic used, the origins are usually dubious and ethicality questionable. Most tactics are in violation of the philosopher Immanuel Kant's categorical imperative—only those actions are acceptable that can be universalized. Most tactics are clearly illegal and unethical. These might range from breaking into a competitors' offices and computer systems to wiretapping, hiring private detectives, and going thorough competitors' trash to find confidential information.

The reason information is collected is intricately linked to aspects of public interest. Of particular concern are cases where corporate intelligence related to national and international security is obtained. In such cases several public interest issues come to the fore. See Figure 9.1 to see how cyberespionage is undertaken.

Some examples of cyberespionage include:

1. Anticompetitive behaviors, including the deliberate removal of competitors
2. Price hikes
3. Entrenchment of a monopoly position

One of the weapons of choice in twenty-first century espionage is the botnet. What required the resources of a nation state in the 1970s and 1980s can now be accomplished by tech-savvy

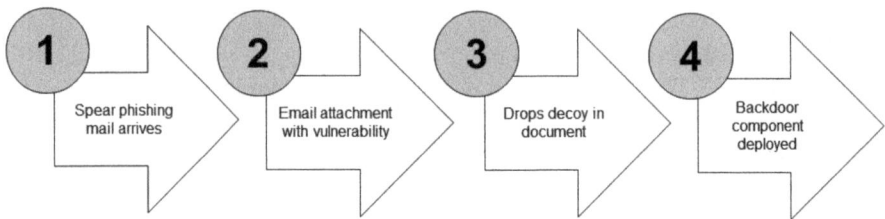

**Figure 9.1.** A typical cyberespionage attack

users located anywhere in the world. A term that is often used is that of "web robots," since these are programs that tend to take over the infected computer. In that sense, a "bot" is a malware, and since a network of computers is usually involved, the term "botnet" is used. "Zombies" is also a term that is often used to describe botnets. This is because typically an infected computer *bids for the master*. An infestation by a bot can enable the following kinds of acts (more details can be found in Schiller and Binkley [6]):

- **Sending.** The computer can send spam, viruses, and spyware.
- **Stealing.** Personal and private information can be stolen and sent back to the malicious user. Such information can include credit card numbers, bank credentials, and other sensitive information.
- **Denial of service.** Cybercriminals can resort to demanding a ransom or a range of other criminal acts.
- **Click fraud.** Bots are used to increase web-advertising billings. This is accomplished by automatic clicking on URLs and online advertisements.

## Cyberterrorism

The convergence of terrorism and cyberspace is known as **cyberterrorism**. The term was first coined by Barry Collin in 1980. The aim of cyberterrorism is to disrupt critical national infrastructure or to intimidate a government or political establishments or civilian enterprise. The attack should cause violence, vandalism, or enough harm to generate fear—for example, explosions, plane crashes, or severe economic loss. The core characteristics of cyberterrorism include

1. It is pursued to create destruction of a nation's critical infrastructure.
2. It can be executed by various means and involves computers and digital technology as the main elements.
3. It can affect government, citizens, or other resources of high economic value.
4. The act can be motivated by religious, social, or political reasons.

A large number of cyberterrorism acts conducted for social and political reasons have come to light over the past few years. In 1996, a hacker associated with a white supremacist group temporarily disabled an Internet service provider and its record-keeping system in Massachusetts. Though the provider attempted to stop the hacker from sending out hate messages, the hacker signed off with the threat, "You have yet to see true electronic terrorism. This is a promise."[1] Another example is from during the Kosovo conflict in 1999, when a group of hackers attacked a NATO computer with email bombs and denial of service attacks. The attack was a way to protest against NATO bombings. In 2013, hackers supported by the Syrian government attacked an Australian Internet company that manages Twitter and *Huffington Post* sites. Later the Syrian Electronic Army claimed responsibility for the attack.[2]

Terrorists groups are increasingly using the Internet to spread their messages and to coordinate their actions and communicate their plans. There are hardly any networks that are fully prepared against all possibilities of threat. In comparison to traditional terrorism, cyberterrorism offers several advantages to terrorist groups. First, it is cheap, and can be executed from any part of the world. Second, the terrorist can be anonymous. Third, it doesn't put a terrorist's life at risk. Fourth, the impact can be catastrophic, and fifth, white collar crime gets more attention. Furthermore, terrorists face a few hardships when executing cyberterrorism: First, the computer systems may be too complex for them to cause the desired level of damage. Second, some terrorists may be interested in causing actions that lead to loss of live, yet cyberattacks usually do not lead to loss of life. Third, terrorists may not be inclined to try new methods of sabotaging a system

## Ensuring the Safety of Systems

The Federal Bureau of Investigation considers information warfare and cyberterrorism as two different types of threats: while cyberterrorism is a kind of information warfare, an act of information warfare may or may not be cyberterrorism. There are several definitions or connotations.

One of the earliest and broader definitions of information warfare is by Thomas Rona. According to Rona, "The strategic, operation, and tactical level competitions across the spectrum of peace, crisis, crisis escalation, conflict, war, war termination, and reconstitution/restoration, waged between competitors, adversaries or enemies using information means to achieve their objectives" (quoted by Libicki, [5]).

Another definition by Haeni puts information warfare as "Actions taken to achieve information superiority by affecting adversary information, information-based processes, information systems, and computer-based networks, while defending one's own information, information-based processes, information systems, and computer-based networks" [2].

As per Libicki [5], seven different forms of information warfare exist:

1. **Command and control warfare (C2W).** According to the Department of Defense (DoD), C2W is the strategy employed by the military to implement information warfare during war; it integrates physical destruction, and the objective is to degrade or compromise the enemy's command and control structure. This decapitation is carried out to serve the nation's strategic objectives.

---
1 https://calhoun.nps.edu/bitstream/handle/10945/55351/Denning_Dorothy_2000_cyberterrorism.pdf?sequence=1. Accessed September 23, 2017.
2 http://www.thedailystar.net/beta2/news/new-york-times-twitter-hacked-by-syrian-group/. Accessed September 23, 2017.

2. **Intelligence-based warfare (IBW).** The real-time feeding of intelligence into operations to target the battle space is known as IBW. This has more to do with the high-tech technology that sends the real-time signals in the battle space and enables the military to take informed actions.
3. **Electronic warfare (EW).** EW targets communication channels by degrading the physical medium to exchange information. The attack is mainly targeted to compromise the network traffic.
4. **Psychological warfare (PSYW).** PSYW is aimed against the human mind. Such a war could be launched to disrupt the national will, the opposition's commander, the opposition's troops, or even its culture.
5. **Hacker warfare (HW).** This warfare is aimed to exploit the security holes in the information system. Hacker warfare varies considerably and could be launched from almost anywhere in the world. The intent could be to shut down the entire system, steal sensitive data, inject viruses, and so on.
6. **Economic information warfare (EIW).** Information warfare and economic warfare, when executed together, lead to EIW. This can lead to either information blockade or information imperialism. Information blockade assumes the well-being of nations determined by the flow of information. By disrupting the accessibility to information flow, the aim is to cripple economies. In contrast, information imperialism has to do with trade. Usually, companies retain highly technical information in-house and command a high price for the products or service they specialize in. Acquiring and maintaining these positions could be considered a war, if the intention is to keep other nations behind.
7. **Cyberwarfare.** Libicki [5] states that cyberwarfare is by far the most broadly defined area, with many forms, such as cyberterrorism, or information terrorism. When compared with traditional terrorism, information terrorism targets individuals by attacking their data files and the effect of compromises could be catastrophic.

# Cyberstalking

*Stalking* refers to the unwanted attention by an individual or a group and typically results in harassment and intimidation. Unlike traditional stalking, where people are harassed by physical actions, *cyberstalking* makes use of the Internet and other electronic devices for harassment. In September 2012, the Bureau of Justice Statistics (BJS) released a report on stalking cases in the United States. According to this report, the most vulnerable people are young adults and women, and the damage could range from physical abuse, to loss of jobs, vandalism, financial loss, identity theft, and even assault and murder.

Stalking has several implications and could also lead to psychological trauma and personal loss. Cyberstalking, however, has added a new dimension to persecution and makes victims feel a palpable sense of fear. Prosecution of criminals is challenging, as stalkers can fake their identities, or can choose to be anonymous. This makes it difficult for the victim and cyberpolice to track them down. The lack of Internet regulation, and the nascence of computer forensics has given stalkers a kind of free hand. Moreover, the rapid evolution of technology leads to the advent of new cyberstalking practices, making it more difficult for victims and cyberpolice to identify and punish a perpetrator

While cyberstalking can come in many forms, the most common tactics of a cyberstalker include

- Threatening, harassing, or manipulative emails sent out from various accounts
- Hacking of personal information and gaining access to a victim's personal records or accounts
- Creating fake accounts on social networking sites to gain information about victims or to connect or lure victims
- Posting information about victim on different forums that cause victim embarrassment, loss of reputation or financial loss
- Signing up different forums using victims credentials
- Seeking privileges (e.g., loan approvals) using victims' information

In the United States, several states have enacted cyberstalking and cyberharassment laws. One of perils of these cyberstalking laws is exemplified by the Petraeus affair.[3] FBI began investigating Paula Broadwell for sending allegedly harassing emails to Jill Kelley. One of the apparent problems emerging from the Petraeus affair case was the fact that majority of the laws are written for speech rather than online intimidation or harassment. Cyberstalking covers a range of crimes that involve the use of the Internet as one of the primary media to engage in a range of criminal activities—false accusations, monitoring, threats, identity theft, data destruction or manipulation, and exploitation of a minor.

Recently a case from Hyattsville, Maryland, came to light, where a man was found guilty of 73 counts of stalking. It included reckless endangerment, harassment, and violation of a protective order. In this case, the victim of an individual's ex-wife reportedly endured almost 45 days of online harassment. The accused kept sending threatening emails, created false profiles of the victim using her real name, and invited men to come to visit her at home. Oddly the accused denied having committed a crime, but the harassing behavior stopped upon his arrest. It is a known fact that angry spouses or lovers perpetrate most of the crimes. Various surveys also show that in 60% of the cases, the victims are females.

In recent years "revenge porn" has also emerged as a new menace. This is when ex-partners post sexually explicit photos on websites. Various calls are currently being made for stricter laws that are specifically aimed at stopping revenge porn. Another example that highlights the effect of cyberstalking on the victim was when Patrick Macchione was sentenced to 4 years in prison and 15 years of probation for cyberstalking a college classmate. One thing in particular to note from this case is that the perpetrator started the online relationship in a seemingly normal manner. After gathering enough information for the stalking, such as the victim's cell phone number and place of employment, his interactions with her became increasingly harassing. At one point, the victim was receiving 30 to 40 calls in a five-hour work shift and the perpetrator would appear at her place of employment—and in

---

3   The Petraeus affair details can be found at http://investigations.nbcnews.com/_news/2012/11/12/15099730-petraeus-revelation-began-as-cyber-harassment-probe-investigation-ended-4-days-before-election.

one instance even chased her vehicle. His interactions included text messages, Facebook and Twitter harassment, as well as in-person contact, whereby he demanded affection from the victim and threatened violence if his demands weren't met. In this case, the victim continues to fear that her perpetrator will find a way to come after her and becomes anxious if she receives too many messages or text messages in a short period of time.

Even those in the military can become infatuated enough with an individual to stalk them. A member of the US Navy was convicted of cyberstalking a former girlfriend. Notable in his case was the use of GPS technologies to track her location, using her cell phone signal as well as somehow managing to get monitoring software on her computer to view her online activities. At one point, he went as far as to create a fake Facebook profile under another name so that he could continue to observe her Facebook-related activity after she successfully got a restraining order against him. Prior to sentencing, the perpetrator underwent psychiatric evaluation to ensure that he could be held accountable for the crimes.

Sometimes the aftermath of a cyberstalking incident goes beyond just mental and emotional harm and includes violence against the victim. Last summer, an entire family was indicted on cyberstalking and murder charges when they conspired to harm and eventually murdered a woman that was divorcing a family member and pushing for custody of the children involved in the relationship. The family utilized social media websites as well as YouTube in a "campaign to ruin [the victim's] reputation."[4] A false website was also created to attempt to sway public opinion regarding the victim, in the hope of currying favor after their planned murder of her. At one point, additional friends were feeding information back to the family, such as license plate numbers and photos of the victim's home.

## Defense against Cyberstalking

Cases of stalking via social networking sites have gained much public attention. It is not only teenagers who can fall victim, but people of all ages and groups. As reported by *Science Daily*, research based on the 2006 Supplemental Victimization Survey from the National Criminal Crime Victimization Survey shows that approximately 70% of the victims of cyberstalking are women, while female victims only represented 58% in stalking cases. In addition, the average age for stalking victims is 40.8 years old, while cyberstalking victims averaged 38.4 years old.[5] Some other recent high-profile cyberstalking cases are identified in Table 9.2.

The best defense against cyberstalking is to use the Internet and social networking with responsibility. Making sensitive information available for public viewing can increase its vulnerability. Before posting or exchanging any information, think about who really can have access to this information and how the information can be exploited. Some simple measures include

---

4 http://www.valleycentral.com/news/story.aspx?id=932632
5 http://www.sciencedaily.com/releases/2013/02/130212075454.htm

**Table 9.2.** Examples of cyberstalking cases

| Case | Description |
|---|---|
| Sean Michael Vest case from Pensacolian (2017) | Sean Micheal Vest was charged with 15 counts of aggravated stalking and cyberstalking. He is accused of sending harassing text messages and voice calls to a number of women. He also used different Internet browsers to collect and distribute images. |
| The Matusiewicz case (2016) | Lenore Matusiewicz and her children, Delaware optometrist David Matusiewicz and nurse Amy Gonzalez, harassed, spied on, and cyberstalked David's ex-wife, Christine Belford. Eventually Belford was shot. All three were sentenced in Wilmington, Delaware. |
| The James Hobgood case (2016) | James Hobgood pleaded guilty after cyberstalking and harassing a woman who had moved to Arkansas. Hobgood had created publicly accessible social media accounts where he stated that the victim was an escort and an exotic dancer. |

1. Limit online exchange of sensitive information, such as financial details; instead, call the person and provide the details.
2. Never "friend" a 10-minute acquaintance; limit your online "friends" to people you actually know in person.
3. Set your privacy settings to the highest.
4. Never post pictures or images with identifiers, such as school names and so on.
5. Lastly, use caution and seek help if required.

### In Brief

- The behavioral aspect of cybersecurity is a major challenge.
- Many of the security vulnerabilities are exploited by people.
- In order to manage security, it is important to understand the motivation of people in organizations to comply or not comply with a security policy. Such motivation may be because of
    - Extrinsic factors
    - Intrinsic factors
- Cyberterrorism is a new emerging threat.
- Cyberstalking is menace to our society.

# Questions and Exercises

**Discussion questions.** These are based on a few topics from the chapter and are intentionally designed for a difference of opinion. These questions can best be used in a classroom or a seminar setting.

1. Women are more likely to be cyberstalked than men. How would you go about creating an awareness campaign ensuring that this message gets through?
2. "Security is more about people management than technology management." Discuss.
3. What are social engineering attacks? What behavioral issues need to be addressed to ensure that such attacks do not succeed?

**Exercise.** Imagine that an employee in your company has been violating the established security policy, particularly by visiting undesirable websites, online shopping during work time, and so on. What kind of intrinsic and extrinsic motivation factors would you identify to ensure that the employee adheres to the security policy?

See Appendix A (after the Index) for a set of Short Questions for this chapter.

## Case Study: Cyberterrorism—A New Reality

When hackers claiming to support the Syrian regime of Bashar Al-Assad attacked and disabled[6] the website of *Al Jazeera*, the Qatar-based satellite news channel, in September 2012, the act was seen as another act of hacktivism,[7] purporting to promote a specific political agenda over another. Hacktivism has become a very visible form of expressing dissent. Even though there have been numerous incidents reported by the media, the first case of hacktivism[8] was documented in 1989 when a member of the Cult of the Dead Cow[9] hacker collective named Omega coined the term in 1996. However, hacktivism is not the only form of cyberprotest and conflict that has everyone from ICT professionals to governments scrambling for solutions. Individuals, enterprises, and governments alike rely in many instances almost completely on network computing technologies, including cloud computing. The international and ever-evolving nature of the Internet along with inadequate law enforcement and the anonymity the global architecture offers creates opportunities for hackers to attack vulnerable nodes for personal, financial, or political gain.

The Internet is also rapidly becoming the political and advocacy platform of choice, bringing with it both positive and negative consequences. Increasingly sophisticated off-the-shelf technologies and easy access to the Internet are significantly increasing incidents of cyberterrorism, netwars,[10] and cyberwarfare. The following are a few examples.

According to The Israel Electric Company, Israel is attacked 1,000 times a minute by cyberterrorists targeting the country's infrastructure—water, electricity, communications, and other services.[11]

The *New York Times*, quoting military officials, said there was a seventeenfold increase in cyberattacks targeting the US critical infrastructure between 2009 and 2011.[12]

The 2010 Data Breach Investigations Report has data recording more than 900 instances of computer hacking and other data breaches in the past seven years, resulting in some 900

---
6 http://www.aljazeera.com/news/middleeast/2012/09/20129510472158245.html
7 http://en.wikipedia.org/wiki/Hacktivism
8 http://www.counterpunch.org/assange11252006.html
9 http://en.wikipedia.org/wiki/Cult_of_the_Dead_Cow
10 http://en.wikipedia.org/wiki/Netwar
11 http://www.timesofisrael.com/israel-fights-off-1000-cyber-attack-hits-a-minute/
12 http://www.nytimes.com/2012/07/27/us/cyberattacks-are-up-national-security-chief-says.html

million compromised records. In 2012, the same study listed 855 breaches, resulting in 174 million compromised records in 2011 alone, up from 4 million in 2010.[13]

Another study of 49 breaches in 2011 reported that the average organizational cost of a data breach (including detection, internal response, notification, postnotification cost) was $5.5 million. This number was down from $7.2 million in 2010.[14]

*The Telegraph* (London) reported that "India blamed a new 'cyber-jihad' by Pakistani militant groups for the exodus of thousands of people from India's north-eastern minorities from its main southern cities in August after text messages warning them to flee went viral."[15]

There have been recorded instances of nations allegedly engaging in cyberwarfare.[16] The Center for the Study of Technology and Society has identified five methods by which cyberwarfare can be used as a means of military action. These include defacing or disrupting websites to spread propaganda, to conduct espionage and gain access to critical information, to disrupt enemy military operations, and to attack critical infrastructure.[17] In 1999, pro-Serbian hacker groups, including the Black Hand,[18] broke into NATO, US, and UK computers during the Kosovo conflict. In 2000, both pro-Israeli and pro-Palestinian groups created panic[19] for government and financial networks, and in 2001, the world saw hacking with a patriotic flavor when Chinese and US hackers traded attacks on computers in both countries.

One of the first widely documented cases was the cyberattack on the Republic of Georgia in 2007. On April 26, a series of distributed denial of service (DDoS) attacks targeted government, media, and financial networks and Internet infrastructure. Many other servers were hacked and websites changed to display pro-Russian messages. Many of the initial attacks were said to have originated from Russia and, in some cases, allegedly from Russian government computers. The first wave of attacks against Estonian websites fizzled out after the Estonian foreign minister publicly declared that many of the attacks had originated from Russian government computers.[20]

The Estonian Internet infrastructure was subjected to more attacks. On April 30, 2007, attackers utilized so-called robot networks (botnets) from numerous sources around the world. About a week later, there were more DDoS attacks, including one on Estonia's Hansabank, which reported a loss of about $1 million because of the attacks. The attacks continued intermittently for a few weeks before finally dying off in the summer of 2007.

Another incident was the South Ossetia conflict between Russia and Georgia in 2008. This Russian-Georgian conflict is classified as the first cyberspace conflict that was synchronized with traditional combat actions.[21] Just as Russian troops were crossing the border,

---

13  http://www.verizonbusiness.com/Products/security/dbir/?CMP=DMC-SMB_Z_ZZ_ZZ_Z_TV_N_Z041
14  https://www.symantec.com/about/newsroom/press releases/2012/symantec_0320_02
15  http://www.telegraph.co.uk/news/worldnews/asia/india/9490007/India-blames-Pakistan-for-cyber-jihad-designed-to-spark-ethnic-strife.html
16  http://en.wikipedia.org/wiki/Cyberwarfare
17  Center for the Study of Technology and Society, "National Security Special Focus: Cyber Warfare" (Washington, DC: Author, 2001).
18  http://news.bbc.co.uk/2/hi/science/nature/200069.stm
19  http://matthewkalman.blogspot.com/2000/10/middle-east-conflict-spills-into.html
20  http://www.smh.com.au/news/Technology/Estonia-urges-firm-EU-NATO-response-to-new-form-of-warfarecyberattacks/2007/05/16/1178995207414.html
21  http://www.wired.com/dangerroom/2009/03/georgia-blames/

websites for communications, finance, government, and many international organizations in Georgia became inaccessible.[22] These actions included various DDoS attacks that disrupted communications and information networks in Georgia. The attackers also defaced Georgian websites, adding pro-Russian images, supposedly for propaganda purposes.[23] One of the first networks attacked was a popular hacker forum in Georgia. Consequently, pro-Georgian hackers made successful attacks against Russian networks as well.[24]

Although both the Estonian and Georgian attacks were widely believed to be the work of state-sponsored Russian hackers, no proof has ever been found conclusively linking Russian authorities to the incidents.[25]

## The "First Cyberwarfare Weapon": Stuxnet

In June 2010, an Iranian nuclear facility in Natanz was said to have been attacked by a sophisticated, standalone malicious malware[26] that replicated itself in order to spread to other computers. The malware, called Stuxnet,[27] initially spread via Microsoft Windows operating system and targeted industrial software and equipment—in particular, certain specific industrial control systems made by Siemens.[28] In all, versions of Stuxnet targeted five Iranian organizations,[29] all allegedly linked to the Iranian nuclear program,[30] and may have caused significant damage to the Iranian nuclear enrichment program facility located at Natanz.[31] Stuxnet is said to have been in use since 2009 and was first identified in July 2010 by VirusBlokAda, an information-technology security company in Belarus,[32] after it was said to have "accidently spread beyond" its intended target, Natanz, via infected USB sticks. However, some experts have argued that Stuxnet is not a "worm," since it was propagated via removable media—CDs, DVDs, thumbdrives—and did not distribute through self-replication over the Internet.[33]

In any event, the 2010 version of Stuxnet has been called the "largest" and "most sophisticated attack software ever built," and one investigative article said that the event foreshadowed the destructive new face of 21st century warfare, writing that "Stuxnet is the Hiroshima of cyberwar."[34] According to a report by Symantec, data from the early days of the Stuxnet attack showed that Iran, Indonesia, and India accounted for the bulk of the infected computers.[35] The report also said that Stuxnet was the first piece of malware to

---

22 http://www.networkworld.com/community/node/44448
23 http://latimesblogs.latimes.com/technology/2008/08/experts-debate.html
24 https://www.networkworld.com/article/2274800/lan-wan/russian-hacker-militia--mobilizes-to-attack-georgia.html
25 http://www.wired.com/dangerroom/2009/03/georgia-blames/
26 http://en.wikipedia.org/wiki/Malware
27 http://en.wikipedia.org/wiki/Stuxnet
28 http://support.automation.siemens.com/WW/llisapi.dll?func=cslib.csinfo&lang=en&objid=43876783&caller=view
29 http://www.bbc.co.uk/news/technology-12465688
30 http://www.bbc.co.uk/news/technology-11388018
31 http://isis-online.org/uploads/isis-reports/documents/stuxnet_FEP_22Dec2010.pdf
32 https://www.infoworld.com/article/2625529/malware/siemens-confirms-german-customer-hit-by-stuxnet-espionage-worm.html
33 http://spectrum.ieee.org/podcast/computing/embedded-systems/stuxnet-leaks-or-lies
34 http://www.vanityfair.com/culture/features/2011/04/stuxnet-201104
35 http://www.symantec.com/security_response/writeup.jsp?docid=2010-071400-3123-99

Case Study: Cyberterrorism—A New Reality • 243

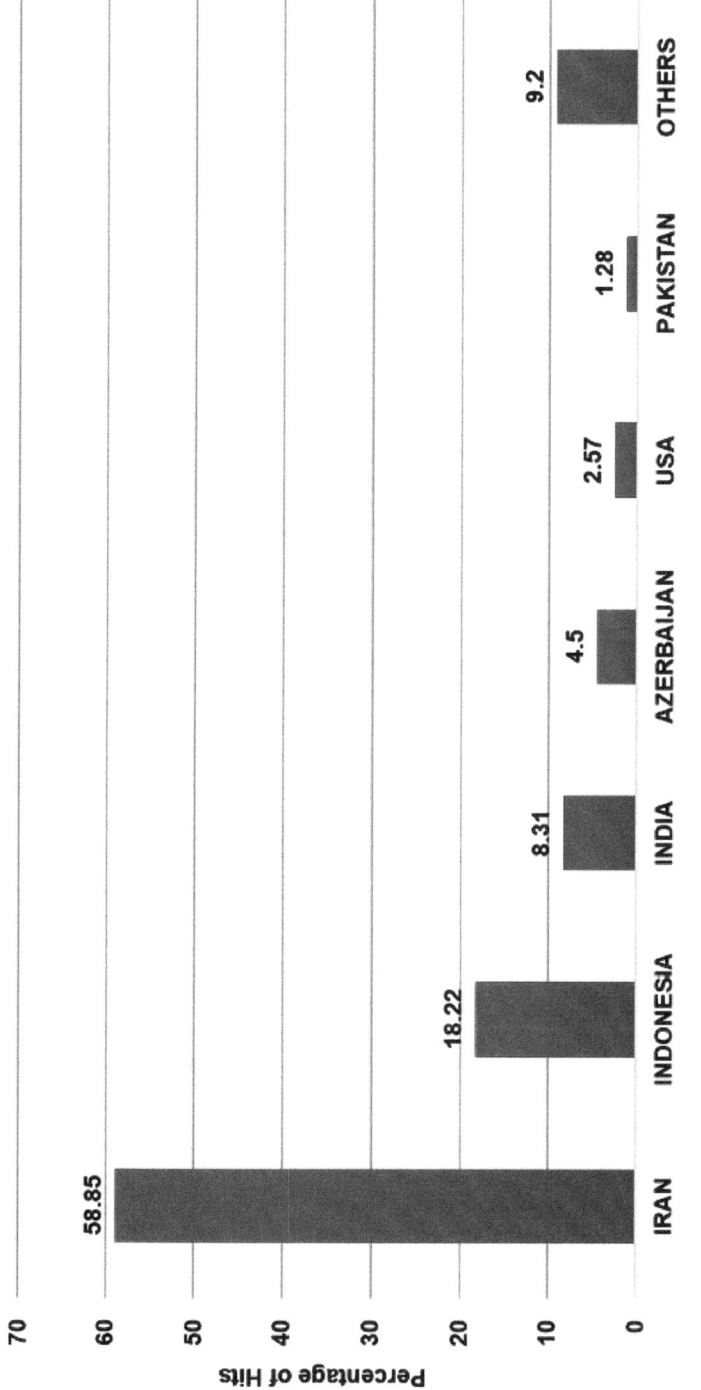

**Figure 9.2.** Percentage of hits from W32.Stuxnet by country

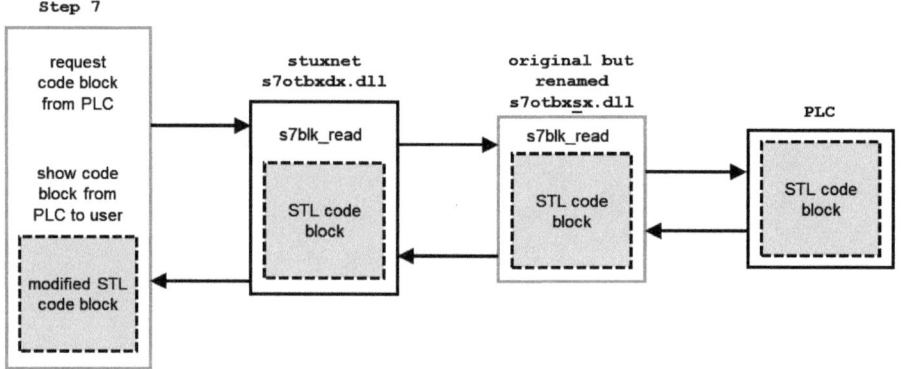

**Figure 9.3.** Overview of Stuxnet

exploit the Microsoft Windows shortcut "LNK/PIF" files' automatic file execution vulnerability[36] in order to spread.

Symantec found that not only did versions of Stuxnet exploit up to four "zero-day"[37] vulnerabilities in the Microsoft Windows operating system, at half a megabyte it was unusually large in size and seemed to have been written in several languages, including portions in C and C++.[38] Another sign of the sophistication was the use of stolen digital certificates from Taiwanese companies, the first from Realtek Semiconductor in January 2010 and the other from JMicron Technology in July 2010. The size, sophistication, and the level of effort has led experts to suggest that the production of the malware was "state-sponsored," and that it is "the first-ever cyberwarfare weapon."[39] The effects of Stuxnet have been likened to a "smart bomb" or "stealth drone," since it sought out a specific target (programmable-logic controllers made by Siemens), masked its presence and effects until after it had done the damage (the operation of the connected motors by changing their rotational speed), and deleted itself from the USB flash drive after the third infection.[40]

Figure 9.3 shows an overview of Stuxnet hijacking communication between Step 7 software and a Siemens programmable-logic controller.[41] As programmed, Stuxnet stopped operating on June 23, 2012, after infecting about 130,000 computers worldwide, with most of them said to be in Iran.

1. How can cyberterrorism, as represented by the Stuxnet, be successfully prevented?

---

36  http://www.securityfocus.com/bid/41732/discuss
37  http://en.wikipedia.org/wiki/Zero-day_attack
38  http://spectrum.ieee.org/podcast/telecom/security/sons-of-stuxnet
39  R Langner, "Security & Privacy," *IEEE*, 2011, ieeexplore.ieee.org.
40  http://www.symantec.com/connect/blogs/stuxnet-breakthrough
41  http://en.wikipedia.org/wiki/File:Stuxnet_modifying_plc.svg

# References

1. Crane, A. 2005. In the company of spies: When competitive intelligence gathering becomes industrial espionage. *Business Horizons* 48(3): 233–240.
2. Haeni, R.E.. 1997. *Firewall penetration testing*. Technical report, The George Washington University Cyberspace Policy Institute.
3. Herath, T., and H.R. Rao. 2009. Encouraging information security behaviors in organizations: Role of penalties, pressures and perceived effectiveness. *Decision Support Systems* 47(2): 154–165.
4. Herath, T., and H.R. Rao. 2009. Protection motivation and deterrence: a framework for security policy compliance in organisations. *European Journal of Information Systems* 18(2): 106–125.
5. Libicki, M.C. 1995. *What is information warfare?* Fort Belvoir, VA: Defense Technical Information Center.
6. Schiller, C., and J.R. Binkley. 2011. *Botnets: The killer web applications*. Rockland, MA: Syngress.
7. Son, J.-Y. 2011. Out of fear or desire? Toward a better understanding of employees' motivation to follow IS security policies. *Information & Management* 48(7): 296–302.
8. Talib, Y., and G. Dhillon. 2015. Employee ISP compliance intentions: An empirical test of empowerment. Paper presented at the International Conference on Information Systems, Fort Worth, TX.

# CHAPTER 10

# Culture and Information System Security

*I was once a member of a mayor's committee on human relations in a large city. My assignment was to estimate what the chances were of non-discriminatory practices being adopted by the different city departments. The first step in this project was to interview the department heads, two of whom were themselves members of minority groups. If one were to believe the words of these officials, it seemed that all of them were more than willing to adopt non-discriminatory labor practices. Yet I felt that, despite what they said, in only one case was there much chance for a change. Why? The answer lay in how they used the silent language of time and space.*

—Edward T. Hall
*The Silent Language* (1959)

Joe Dawson was amazed to find out that any amount of formal control mechanisms, including secure design protocols, risk management, and access control, were good enough only if there was corresponding culture to sustain the security policies. If the people had the wrong attitude or were disgruntled, the organization faced the challenge of managing adequate security.

Joe's efforts to ensure proper information security at SureSteel had caught the attention of a doctoral student, who had requested to follow Joe around in order to undertake an ethnographic study on security organizations. As a byproduct of the research, the doctoral student had written a column in a local newspaper. Joe had given permission to use the company name and so on, thinking that he had nothing to hide. Interestingly, however, the column had generated significant publicity and interest in SureSteel. This came as a blessing in disguise. It was a blessing because SureSteel came to the limelight and Joe Dawson started getting invitations to talk about his experiences. At one such presentation, a member in the audience had pointed Joe to look at the work of The Homeland Security Cultural Bureau. When Joe visited the bureau's website, it was refreshing to note that there was actually a formal program that linked security and culture. Joe read the mission statement with interest:

**HSCB** is protecting the interests of the country's national security by employing efforts to direct and guide the parameters of cultural production.

**PURPOSE:** To protect the interests of the country's national security by employing efforts to direct and guide the parameters of cultural production.

**VISION:** A center of excellence serving as an agent of change to promote innovative thinking about culture and security in a dynamic international environment.

**MISSION:** To provide executive and public awareness of the role that culture can play in both endangering, as well as promoting, a secure nation.

**ACTIVITIES:** To explore issues, conduct studies and analysis, locate and eliminate projects and institutions which undermine national security. To develop and promote a cultural agenda which cultivates a positive image of America, cultural initiatives in the homeland and abroad. To support good cultural initiatives, consult cultural institutions, and provide executive education through a variety of activities including: workshops, conferences, table-top exercises, publications, outreach programs, and promote dialogue.[1]

The mission of HSCB resonated with what Joe had always considered to be important—a mechanism for promoting and ensuring culture as a means for ensuring security. Clearly such an approach could be used to protect information resources of a firm. Given that Joe was partially academic in orientation, he decided to see if there had been any research in this area and if he could find some model or framework that would allow him to think about the range of security culture issues.

Since Joe was a member of ACM, the ACM digital library was an obvious place to start looking. What Joe found was absolutely amazing. Although various scholars had considered culture to be important, there was hardly any systematic research to investigate the nature and scope of the problem domain, specifically with respect to information security.

As Joe Dawson pondered over the range of security issues he had attempted to understand over past several weeks, it seemed clear that management of information system security went beyond implementing technical controls. Besides, security management could also not be accomplished just by having a policy or other related rules. When Joe had begun his journey, attempting to understand the nuts and bolts of security, he had stumbled across an address given by Gail Thackeray, an Arizona-based cybercrime cop. Joe did not exactly remember where he had seen the article, but searching for the name in Google helped him locate the article at www.findwealth.com. In the article Thackeray had been quoted as saying: "If you want to protect your stuff from people who do not share your values, you need to do a better job. You need to do it in combination with law enforcement around the country. You need better ways to communicate with the industry and law enforcement. It is only going to get worse."

Joe saw an important message in what Thackeray was saying. Clearly there was a need to work with law enforcement and other agencies to report suspected criminals. It was also equally important, if not more so, to develop a shared vision and culture.

---

1  https://www.dhs.gov/xlibrary/assets/hsac_ctfreport_200701.pdf

Some very fundamental questions came to Joe's mind: What would a shared culture for information system security be? How could he facilitate development of such a culture? How could he tell if his company had a "good" security culture? Obviously these were rather difficult issues to deal with. Joe felt that he had to do some research.

Joe started out by going to the www.cio.com site and simply searched for "security culture." To his surprise, there were practically no reports or news items on the subject matter. Clearly whenever Joe talked to colleagues and others associated with security culture, values and norms seemed to pop up in the discussions. Joe wondered why no one was writing anything about the subject. *Maybe there's something in the ACM digital library*, Joe thought. A search did not reveal much, apart from a paper by Ioannis Koskosas and Ray Paul of Brunel University, England. The paper, titled "The Interrelationship and Effect of Culture Communication in Setting Internet" and presented at the 2004 Sixth International Conference on Electronic Commerce, highlighted the importance of socioorganizational factors and put forward the following conclusion: "A major conclusion with regard to security is that socio-organizational perspectives such as culture and risk communication play an important role in the process of goal setting....Failure to recognize and improve such socio-organizational perspectives may lead to an inefficient process of goal setting, whereas security risks with regard to the management information through the internet banking channel may arise."[2]

Reading this paper left Joe even more confused. What did the authors mean by "culture"? How could he tell if it was the right culture? These questions still bothered Joe. *Maybe*, Joe thought, *the answer is in understanding what culture is and how a right kind of an environment can be established.* Perhaps this kind of research would have been done in the management field. *It could be worthwhile exploring that literature*, Joe considered.

---

In Chapter 1 we argued that the informal system is the natural means to sustain the formal system. The formal system, as noted previously, is constituted of the rules and procedures that cannot work on their own unless people adopt and accept them. Such adoption is essentially a social process. People interact with technical systems and the prevalent rules and adjust their own beliefs so as to ensure that the end purpose is achieved. To a large extent such institutionalization occurs through informal communications. Individuals and groups share experiences and create meanings associated with their actions. Most commonly we refer to such shared patterns of behavior as *culture*. In many ways it is the culture that binds an organization together.

Security of informal systems is thus no more than ensuring that the integrity of the belief systems stays intact. Although it may be difficult to pinpoint and draw clear links between problems in the informal systems and security, there are numerous instances where in fact it is the softer issues that have had an adverse impact on the security of the systems. Many researchers have termed these as "pragmatic"[3] issues. The word *pragmatics* is an inter-

---

2 http://dl.acm.org/citation.cfm?id=1052264

3 Pragmatics is also one of the branches of Semiotics, the Theory of Signs. Semiotics as a field was made popular by the works of Umberto Eco (see *A theory of semiotics*, University of Indiana Press, 1976), who besides studying signs at a theoretical level has also implicitly applied the concepts in numerous popular pieces such as *Foucault's pendulum* (Houghton Mifflin Harcourt, 2007) and *The name of the rose* (Random House, 2004).

esting one. Although it connotes a reasonable, sensible, and intelligent behavior, which most organizations and institutions take for granted, a majority of information system security problems seem to arise because of inconsistencies in the behavior itself. Therefore, in terms of managing information system security, it is important that we focus our attention on maintaining the behavior, values, and integrity of the people.

This chapter relates such inconsistent patterns of behavior, lack of learning, and negative human influences with the emergent security concerns. Implicitly the focus is on maintaining the integrity of the norm structures, prevalent organizational culture, and communication patterns.

## Understanding the Concept of Security Culture

Culture is an elusive concept. It cannot be seen or touched. It can only be felt and sensed. Various scholars, however, have defined culture as the system of shared beliefs, values, customs, behaviors, and artifacts that members of the society or organization use to cope with their world and with one another. Edgar Schein [10], who has been considered the thought leader in the area of organizational culture, suggests culture to be "the pattern of basic assumptions that a given group has invented, discovered, or developed in learning to cope with its problems of external adaptation and internal integration, and that have worked well enough to be considered valid, and, therefore to be taught to new members as the correct way to perceive, think, and feel in relation to those problems."

In a recent article in *CIO* magazine,[4] the author argues culture is perhaps the single most reason why certain companies are more successful than others. Quoting the case of Zappos, the author notes how growing from 70 to 5,000 people and maintaining a culture that focused on providing great service to the customers allowed the company to be extremely successful. Zappos boasts of a "work hard, play hard" culture and makes every attempt to nurture the same ethos. As a matter of fact, the Zappos family core values[5] are explicitly stated as

1. Deliver WOW Through Service
2. Embrace and Drive Change
3. Create Fun and A Little Weirdness
4. Be Adventurous, Creative, and Open-Minded
5. Pursue Growth and Learning
6. Build Open and Honest Relationships With Communication
7. Build a Positive Team and Family Spirit
8. Do More With Less
9. Be Passionate and Determined
10. Be Humble

At the same time culture can also reflect negatively about a company. Consider the case of Wells Fargo. The bank explicitly states "ethics" and "what's right for customers," as two core values. Yet the enterprise did not hesitate to create nearly two million ghost accounts and

---

4 https://www.cio.com/article/3187088/leadership-management/how-company-culture-can-make-or-break-your-business.html
5 http://www.zappos.com/core-values

defraud its customers. In the business and academic literature, it has often been contended that there is a strong correlation between culture and organizational performance. Culture, however, is a word that has many meanings. For anthropologists, culture has been referred to as a way of life, a sum of behavioral patterns, attitudes, and material things. Organizational theorists have considered culture to be a set of norms and values that an organization or a group shares. In a business setting, it makes sense to consider culture in terms of shared norms and values essentially because such a shared understanding, at least intuitively, is bound to motivate employees. A strong culture has in fact been linked to enhanced coordination and control, improved goal alignment, and increased employee effort [8], [5].

Culture has been considered as the single most important factor leading to the success or failure of a firm. A proper mix of values and norms makes all the difference. Years of research into management of IS security has also established a clear link between good management practices and IS security—that is, if coordination and control mechanisms, goal alignment, and employee morale are well-established and good, these are a consequence of "good management practices." And organizations that have good management practices are essentially secure organizations—or, the probability of occurrence of adverse events in such organizations is minimal at best.

Developing and sustaining a security culture is important as well. Consider an organization where new IT systems are being implemented. It is rather easy for a technologist to forget the social context, which would justify the use of technology in the first place. Often, problem domains are considered without an appreciation of the context, and hence solutions get developed in isolation of complete understanding of the problems. This results in consequences at two levels. First, the computer-based systems do not serve the purpose they were originally conceived for. Second, there is a general discordance between the rules imposed by the computer-based systems and the corporate objectives. Security problems do not necessarily arise because of failure of technical controls, but more because of a lack of integrity of rules at different levels and the controls. Lack of institutionalization of control structures also results in security problems. Most research- and practice-based work tends to suggest that technical controls are perhaps the best means to ensure security. While the role of technical controls cannot be underestimated, true security can only be achieved if the technical controls have been adequately institutionalized.

Even the underlying principles for security evaluation criteria such as TCSEC and now Common Criteria (discussed in Chapter 7) ignore the informal aspects of the evaluation process. These are founded on the principle that security controls are verified against the formal model of the security policy. This may be adequate in certain situations, but as a methodology, it falls short of providing complete assurance. Among the earlier researchers to recognize this problem was Richard Baskerville, when he noted that security controls are often an afterthought and suggested that security should be considered at the logical design phase of systems development [1]. What in fact is needed is to consider security at the requirements analysis stage of a computer-based information systems development process. In considering security in such a way, we are able to develop a culture where security ends up being considered integral to all functions of the organization.

Security culture is the totality of patterns of behavior that come together to ensure protection of information resources of a firm. Once well-formed, security culture acts as the glue that brings together the expectations of different stakeholders. A lack of security culture results in a loss of integrity of the whole organization and even questions the technical adequacy of the controls. This is because it is the people who ensure proper functioning of the controls. Security failures, more often than not, can be explained by the breakdown of communications. Since communication is often considered only in terms of people talking or sending memos, the nonverbal aspects are usually ignored. However, proper security culture can only be sustained if nonverbal, technical, and verbal aspects are dually understood.

As has been argued by Vroom and Von Solms [13] that the utopian view of IS security is where employees voluntarily follow all guidelines and that these are their second nature. For example, it is second nature to change the system password every three weeks, or to back up data every month. However, as Schien [10] notes, culture really exists at three levels. First are the artifacts that are visible. These may be locked doors, obvious firewalls, and well-displayed policies (Figure 10.1).

At the second level are the espoused values and norms. These are not obvious to outsiders and only partially visible in form of good communication, knowledge of policies, audit trails, and so on. The third level is that of tacit assumptions. These are largely hidden. These basic assumptions suggest the underlying beliefs and values of the people.

All levels influence each other. A change in basic assumptions means fundamental changes in the values of the firm. This would have an affect on how outsiders will view the firm. For example, if the firm has a "go-getter at whatever cost" attitude, and employees are forced to perform and achieve specified targets or face consequences, it would mean that a range of compliance issues might be ignored or overlooked. Eventually certain artifacts may also emanate the prevalent culture. Such an impact was evident in the case of Kidder

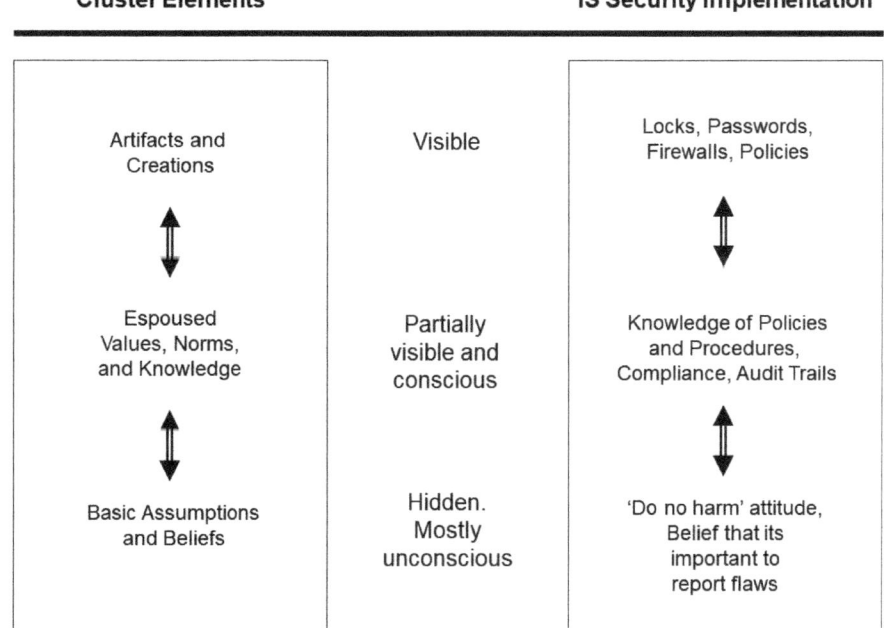

**Figure 10.1.** Elements of culture and security interpretations

Peabody and the manner in which employee Joseph Jett took advantage of the situation (see Box 10.1 for details).

So, how can one institutionalize controls and build a strong security oriented culture? The following six steps set the tone for a defining a more vigilant and accountable workforce:

- **Step 1: Rethink the structure of the C-suite.** The systematic position and structure of the information security roles sends a silent message to the rest of the organization. There was a time when IT typically reported to finance or accounting, and cybersecurity would be considered part of the IT department. While making logical sense to place cybersecurity within IT and a chief information security officer (CSIO) reporting to the chief information officer (CIO), it is not a stretch if the CSIO reported directly to the CEO. This gives a clear indication that cybersecurity is important and is not just relegated to the IT department.
- **Step 2: Prioritize the security literacy of end users.** Time and again various surveys have suggested that human error is the second most reason for security vulnerabilities. Yet many companies only play lip service to training and awareness programs. Many times the training and awareness programs do not necessarily relate to the task at hand. It is important that relevant training is imparted on an ongoing basis and mock exercises conducted. Adequate funding needs to be provided for training as well.
- **Step 3: Establish a security metrics.** As the age-old adage goes—*if you can measure it, you can't manage it.* It is important to establish security metrics. All cybersecurity decision-making should be based on facts rather than feelings. Deviations from industry standards should dictate how priorities are identified and remedial actions taken.
- **Step 4: Link business and technology processes.** Alignment between IT and business has been an ongoing challenge. But in terms of ensuring security, the alignment of IT and business processes is a necessity. No organization can expect risk and compliance management, vendor selection, security training, and so on to be imposed onto the business by an IT department. These should be treated as business issues first. All business leaders need to be actively involved in shaping policies regarding all aspects of security.
- **Step 5: Define an outlook for security spending.** Over the years, IT spending has always been considered as an expense. The evolving nature of business and the growing dependence on data and its protection demands that any investment in security should be considered as value addition. Security investments do not represent and expense. Security investments ensure that assets remain protected and the business flourishes. Security departments should clearly link spending with the business benefits.
- **Step 6: Ensure accountability.** Responsibility and accountability aspects are structural in nature. And there is no doubt that proper structures should be established to ensure that lines of communication are proper and well-established. However, another aspect that needs consideration is that of incentivizing accountability. This means employees who offer new insights, take security seriously, and proactively provide solutions to security problems are adequately incentivized.

## Box 10.1. The case of Kidder Peabody

This case of Kidder Peabody and how Joseph Jett defrauded the company shocked nearly everyone when it came to light. Over the course of more than two-and-a-half years, Joseph Jett was able to exploit the Kidder trading and accounting systems to fabricate profits of approximately $339 million. Joseph Jett was eventually removed from the services of Kidder. The US Securities and Exchange Commission claimed that Jett engaged in more than 1,000 violations in creating millions of dollars in phony profits so as to earn millions in bonuses. During the course of Jett's involvement with Kidder, he amassed a personal fortune of around $5.5 million and earned upward of $9 million in salary and bonuses. In a single year he had made nearly 80% of the firm's entire annual profit of $439 million.

Prior to joining Kidder, Jett was aware of the manner in which the brokerage business was conducted and the specific conditions at Kidder. Jett realized the shortcomings of the accounting system and began tinkering with it. The management overlooked Jett's actions, especially because he seemed to be performing very well and was adding to the firms' profitability. Jett had been hired to perform arbitrage between treasury bonds and "strips" ("separate trading of registered interest and principal of securities"). Kidder relied heavily on expert systems to perform and value transactions in the bond markets. Based on the valuation of the transactions, Kidder systems automatically updated the firm's inventory and profit and loss statements. Jett found out that by entering forward transactions on the reconstituted strips, he could indefinitely postpone the time when actual losses could be recognized in a profit and loss statement. Jett was able to do this by racking up larger positions and reconstituting the strips. This resulted in Jett's trading profits touching a $32 million record, previously unheard of in dealing with strips. Jett's personal bonus was $2 million. The following year Jett reported a $151 million profit and earned $12 million in bonus. When included $47 billion worth of strips and $42 billion worth of reconstituted strips, senior management decided to look deeper into the dealings.

The factors presented in this scenario clearly suggest that basic safeguards had not been instituted at Kidder. Although the junior traders at Kidder were aware of Jett's activities, the senior management did not make any effort to access this information. These factors certainly influenced Jett's beliefs and his intentions regarding the advantages and disadvantages of engaging in the illicit activities. As a consequence, Jett manipulated the accounting information system and deceived the senior management at Kidder.

As is typical of many merchant banks, bonuses earned are intricately linked with the profitability of the concern. Kidder offered substantial bonuses for individual contributions to company profits. Even within the parent company, General Electric, the CEO told his employees that he wanted to create a "cadre of professions" who could perform and be more marketable. This resulted in the employees being subjected to intense pressure to perform and having a focus on serving their self-interest. As a consequence, General Electric did not necessarily afford a culture of loyalty and truthfulness. The employees were inadvertently getting the silent message that they should "look after themselves and win at any cost."

Kidder Peabody and the parent company General Electric were determined to court political and business acclaim by recruiting a large number of people from ethnic minorities. The bank considered this to be a means of being socially responsible and perhaps gaining esteem from others by lending a helping hand to certain underprivileged sections of the society. There are claims that Jett had falsified information on his CV, thus making it extremely impressive. The personnel department at Kidder took this information on face value and did not make any attempt to verify it. It follows therefore that Jett's risk-taking character and involvement in unethical deeds may have influenced his beliefs.

## Silent Messages and IS Security

Culture is shared and facilitates mutual understanding, but can only be understood in terms of many subtle and silent messages. Thus culture can be studied by analyzing the communication processes. This also means that culturally determined patterns of behavior are messages that can be communicated. Seminal work to study culture and understand its range of silent messages was originally undertaken by Edward T. Hall [6]. Hall considers culture as communication and communication as culture. Consequently, culture is concerned more with messages than with the manner in which they are controlled. Based on extensive research spanning several decades, Hall synthesizes the study of culture along 10 dimensions, referred to as *streams*. Each cultural stream interacts with each other to produce a *web of a culture* (see Figure 10.2). The resultant interactions among the streams result in the explication of patterns of behavior. The 10 streams of culture are interaction, association, subsistence, bisexuality, territoriality, temporality, learning, play, defense, exploitation. All streams of the culture may not be relevant for a given situation and hence may not be used in any cultural evaluation. The web of culture is a useful tool to address security issues that might emerge in any given setting.

### Interaction

According to Hall, interaction has its basis in the underlying irritability of all living beings. One of the most highly elaborated forms of interaction is speech, which is reinforced by the tone, voice, and gesture. Interaction lies at the hub of the "universe of culture" and

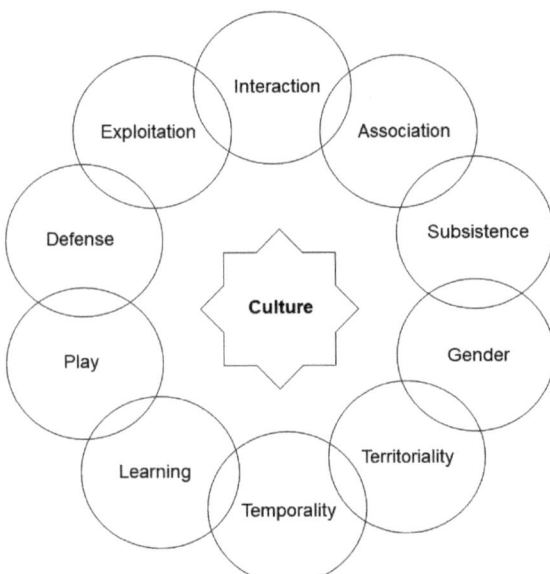

**Figure 10.2.** Web of culture

everything grows from it. A typical example in the domain of information systems can be drawn from the interaction between the information manager and the users. This interaction occurs both at the formal and informal levels—formally through the documentation of profiles and informally through pragmatic monitoring mechanisms.

The introduction of any IT-based system usually results in new communication patterns between different parts of the organization. At times there may be a risk of the technology emerging as the new supervisor. This affects the patterns of interaction between different roles within the organization. A change in the status quo may also be observed. This results in new ways of doing work and questions the ability of the organization to develop a shared vision and consensus about the patterns of behavior. This results in the risk of communication breakdowns. Employees may end up being extremely unhappy and show resentment. This may create a perfect environment for a possible abuse.

## Association

Kitiyadisai [7] describes *association* in a business setting as that where an information manager acquires an important role of supplying relevant information and managing the information systems for the users. The prestige of the information systems group increases as their work gets recognized by the public. An association of this kind facilitates adaptive human behavior

The introduction of new IT-based systems changes the associations of individuals and groups within the organization. This is especially the case when systems are introduced in a rather authoritative manner. It results in severe "organizational" and "territoriality" implications. Typically managers may force a set of objectives onto organizational members who may have to reconsider and align their ideas with the "authoritarian" corporate objectives. In a highly professional and specialist environment, such as a hospital or a university, it results in a fragmented organizational culture. A typical response is when organizational members show allegiance to their professions (e.g., medial associations) rather than their respective organizations. The mismatch between corporate objectives and professional practices leads to divergent viewpoints. Hence concerns arise about developing and sustaining a security culture. It is important that the organization has a vision for security; otherwise corporate policies and procedures are difficult to realize.

With new social media–based communication technologies, association among individuals, between individuals and corporates, and among corporates themselves are changing. The resultant new associations have significant implications for security and privacy. KLM Royal Ditch Airlines has, for instance, adopted Facebook Messenger to communicate with customers. Passengers can now chat with KLM support staff. Establishing new communication channels was not easy. There are some serious security and privacy issues that need consideration (see Box 10.2).

> **Box 10.2. Messenger use by KLM for new associations**
>
> KLM has built in a piece of code on KLM.com that makes it possible for Facebook to show the Messenger plugin to the relevant people on KLM's website. KLM doesn't receive any data from you—only Facebook is able to match whether you are using Messenger. The Messenger plugin on KLM.com leverages Facebook cookies and other similar technologies. Only passengers who have logged in on Facebook or Messenger on the same device in the past 90 days will see the Messenger plugin. If you have deleted your cookies or logged out of Facebook, you won't see the plugin either.
>
> *Extracted from https://messenger.klm.com.*

## Subsistence

*Subsistence* relates to physical livelihood, eating, working for a living, and income (indirectly). For example, when a company tells a new middle manager of his status, subsistence refers to access to management dining room and washing facilities, receipt of a fairly good salary, and so on. IT-based systems adversely affect the subsistence issues related to different stakeholder groups. Any structural change prior to or after the introduction of IT-based systems questions the traditional ways of working. Such changes get queried by different groups. Since new structures may result in different reporting mechanisms, there is usually a feeling of discontent among employees. Such occurrences can potentially lead to rancor and conflict within an organization. This may lead to a situation where a complex interplay among different factors may lead to the occurrence of some adverse consequences.

## Bisexuality/Gender

This refers to differentiation of sexes, marriage, and family. The concept of *bisexuality*[6] is exemplified in an organization by a predominantly male middle management displaying machismo. Bisexuality has implications for the manner in which men and women deal with technology. A 1994 study found that in a group of fourth through sixth graders in school who had been defined as ardent computer users, the ratio of girls to boys using computers was 1:4. This gap continued to increase through to high school [9]. Part of the reason is the differing expectations from boys and girls with respect to their behaviors, attitudes, and perceptions.

---

6 E. T. Hall (1959) uses the term "bisexuality" to refer to all aspects of gender.

> ### Box 10.3. Gender differences
>
> Internet auction fraud was by far the most reported offence. Investment fraud, business fraud, confidence fraud, and identity theft round out the top seven categories of complaints referred to law enforcement during the year. Among those individuals who reported a dollar loss, the highest median dollar losses were found among Nigerian letter fraud, identity theft, and check fraud complainants.
> - Among perpetrators, nearly four in five (79%) are male
> - Among complainants, 71% are male, half are between the ages of 30 and 50 (the average age is 39.4).
> - The amount lost by complainants tends to be related to a number of factors. Males tend to lose more than females. This may be a function of both online purchasing differences by gender, and the type of fraud the individual finds himself or herself involved with. While there isn't a strong relationship between age and loss, the proportion of individuals losing at least $5,000 is higher for those 60 years and older than it is for any other age category.
> - Electronic mail (email) and web pages are the two primary mechanisms by which the fraudulent contact took place.
>
> *Extracted from the IFCC 2002 Internet Fraud Report.*

## Territoriality/Location

*Territoriality* refers to division of space, where things go, where to do things, and ownership. Space (or territoriality) meshes very subtly with the rest of the culture in many different ways. For example, status is indicated by the distance one sits from the head of the table on formal occasions.

IT-based systems can create many artificial boundaries within an organization. Such boundaries do not necessarily map on to the organizational structures. In such a situation there are concerns about the ownership of information and privacy of personal data. Problems with the ownership of systems and information reflect concerns about structures of authority and responsibility. Hence there may be problems with consistent territory objectives (i.e., areas of operation). Failure to come to grips with the territory issues can be detrimental to the organization, since there can be no accountability in the case of an incident.

## Temporality/Time

*Temporality* refers to division of time, when to do things, sequence, duration, and space. It is intertwined with life in many different ways. In a business setting, examples of temporality can be found in flexible working hours, being "on call," "who waits for whom," and so on.

IT-based systems usually provide comprehensive management information and typically computerize paper-based systems. Technically the system may be very sound in performing basic administrative tasks. However, it can be restrictive and inappropriate as well. This is usually the function of how the system gets specified and the extent to which formalisms have been incorporated into the technical system. Hence it may not serve the needs of many users. In this regard the users end up seeking independent advice on IT use, which defeats the core objective of any security policy.

## Learning

*Learning* is "one of the basic activities of life, and educators might have a better grasp of their art if they would take a leaf out of the book of the early pioneers in descriptive linguistics and learn about their subject by studying the acquired context in which other people learn" [6]. In an organization, management development programs and short courses are typical examples.

IT-based systems provide good training to those unfamiliar with the core operations of an organization. The users who feel that their needs have been met through IT have to establish a trade-off between ease of use and access. Companies need to resolve how such a balance can be achieved. Once a system is developed, access rights need to be developed, communicated, and integrity ensured between the technical system and the bureaucratic organization.

## Play

In the course of evolution, Hall considers *play* to be a recent and a not too well-understood addition to living processes. *Play* and *defense* are often closely related, since humor is often used to hide or protect vulnerabilities. In Western economies *play* is often associated with competition. *Play* also seems to have a bearing on the security of the enterprise, besides being a means to ensure security and increase awareness.

Many organizations use *play* as a means to prepare for possible disasters. For instance, Virginia Department of Emergency Management runs a Tornado Preparedness Drill (Figure 10.3), which is a means to make citizens familiar with a range of issues related to tornados. Such drills and games have often been used to inculcate a culture within organizations.

In recent years, *play* has extensively been used to become familiar with security breaches and related problems. In San Antonio, Texas, for example, an exercise termed "Dark Screen" was initiated. The intent was to bring together representatives from the private sector,

**Figure 10.3.** Playing for crisis management (Source: wwwvaemergency.com)

federal, state, and local government agencies, and help each other in identifying and testing resources for prevention of cybersecurity incidents (for details, see White et al. [14]).

## Defense/Security

*Defense* is considered to be an extremely important element of any culture. Over the years, people have elaborated their defense techniques with astounding ingenuity. Different organizational cultures treat defense principles in different ways, which adversely affects the protective mechanisms in place. A good defense system would increase the probability of being informed of any new development and intelligence by the computer-based systems of an organization.

IT-based systems allow for different levels of where password control can be established. Although it may be technically possible to delineate levels, it is usually not possible to maintain integrity between system-access levels and organizational structures. This is largely because of influences of different interest groups and disruption of power structures. This is an extremely important security issue and cannot be resolved on its own, unless various operational, organizational, and cultural issues are adequately understood.

## Exploitation

Hall draws an analogy with the living systems and points out that "in order to exploit the environment all organisms adapt their bodies to meet specialised environmental conditions." Similarly organizations need to adapt to the wider context in which they operate. Hence companies that are able to use their tools, techniques, materials, and skills better will be more successful in a competitive environment.

Today IT-based systems are increasingly becoming interconnected. There usually are aspirations to integrate various systems, which themselves transcend organizations and national boundaries. It seems that most organizations do not have the competence to manage such a complex information infrastructure. As a consequence, the inability to deal with potential threats poses serious challenges. In recent years, infections by the Slammer, Blaster, and SoBig worms are cases in point. The 1988 Morris Worm is also a testament to the destruction that can be caused in the interconnected world and the apparent helplessness over the past two decades in curtailing such exploitations.

# Leadership and Security Culture

The importance of leadership and governance structures in managing security gained prominence with the introduction of Sarbanes-Oxley (SOX) legislation in the US (and its corresponding legislation in other countries). SOX made senior management responsible for ensuring security. Failure to comply has the potential for criminal fraud liability. While information security is not explicitly discussed in the text of the act, the reality of modern businesses is that most business processes are dependent on technology and technical controls mirror management control. Hence responsibility and accountability become rather important elements in management of security.

Unfortunately, governance boards are not necessarily cyberaware. There is lack of awareness of critical cybersecurity issues. An *IT Governance* "Boardroom Cyber Security Watch Survey" found that

1. 32% of respondents did not receive any regular report on the cyberdefenses. This means that a large proportion of board members are not aware of the range of cyberdefenses.
2. 21% of the respondents felt that reports to the board do not include the complete cybersecurity picture. Another 28% were unsure if any information was provided. This suggests issues with the quality of reports.
3. 30% of the respondents feel that the board lacks the necessary cybersecurity knowledge and skills. This means that there is a general lack of competence amongst board members.

There are some easy and simple steps that need to be undertaken. Cybersecurity professionals need to learn how they should communicate with the senior leadership. Remaining engrossed in the technical jargon prevents the message from getting communicated. Usually security professionals stress how well their technical controls are functioning. From a senior management perspective, the functioning of the controls is less important than how security has enabled the delivery of business benefits.

In a very interesting paper published in 2015 in the *Journal of Information System Security* [16], the authors found several issues that are pertinent to the role of leadership in promoting a security culture within the organization (Figure 10.4):

1. Management commitment to security reflects on employee commitment as well. This means that if the leadership is less committed to security practices, employees are going to mirror that behavior.
2. Lack of communication by the leadership on important security issues results in (a) a lack of motivation in the employees and (b) reduced awareness of the necessary controls.
3. Lack of motivation and awareness on part of the employees results in ill-defined responsibility of the employees, which results in limited knowledge of security practices.
4. A combination of reduced awareness, lack of motivation, misplaced responsibility, and reduced knowledge results in a negligent employee, which is a cause of great concern to the organization.

## Institutionalizing "Obedient" Security Culture[7]

An organization's employees are considered a corporate resource and asset, requiring constant care and management in their daily activities. For the majority of these employees, their first priority at work is to get the job done. As a result of the pressure related to an employee's job, any tasks that are not deemed essential to "getting things done" are often overlooked or regarded as hindrances. Therefore if employees see information security practices as an obstacle to their work, they might not take on proper responsibilities and, worse, might go out of their way to find a "workaround" to any security measure they do not consider necessary. Consequently, it is often the behavior and actions of employees that

---

[7] Part of this section is drawn from [11]. Used with permission.

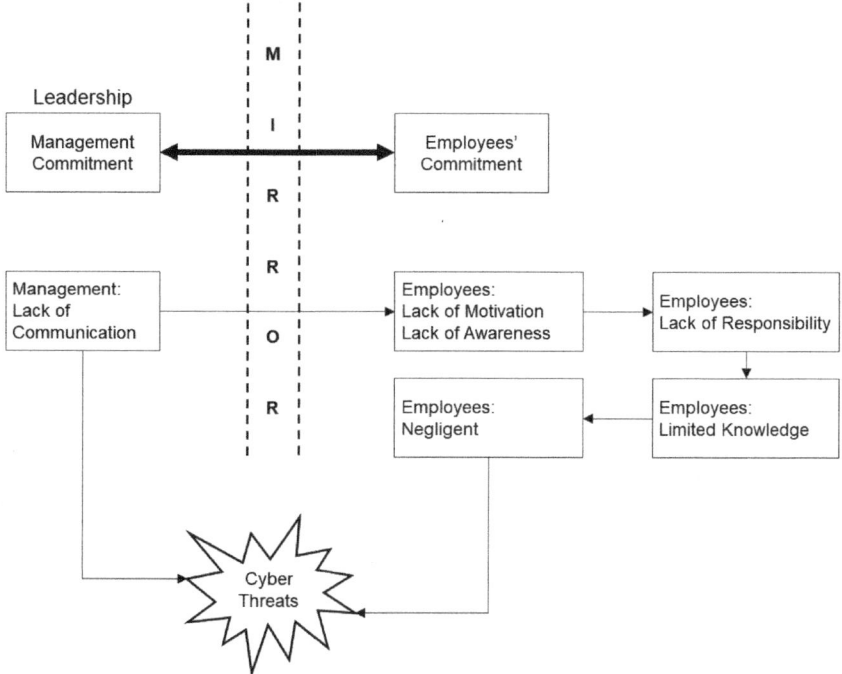

**Figure 10.4.** Role of leadership in cybersecurity [16] (used with permission from the *Journal of Information System Security*)

are the weakest link in the information security chain in an organization. Furthermore, given that even seemingly innocuous information in the hands of a malicious person could be exploited to obtain the key to an organization's most valued secrets, employees should be alert to possible threats to an organization's information. Many information technology professionals have the mistaken belief that their organizations are impervious to attack, as they have deployed security features such as authentication, intrusion detection, and firewalls. Likewise, as many aspects of information security involve technology, many employees believe that security issues are being handled by firewalls and other security technologies deployed by the information technology professionals in their organization. However, the reality is that most employees are the "front line" required to protect the overall security of the organization.

The human part of any information security solution is the most essential. Almost all information security solutions rely on the human element to a large degree, and employees continue to be the most severe threat to information security. Employees must learn to fully appreciate that a severe loss of sensitive corporate information could jeopardize not only the organization but their own personal information and jobs as well. It is vital that employees realize the importance of protecting information assets and the role that they should be playing in the security of information. Employees who do not fully understand their roles are very often apathetic to information security and may be enticed to overlook or simply ignore their security responsibilities. An information security program should be more than just implementing an assortment of technical controls. It should also address the behavior and resulting actions of employees.

It is the corporate culture of an organization that largely influences the behavior of employees. Therefore, in order to safeguard its information assets, it is imperative for organizations to stimulate a corporate culture that promotes employee loyalty, high morale, and job satisfaction. Through the corporate culture, employees should be aware of the need for protecting information and of the ways inappropriate actions could affect the organization's success. It is vital to ensure that employees are committed to and knowledgeable about their roles and responsibilities with regard to information assets. As part of its corporate governance duties, senior management is accountable for the protection of all assets in an organization. Information is one of the most important assets that most organizations possess. It is vital, therefore, for senior management to provide guidance and direction toward the protection of information assets. Further, it is imperative for senior management to create, enforce, and commit to a sound security program.

To describe this relationship between the corporate governance duties of senior management to protect information assets, the requirement to change the behavior and attitudes of employees through the corporate culture, and instilling information security practices, Thomson and von Solms [12] defined the term "corporate information security obedience." the definition of "corporate information security obedience" is "de facto employee behavior complying with the vision of the Board of Directors and Executive Management, as defined in the Corporate Information Security Policy." Therefore, if corporate information security obedience is evident in an organization, the actions of employees must comply with that which is required by senior management in terms of information security. However, they will comply not because of incentives or consequences if they do not, but rather because they believe that the protection of information assets should be part of their daily activities.

Therefore, to be genuinely effective, information security needs to become part of the way every employee conducts his or her daily business. Information security should become an organizational lifestyle that is driven from the top, and employees should become conscious of their information security responsibilities. In other words, the corporate culture in an organization should evolve to become an information security obedient corporate culture and ensure employees are no longer a significant threat to information assets. In order for this to be achieved, it is necessary for new knowledge to be created in an organization.

## Security Culture Framework

In the previous section we looked at the range of silent messages that come together to manifest some form of culture. The 10 cultural streams interact with each other to offer some 100 different kinds of messages, which tell a story of institutions, their structure, and social and psychological makeup of individuals. A more parsimonious organization of the cultural streams is to consider individual values and silent messages along two dimensions [3]: one differentiating flexibility, discretion, and dynamism from stability, order, and control; and the other differentiating internal orientation, integration, and unity from external orientation, differentiation, and rivalry (Figure 10.5).

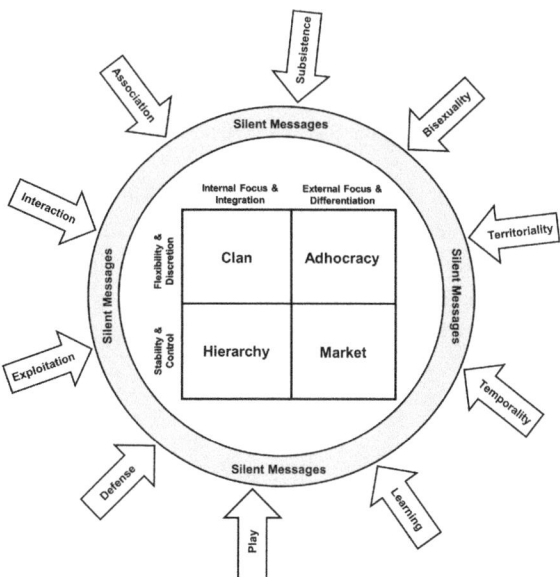

**Figure 10.5.** Types of culture

Flexibility, discretion, and dynamism are typically found in situations (and organizations) that are more innovative and research oriented. Management consulting firms, research and development labs, and the like are typical examples. The business context in these organizations demands them to be more flexible and less bureaucratic, allowing employees to have more discretion. In contrast, there are contexts that demand stability, order, and control. Institutions such as the military, some government agencies, and companies dealing with mission critical projects generally fall in this category.

The dimension representing internal orientation, integration, and unity typifies organizations. Such organizations strive to maintain internal consistency and aspire to present a unified view to the outside world. Integration and interconnectedness of activities and processes are central. In contrast, the external orientation focuses on differentiation and rivalry. This is typical of more market-oriented environments.

The two dimensions present a fourfold classification of culture. Each of the classes defines the core values for the specific culture type. The culture classes provide a basis for fine-tuning security implementations and developing a context-specific culture. The four classes are

- Adhocracy culture
- Hierarchy culture
- Clan culture
- Market culture

## Adhocracy Culture
This culture is typically found in organizations that undertake community based work or are involved in special projects (*viz.* research and development). External focus, flexibility, and

discretion are the hallmarks. Typical examples of organizations with a dominant adhocracy culture are nonprofit firms and consulting companies. Members of these organizations usually tend to take risks in their efforts to try new things. Commitment to creativity and innovation tends to hold such organizations together. Adhocracy also encourages individual initiative and freedom.

Security controls by nature are restrictive. Adhocracy culture signifies trust among individuals working together. Therefore excessive process and structural controls tend to be at odds with the dominant culture. This does not mean that there should not be any controls. Rather process integrity and focus on individual ethics becomes more important. Unnecessary bureaucratic controls are actually more dangerous and may have a detrimental impact on security relative to fewer controls.

## Hierarchy Culture

The key characteristic of the hierarchy culture is focus on internal stability and control. It represents a formalized and a structured organization. Various procedures and rules are laid out, which determine how people function. Since most of the rules and procedures for getting work done have been spelled out, management per se is the administrative task of coordination—being efficient and yet ensuring integrity is the cornerstone. Organizations such as the military and government agencies are typically hierarchical in nature.

Most of the early work in security was undertaken with hierarchy culture guiding the development of tools, policies, and practices. Most access control techniques, the setting up of privileges requires a hierarchical orientation. This is largely because the user requirements were derived from a hierarchical organization, such as the military. These systems are generally very well-defined and are complete, albeit for the particular culture they aspire to represent. For instance, the US Department of Defense has been using the Trusted Computer System Evaluation Criteria for years, and clearly they are valid and complete. So are the Bell La Padula and Denning Models for confidentiality of access control (see Chapter 3). Similarly, the validity and completeness of other models such as Rushby's Separation Model and the Biba Model for integrity have also been established. However, their validity exists not because of the completeness of their internal working and their derivations through axioms, but because the reality they are modeling is well defined (i.e., the military organization). The military, to a large extent, represents a culture of trust among its members and a system of clear roles and responsibilities. Hence the classifications of security within the models do not represent the constructs of the models, but instead reflect the very organization they are modeling. A challenge, however, exists when these models are applied in alternative cultural settings. Obviously in the commercial environment the formal models for managing security fall short of maintaining their completeness and validity.

## Clan Culture

This culture is generally found in smaller organizations that have internal consistency, flexibility, sensitivity, and concern for people as its primary objectives. Clan culture tends to place more orientation on mutual trust and a sense of obligation. Loyalty to the group and commitment to tradition is considered important. A lot of importance is placed on long-term benefit of developing individuals and ensuring cohesion and high morale. Teamwork, consensus, and participation are encouraged.

While the tenets of clan culture are much sought after, even in large organizations, it is often difficult to ensure compete compliance. In large organizations, the best means to achieve some of the objectives is perhaps through an ethics program that encourages people to be "good" and loyal to the company. In smaller organizations it is easier to uphold this culture, essentially because of factors such as shame. A detailed discussion on this aspect can be found in the book entitled *Crime, Shame, and Reintegration* [2].

## Market Culture

Market culture poses an interesting challenge on the organization. While outward orientation and customer impact are the cornerstones, some level of stability and control is also desired. The conflicting objectives of stability and process clarity, outward orientation of procuring more orders or aspiring to reach out, often play out in rather interesting ways. The operations-oriented people generally strive for more elegance in procedures, which tends to be at odds with the more marketing-oriented individuals. People are both mission- and goal-oriented, while leaders demand excellence. Emphasis on success in accomplishing the mission holds the organization together. Market culture organizations tend to focus on achieving long-term measurable goals and targets.

In this culture, security is treated more as a hindrance to the day to day operations of the firm. It is usually a challenge to get a more market-oriented culture to comply with regulations, procedures, and controls. To a large extent regulation and control is at odds with the adventurism symbolized by a market culture. Nevertheless security needs to be maintained, and it's important to understand the culture so that adequate controls are established.

---

**Box 10.4. Security culture at Harrah's**

Harrah's Entertainment owns hotels and casinos in a number of cities. The hotel has been using technology for a number of years to better serve its customers. The implementation of the Harrah's Customer Relationship Management System has been well-reported. In 1999 Harrah's had expanded their Total Rewards program and combined the information they already had with customer reservation data. As a result, it became possible for the marketing department to tweak Harrah's products. It also became possible for service representatives to access data in real time, thus being able to reward valued customers with a range of promotions—room upgrades, event tickets, complimentary meals, and so on.

Casino hotels generally operate in a very hierarchical culture, with rather strict rules for accessing information. At the same time, however, Harrah's CRM system is accessed by at least 12,000 of the 47,000 employees. Employees of the Las Vegas property handle $10–15 million in cash every day. The CRM system has a lot of sensitive data that keep track of how customers gamble and what they do during their visit to the casino.

As the CIO put it, "There's an implicit trust that we have with our employees." There are, however, a lot of checks and balances as well, which "force" employees to be honest. Harrah's indeed represents a dual culture that coexists—hierarchy and adhocracy.

Harrah's uses the following checklist to manage security:

- Physical surveillance—Close circuit cameras for physical protection of the premises
- Security badges—A must for employees to wear
- User account monitoring—Employee accounts are closed within a day of their leaving the company. Every quarter personnel files are compared to spot discrepancies.
- Daily log reviews—Each day changes to customer credit limits and so on are reviewed.
- Checks and balances—There are at least three employees when gaming table chips are replenished.
- Location-specific access limitation—The nature of access is specific to the location. For instance, room restaurant computers do not have access to gaming floor computers.
- Strict data center access—Sophisticated access control mechanisms have been implemented.
- Limited access to the production system—Access is monitored for any changes. IT staff are issued a temporary password. Proper authorization is necessary.

Research has shown that organizations tend to represent all four kinds of cultures—some are strong in one while weak in others [15]. The intention of classifying cultures is to provide an understanding of how these might manifest themselves in any organizational situation and what could be done to adequately manage situations.

A series of radar maps can be created for each of the dimensions of security. In this book a number of dimensions have been discussed and presented—confidentiality, integrity, availability, responsibility, integrity of roles, trust, and ethicality. Chapter 16 provides a summary. Each of these dimensions can be evaluated in terms of the four cultures. A typical radar map can help a manager pinpoint where they might be with respect to each of the dimensions.

Consider the situation in a medical clinic. There may be just two doctors in this clinic with a few staff. With respect to maintaining confidentiality, organizational members can define what their ideal type might be. This could be based on patient expectations and on other regulatory requirements. Given this setting, it may make sense to conduct business as a family, hence suggesting the importance of a clan culture. However, at a certain point in time there may be less of a clan culture and more hierarchical culture. This may be perhaps because none of the organizational members know each other well (as depicted in Figure 10.6 for the year 2013). A conscious decision may, however, be taken to improve—representing the ideal type aspired for in Figure 10.6. A map drawn in subsequent years (e.g., 2017) could help in understanding where progress has been made and what other aspects need to be considered. Similar maps can be drawn for other dimensions of security. The maps are snapshots of culture, which become guiding frameworks for organizational policy setting.

## OECD Principles for Security Culture

As discussed elsewhere (Chapter 7), the Organization for Economic Cooperation and Development (OECD) came into being pursuant to article 1 of the convention signed in Paris on December 14, 1960. OECD was established to further cooperation and interaction among member countries. Such cooperation was for economic development to further

OECD Principles for Security Culture • 267

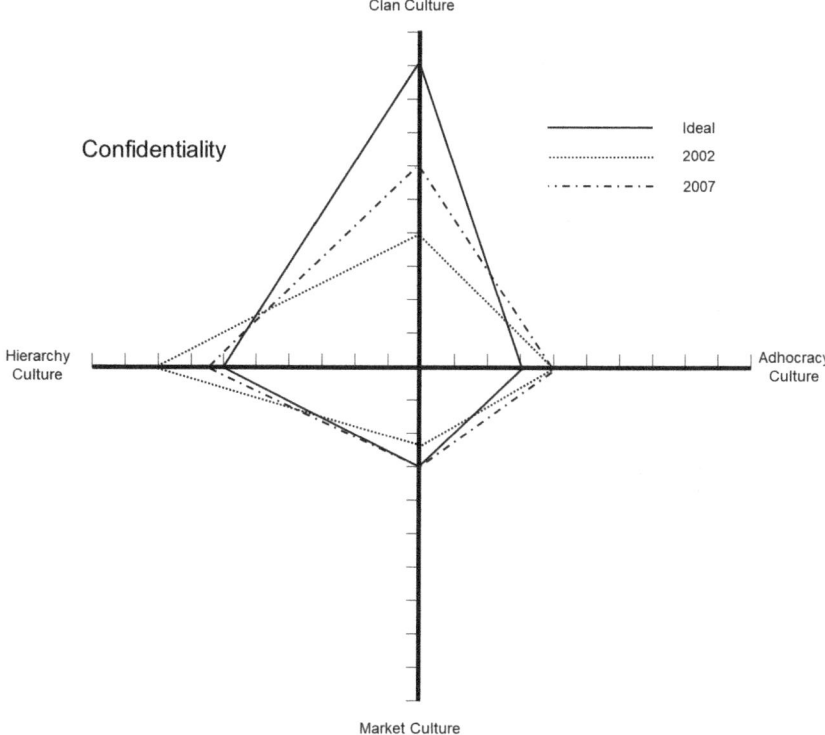

**Figure 10.6.** Radar map for confidentiality culture in a medical clinic

trade and improvement in living standards. Austria, Belgium, Canada, Denmark, France, Germany, Greece, Iceland, Ireland, Italy, Luxembourg, the Netherlands, Norway, Portugal, Spain, Sweden, Switzerland, Turkey, the United Kingdom, and the United States were among the original members.

In 2002 OECD identified and adopted the following nine principles for IS Security culture.

1. **Awareness. Participants should be aware of the need for security of information systems and networks and what they can do to enhance security.** Clearly being aware that risks exist is perhaps the cheapest of the security controls. As has been discussed elsewhere in this book, vulnerabilities can be both internal or external to the organization. Organizational members should appreciate that security breaches can significantly impact networks and systems directly under their control. Potential harm can also be caused because of interconnectivity of the systems and the interdependence of/on third parties.
2. **Responsibility. All participants are responsible for the security of information systems and networks.** Individual responsibility is an important aspect for developing a security culture. There should be clear cut accountability and

attribution of blame. Organizational members need to regularly review their own policies, practices, measures, and procedures and assess their appropriateness.
3. **Response. Participants should act in a timely and cooperative manner to prevent, detect, and respond to security incidents.** All organizational members should share information about current and potential threats. Informal mechanisms need to be established for such sharing and response strategies to be permeated in the organization (see Figure 10.7 for an example). Where possible, such sharing should take place across companies.
4. **Ethics. Participants should respect the legitimate interests of others.** All organizational members need to recognize that any or all of their actions could harm others. The recognition helps in ensuring that legitimate interests of others get respected. Ethical conduct of this type is important and organizational members need to work toward developing and adopting best practices in this regards.
5. **Democracy. The security of information systems and networks should be compatible with essential values of a democratic society.** Any security measure implemented should be in consort with tenets of a democratic society. This means that there should be freedom to exchange thoughts and ideas, besides a free flow of information, but protecting confidentiality of information and communication.
6. **Risk assessment. Participants should conduct risk assessments.** Risk assessment needs to be properly understood and should be sufficiently broad based to cover a range of issues—technological, human factors, policies, third party, and so on. A proper risk assessment allows for identifying a sufficient level of risk to be understood and hence adequately managed. Since most systems are interconnected and interdependent, any risk assessment should also consider threats that might originate elsewhere.
7. **Security design and implementation. Participants should incorporate security as an essential element of information systems and networks.** Security should not be considered as an afterthought, but be well-integrated into all system design and development phases. Usually both technical and nontechnical safeguards are required. Developing a culture that considers security in all phases ensures that due consideration has been given to all products, services, systems, and networks.
8. **Security management. Participants should adopt a comprehensive approach to security management.** Security management is an evolutionary process. It needs to be dynamically structured and be proactive. Network security policies, practices, and measures should be reviewed and integrated into a coherent system of security. Requirements of security management are a function of the role of participants, risks involved, and system requirements.
9. **Reassessment. Participants should review and reassess the security of information systems and networks, and make appropriate modifications to security policies, practices, measures, and procedures.** New threats and vulnerabilities continuously emerge and become known. It is the responsibility of all organizational members to review and reassess controls, and assess how these address the emergent risks.

OECD Principles for Security Culture • 269

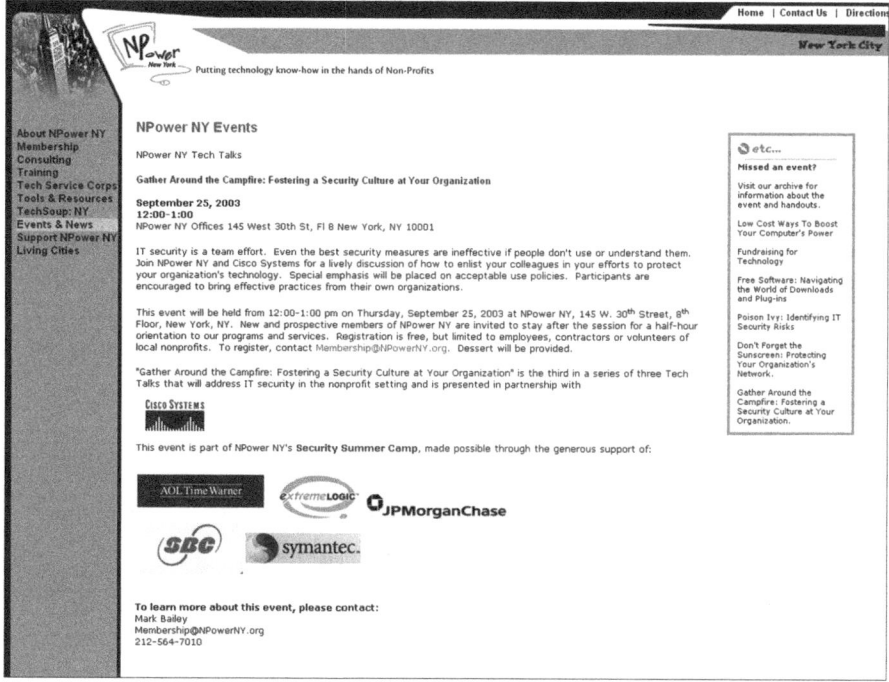

**Figure 10.7.** A call for an informal security culture developing campfire

# Concluding Remarks

In this chapter we have introduced the concept of security culture and have emphasized the importance of understanding different kinds of culture. The relationship of culture to IS security management has also been presented. Ten cultural streams were introduced. These cultural streams come together to manifest four classes of culture:

1. Clan culture
2. Hierarchy culture
3. Market culture
4. Adhocracy culture

The four culture types may coexist, but one type may be more dominant than the other. This would depend on the nature and scope of the organization. The four culture types also form the basis for assessing the real and idea security culture that might exist at any given time.

### In Brief

- Computer systems do not become vulnerable only because adequate technical controls have not been implemented, but because there is discordance between the organizational vision, its policy, the formal systems, and the technical structures.
- Security culture is the totality of patterns of behavior in an organization that contribute to the protection of information of all kinds.
- The prevalence of a security culture acts as a glue that binds together the actions of different stakeholders in an organization.
- Security failures, more often than not, can be explained by the breakdown of communications.
- Culture is shared and facilitates mutual understanding, but can only be understood in terms of many subtle and silent messages.
- Culture is concerned more with messages than with the manner in which they are controlled. E. T. Hall identifies 10 streams of culture and argues that these interact with each other to afford patterns of behavior.
- The 10 streams of culture are interaction, association, subsistence, bisexuality, territoriality, temporality, learning, play, defense, and exploitation.
- Consequently, culture is concerned more with messages than with the manner in which they are controlled. Hall identifies 10 streams of culture and argues that these interact with each other to afford patterns of behavior.
  - The 10 cultural streams manifest themselves in four kinds of culture. The four classes are adhocracy culture, hierarchy culture, clan culture, and market culture.

## Questions and Exercises

**Discussion questions.** These are based on a few topics from the chapter and are intentionally designed for a difference of opinion. These questions can best be used in a classroom or a seminar setting.

1. In an address to a ballroom full of the nation's top chief information officers today, Gail Thackeray, Special Counsel on Technology Crimes for the Arizona Attorney General's Office, said CIOs need to "do a better job" protecting the country's infrastructure.[8] Thackeray advised, "If you want to protect your stuff from people who do not share your values, you need to do a better job. You need to do it in combination with law enforcement around the country.... You need better ways to communicate with the industry and law enforcement. It is only going to get worse." Comment.

2. People who tend to pose the greatest IS security risks are those who have low self-esteem and strongly desire the approval of their peers. People who lay more emphasis on associations and friendships relative to maintaining the organization's value system can cause serious damage to the security. Discuss.

---

8 Extracted in part from *CIO* magazine, October 17, 2000.

**Exercise.** Undertake research to find how security culture is developed and maintained in non-IT-based environments. How can lessons from these implementations be used for developing and sustaining IS security culture?

See Appendix A (after the Index) for a set of Short Questions for this chapter.

## Case Study: The T-Mobile Hack

Recently a hacker was able to access the wireless carrier T-Mobile's network and read the private emails and personal files of hundreds of customers. The government revealed that emails and other sensitive communications from the Secret Service agent investigating the incident were among those obtained by the hacker. The agent, Peter Caviccha, was apparently using his own personal handheld computer to view emails while travelling. His supervisors were apparently aware of this, since they frequently directed emails to his personal email address while he was travelling. The agency had a policy in place that disallowed agents from storing work-related files on personal computers. The hacker was able to view the names and Social Security numbers of 400 T-Mobile customers, and they were notified in writing regarding the security breach. T-Mobile has more than 63 million customers in the United States. The network attack occurred over a seven-month period and was discovered during a broad investigation by the Secret Service while studying several underground hacker organizations. Nicolas Lee Jacobson, a computer engineer from Santa Anna, California, was charged in the cyberattack after investigators traced his online activities to a hotel where he was staying. He was later arrested and then been released on a $25,000 bond.

1. What role did the culture displayed by Caviccha and his supervisors play in the case?
2. Were additional charges warranted against the agent if agency rules were violated?
3. Which type of culture, hierarchical or hybrid, was displayed, and did the type of culture play a role in charges against the agent?
4. What additional security measures are required to prevent eavesdropping on wireless communications?

## References

1. Baskerville, R. 1993. Information systems security design methods: Implications for information systems development. *ACM Computing Surveys*, 25(4): 375–414.
2. Braithwaite, J. 1989. *Crime, shame and reintegration*. Cambridge, UK: Cambridge University Press.
3. Cameron, K.S., and R.E. Quinn. 1999. *Diagnosing and changing organizational culture: Based on the competing values framework*. Reading, MA: Addison-Wesley.
4. Eco, U. 1976. *A theory of semiotics*. Bloomington: University of Indiana Press.
5. Gordon, G.G., and N. DiTomaso. 1992. Predicting corporate performance from organizational culture. *Journal of Management Studies*, 29: 783–799.
6. Hall, E.T. 1959. *The silent language*, 2nd ed. New York: Anchor Books.

7. Kitiyadisai, K. 1991. *Relevance and information systems, in London School of Economics*. London: University of London.
8. Kotter, J.P., and J.L. Heskett. 1992. *Corporate culture and performance*. New York: Free Press.
9. Sakamoto, A. 1994. Video game use and the development of socio-cognitive abilities in children: Three Surveys of elementary school students. *Journal of Applied Social Psychology*, 24(1): 21–42.
10. Schein, E. 1999. *The corporate culture survival guide*. San Francisco: Jossey-Bass.
11. Thomson, K.-L. 2010. Information security conscience: A precondition to an information security culture? *Journal of Information System Security*, 6(4): 3–19.
12. Thomson, K.-L., and R. von Solms. 2005. Information security obedience: A definition. *Computers & Security*, 24(1): 69–75.
13. Vroom, C., and R.V. Solms. 2004. Towards information security behavioural compliance. *Computers & Security*, 23(3): 191–198.
14. White, G.B., G. Dietrich, and T. Goles. 2004. Cyber-security exercises: Testing an organization's ability to prevent, detect, and respond to cyber-security events. Presented at the 37th Hawaii International Conference on System Sciences, Hawaii.
15. Yeung, A.K.O., J.W. Brockbank, and D.O. Ulrich. 1991. Organizational culture and human resource practices: An empirical assessment. *Research in Organizational Change and Development*, 5: 59–81.
16. Zainudin, D., and A. Ur-Rahman. 2015. The impact of the leadership role on human failures in the face of cyber-threats. *Journal of Information System Security*, **11**(2): 89–109.

# CHAPTER 11

# Ethical and Professional Issues in IS Security Management

---

> *I think computer viruses should count as life. I think it says something about human nature that the only form of life we have created so far is purely destructive. We've created life in our own image.*
>
> —Stephen Hawking

Joe Dawson had been on a journey. A journey to understand various aspects of cybersecurity and learn how some of the principles could be applied to his organization. One thing was sure. Managing security in an organization was far more complex than he had previously thought. While understanding and trying to manage cybersecurity, Joe Dawson was facing another problem. There was increased pressure to "make in the USA." The steel industry was particularly affected following Donald Trump's blasting of GM when he said, "General Motors is sending a Mexican-made model of the Chevy Cruze to US car dealers—tax-free across the border. Make in the USA or pay big border tax!"[1]

For Joe, it was easier said than done. The US steel industry, while a backbone of American manufacturing, was hurting by an unprecedented surge in cheap subsidized imports. While Joe sat at his coffee table thumbing through the recent edition of *The Economist*, an article that caught his attention was titled "How to manage the computer security threat." As Joe read through the article, a particular quote caught his eye, which resonated with his experiences in the steel industry: "One reason computer security is so bad today is that few people were taking it seriously yesterday."[2]

This was so very true. It was an ethical dilemma of sorts. And the steel industry was in the state it was because of complacency in the past. Joe, however, had more pressing things

---

1 https://www.cnbc.com/2017/01/03/trump-blasts-general-motors-make-chevy-cruze-model-in-us-or-pay-big-border-tax.html. Accessed September 24, 2017.
2 https://www.economist.com/news/leaders/21720279-incentives-software-firms-take-security-seriously-are-too-weak-how-manage. Accessed September 24, 2017.

to address. His network manager had noticed some activity of sites that were not approved. Some employees were visiting dating sites and had downloaded some apps. He had to focus a different set of ethical dilemmas, at least for now.

---

Ethics are the moral principles that guide individual behavior. In order for a sound cybersecurity strategy, the recognition of ethical principles is important. If the ethical standards are unclear, cybersecurity professionals are indistinguishable from the black-hat criminals. There is no simple way to ensure a suitable ethical environment. The study of cybersecurity ethics is complex, and there are a number of approaches and schools of thought. The cybersecurity landscape evolves every few years. While it represents a booming industry, there is a shortfall of qualified skilled graduates. And organizations are desperate to fill in the job openings.

A Forbes[3] survey reports that the cybersecurity market is expected to grow from $75 billion in 2015 to $170 billion by 2010. It is estimated that there are 209,000 cybersecurity jobs in the United States that have remained unfilled. The figure represents a 74% increase in posting since 2010. Globally, Cisco estimates that there are nearly one million cybersecurity job openings. And it is expected to rise to six million by 2019. The increased demand for cybersecurity professionals does not come without issues. One of the biggest challenges is in finding suitable candidates to fill the positions.

In any recruitment process, there are typically three elements that need consideration:

1. *Qualifications* can be judged from courses taken and degrees and diploma's completed.
2. *Skills* are usually evaluated based on credentials and certifications that an individual might hold.
3. *Experience* can also be evaluated, based on prior projects completed.

The most challenging aspect in recruiting a cybersecurity professional, however, is the person's ethical stance. A company may get an individual who is well-qualified and has the requisite skill set, but may be unethical. Getting an individual to adhere to a code of conduct may be easy, but the proper enculturation of ethical practices takes time.

## Coping with Unemployment and Underemployment

One of the emergent challenges faced by businesses is that of globalization, which has resulted in a skewed job market. There is no doubt that certain kinds of jobs have been created. But, there are are a large number of professions, were technology has resulted in job losses. The United States, for instance, has lost nearly 5.6 million manufacturing jobs between 2000 and 2010. While outsourcing and offshoring may have played a role, a majority of these job losses can be attributed to technological change and automation. Estimates

---

3 https://www.forbes.com/sites/stevemorgan/2016/01/02/one-million-cybersecurity-job-openings-in-2016/#70b00d3827ea

suggest that Chinese imports account for up to 2.4 million lost American jobs.[4] These changes in the nature of work have two distinct consequences:

1. Increased unemployment
2. Increased technological reliance

Both increased unemployment and technological reliance significantly contribute to cybersecurity threats. In related research on psychological well-being following unemployment, researchers have argued that unemployment causes tension, rancor, and conflict among individuals and the communities they live in. The reduced sense of belonging causes people to feel alienated [1]. The "I don't care" attitude is detrimental to the ethical use of technology and the protection of information resources of a firm. The consequences of unemployment and underemployment are known to motivate individuals to circumvent controls and find shortcuts for monetary gain. The solution to economic consequences of unemployment and underemployment are not as straightforward as increasing opportunities for employment. Rather, these problems require a multipronged approach, including establishing reeducation and retraining opportunities.

Increased technological reliance is not without problems. Robotic systems are inherently complex. Security researchers have long argued that when technological complexity is high [2] [3], there is an increased need to design security into the systems. This is particularly important when dealing with teleoperated systems. In robotic systems, it is possible to orchestrate an attack that exploits intermediate network nodes, which disrupts operations from a distance. In most cases such exploits manipulate the authorization process between an operator and a robot. In the security literature such attacks have typically been referred to as "man in the middle" attacks. Such attacks allow the adversary to gain control of the "reach-back" process. The remote robot and the operator are lead to believe that they are talking to each other securely. Attacks on complex intertwined technologies are pretty dangerous, since the adversary becomes part of the exploitation process. There are four classes of attacks that are possible:

1. Replay attacks—Where an adversary, on receiving a message, replays it to the recipient
2. Message dropping—Where an adversary drops messages received
3. Delay and drop attack—Where an adversary randomly delays or drops messages from human operators
4. Message spoofing—Where an adversary modifies or creates a false message, which is sent to a human operator

## Intellectual Property and Crime

Intellectual property is an area of ethics that has received a lot of attention recently, with good reason. There have been known cases where intellectual property theft has occurred because of outsourcing. In the early 1990s, when many businesses jumped onto the bandwagon of

---

4 https://www.ft.com/content/dec677c0-b7e6-11e6-ba85-95d1533d9a62

outsourcing, there were many cases where customer data was either stolen intentionally or was lost because of broken business processes.

The unethical use of IS and IT is pervasive. It also seems to be growing at least as rapidly as the technology. In the United States, most states have laws against software piracy and unauthorized system access. Similar laws exist in other countries as well. Nevertheless, many users pay little or no attention to these laws. In the 1990s, much was discussed and talked about with respect to software piracy. As a matter of fact, in a 1997 estimate, the Software Publishers Association (SPA) estimated that global revenue losses due to piracy in the business application software market approached $11.4 billion worldwide. Studies such as those by Moore and Dhillon [4], who examined the extent of software piracy in Hong Kong, serve to lend credence to the belief that the violation of intellectual property laws occurs all over the world on an unprecedented scale. The problem of software piracy has, however, evolved. There was a time when pirated software was sold on CDs in shady joints around the world. Today much of the piracy takes place through peer to peer networks, making it extremely difficult to detect. While many do not acknowledge the ill effects of something like software piracy (or movie piracy), the illicit sale is generating millions of dollars in sales, which is contributing to the underground dark markets. It is a known fact that the Asian software piracy market is deeply involved with the Japanese Yakuza. There are also reports that the drug trade, originating from Afghanistan, has links to underground markets and money laundering.

With advances in IT, today money-laundering activities have also become sophisticated. While estimating the amount of money laundering is problematic, because it is a concealed activity, some researchers have put forward efforts to quantify money laundering. Schneider and Windischbauer estimate that money-laundering turnover on a global scale was $595 billion in 2001 and rose to $790 billion in 2006 [5]. Agarawal and Agarawal state that the amount of globally laundered money in 2005 was more than $2.0–2.5 trillion, which was about 5–6% of the world GDP [6]; this increased dramatically between 2005 to 2009, by about 33% [7]. A 2015 report notes that an estimated $2 trillion in illicit capital finds its way into various financial institutions [8].

In recent years, crypto currencies have redefined the landscape for money laundering. This is because of lack of regulation and the democratized trust enabled through block chain technologies. Peer-to-peer currency systems are fast evolving and offer an alternative to the more traditional Fiat money, which is backed by nation states. Yet, because of the distributed nature of the currencies, there are concerns that money launderers use them systematically.

## IT and White Collar Crime

Money laundering is a form of white-collar crime. However, in the literature, the definition of white-collar crimes has remained elusive. This is because new revelations and emergent challenges have surpassed traditional boundaries. The concept has often been related to abuse of power, drug trafficking, fraud, exploitation, concealment, and financial and psychological damage. And it has become difficult to demarcate mafia-like organized crimes from political and economic corruption. This has resulted in spillover between the "good" and the "bad" guys.

The Bank of Credit and Commerce International (BCCI) epitomized how white-collar crime and money laundering could be carried about publically without any significant intervention from the regulators. BCCI was founded in 1972. It was registered in Luxembourg and had headquarters in London and Karachi. It operated in 78 countries and had assets in excess of $20 billion. In 1991 regulators from seven countries finally raided and locked down the offices. Investigators found that BCCI was deliberately founded to avoid centralized regulatory review. BCCI was known to have laundered money for people like Saddam Hussein and Manuel Noriega, among others.

The criminology literature has largely tried to understand white-collar crime by abstracting the individual from the context and labeling such people as "deviant." Such individuals have been characterized as greedy who typically lack self-control. An exclusive focus on individuals unfortunately draws attention away from institutional and normative structures that provide an opportunity. The prospect may emerge from the fallacies of the modern economic system or the manner in which the monetary system has evolved, or the innovations abetted by modern technologies.

Fiat money invokes a level of trust between the individual and the banking system. In many countries, currency bills note statements like the following (taken from Hong Kong dollar currency bills): "The Hong Kong and Shanghai Banking Corporation Limited promises to the pay the bearer on demand at its offices here Twenty Hong Kong Dollars." Such a demand could be made because the currency was linked to either gold or silver. In the US the currency bills note something to this effect: "This note is legal tender for all debts, public and private." Today, however, the gold standard does not exist, and most exchanges are based on the trust we place in the current banking systems and nation states. The emergence of democratized trust orchestrated by bitcoin and block chain technologies challenges the Fiat currencies. Yet the lack of regulation poses interesting challenges in terms of white-collar crime, money laundering, and the role of centralized clearance houses. It can be argued that much of the increased incidents of cybercrime, white collar crime, and money laundering are a result of the changing nature of capitalism, where increasingly profits are generated from speculation, insider trading, and circumventing exchange controls. Coupled with this is the general erosion of moral values—the Enron fiasco was a case in point where a majority of the executives were business graduates from elite schools.

# Managing Ethical Behavior

There is no doubt that with each and every advancing technology, there are corresponding opportunities for illegal use and unethical behavior. So what are some tangible ways in which ethical behavior can be inculcated? We discuss several options in the following sections.

## Monitoring for and Preventing Deviant Behavior

Since companies want to avoid deviant behavior from their employees, it is in their best interests to reinforce positive beliefs and attitudes, while at the same time modifying undesirable ones. However, before a business can begin to reinforce these positive aspects in its employees, it must first determine which behaviors and actions are considered desirable. The business can then consider what kinds of beliefs and attitudes will help achieve the specified behavioral outcomes. At the same time, it would also be useful to determine the range of undesirable behaviors. This will allow the company to take proactive steps to reduce or eliminate certain deviant beliefs and attitudes before they can lead to adverse consequences. Such an orientation would allow a company to develop adequate methods and procedures for promoting certain kinds of behavior and for monitoring negative changes in employee attitude.

Monitoring employee behavior is important for determining how well its workers are conforming to the desired "code of conduct," and to deal with any deviations before they become serious. A company, however, should not be content with fixing problems as they occur. There should also be an emphasis to prevent any problems from occurring in the first place. To a large extent, this can be accomplished by instilling a company's values and beliefs in its employees. Businesses and other organizations have a duty to themselves, and to their shareholders, to make absolutely certain that all of their employees fully understand the organization's goals and objectives. Achieving this should begin with the development and publicizing of such things as "mission statements" and "company credos."

## Communicating Appropriate Behavior and Attitudes

Many companies conduct periodic (e.g., annual) briefings for their employees on such topics as sexual harassment, personal responsibility, and employee excellence. Companies use such opportunities to try and communicate an organization's attitude on various subjects. Similarly, such seminars could also be used as a platform to communicate about appropriate and expected behavior in different situations. This may be accomplished through lectures, case examples, or by working test cases in a team environment. Such briefings and classes demonstrate to their employees how they are supposed to act, and that there may be consequences for not following the given norms of behavior. Certain companies have been proactive in conducting such training sessions, albeit following a fraud or some illegal activity. For example, some years ago, six employees of Rockwell International were accused of illegal actions regarding charging to government contracts. Even though the employees were at fault, Rockwell was still held accountable for its employees' actions. As part of the settlement of the case, Rockwell agreed to have every single employee of the corporation attend annual "ethics training" sessions.

In order for such campaigns to be effective in influencing the behavior of employees, company management must sincerely demonstrate to the employees that they are not to engage in unethical business practices. Simply issuing a mission statement or a credo, without following it up with training and publicity to reinforce the message, will not be enough to prevent employee transgressions. Further, in order for the campaign to be effective, upper management must be highly visible in the conduct of the campaign. Management must make employees at all levels really accept (in other words, internalize) the code of conduct they want the employees to follow.

## Making the Employees Believe

The beliefs and attitudes conveyed by the company to its employees must also be consistent, and should be continuously communicated to the employees over a long period of time. Exposing the employees to the company's moral and ethical beliefs only sporadically is not enough. They will not internalize them, and are not likely to follow them. The same message should always be present, day in and day out, in the work environment. Management also needs to make certain that it does not allow its desire to succeed to take precedence over the company's own beliefs and attitudes regarding personal and business ethics.

Clearly ethical training and "wheel alignment" campaigns can have the desired effect on the bulk of the employees of an organization. Most of the employees can be endowed with a strong enough set of beliefs and morals to act as effective controls over the way they conduct themselves within a company. It can be safe to assume that most employees will perform their duties according to the company credos, and will not try to subvert controls by engaging in illegal activities. However, there will always be a small percentage of employees who remain immune to the ethical training and to the attitudes surrounding them. These people present a real threat to the company. Even when a system of controls is in place, a company can be severely damaged, or even destroyed, due to the unethical or illegal actions of a single employee. It is therefore important to devise methods to identify such individuals and keep tabs on them.

Some employees might also evade normative controls because of cultural idiosyncrasies. Employees steeped in certain cultures might find it extremely difficult to admit mistakes, and could resort to unethical or illegal activities to try and hide such mistakes. Employees from other cultures might put faith in personal relationships at the expense of company procedures, and, for example, ignore competitive bidding requirements in the awarding of contracts.

# Codes of Conduct

Various professional bodies have established well-formed codes of conduct, which guide members to perform cybersecurity tasks in an utmost professional manner. In the following sections, some of the major codes are presented.

## ACM Code of Ethics and Professional Practice[5]

Association for Computing Machinery (ACM) is a global organization, which brings together computing educators, researchers and professionals, who engage in a dialogue related to all aspects of computing. ACM has a well-established code of ethics and professional practice, which was established in 1999. A new task force is in place to update the code. It will be released in 2018. ACM suggests that software engineers should commit themselves to making the analysis, specification, design, development, testing and maintenance of software a beneficial and respected profession. In accordance with their commitment to the health, safety and welfare of the public, software engineers shall adhere to the following Eight Principles:

- PUBLIC—Software engineers shall act consistently with the public interest.
- CLIENT AND EMPLOYER—Software engineers shall act in a manner that is in the best interests of their client and employer, consistent with the public interest.

---
5 Source: http://www.acm.org/about/se-code#short

- PRODUCT—Software engineers shall ensure that their products and related modifications meet the highest professional standards possible.
- JUDGMENT—Software engineers shall maintain integrity and independence in their professional judgment.
- MANAGEMENT—Software engineering managers and leaders shall subscribe to, and promote an ethical approach to the management of software development and maintenance.
- PROFESSION—Software engineers shall advance the integrity and reputation of the profession, consistent with the public interest.
- COLLEAGUES—Software engineers shall be fair to, and supportive of their colleagues.
- SELF—Software engineers shall participate in lifelong learning regarding the practice of their profession and shall promote an ethical approach to the practice of the profession.

## SANS IT Code of Ethics[6]

SANS is largely an information security training organization. The SANS IT code of ethics was released in 2004. The code is divided into three parts.

**I will strive to know myself and be honest about my capability.**

1. I will strive for technical excellence in the IT profession by maintaining and enhancing my own knowledge and skills. I acknowledge that there are many free resources available on the Internet and affordable books and that the lack of my employer's training budget is not an excuse nor limits my ability to stay current in IT.
2. When possible I will demonstrate my performance capability with my skills via projects, leadership, and/or accredited educational programs and will encourage others to do so as well.
3. I will not hesitate to seek assistance or guidance when faced with a task beyond my abilities or experience. I will embrace other professionals' advice and learn from their experiences and mistakes. I will treat this as an opportunity to learn new techniques and approaches. When the situation arises that my assistance is called upon, I will respond willingly to share my knowledge with others.
4. I will strive to convey any knowledge (specialist or otherwise) that I have gained to others so everyone gains the benefit of each other's knowledge.
5. I will teach the willing and empower others with Industry Best Practices (IBP). I will offer my knowledge to show others how to become security professionals in their own right. I will strive to be perceived as and be an honest and trustworthy employee.
6. I will not advance private interests at the expense of end users, colleagues, or my employer.
7. I will not abuse my power. I will use my technical knowledge, user rights, and permissions only to fulfill my responsibilities to my employer.

---

6 Source: https://www.sans.org/security-resources/ethics

8. I will avoid and be alert to any circumstances or actions that might lead to conflicts of interest or the perception of conflicts of interest. If such circumstance occurs, I will notify my employer or business partners.
9. I will not steal property, time or resources.
10. I will reject bribery or kickbacks and will report such illegal activity.
11. I will report on the illegal activities of myself and others without respect to the punishments involved. I will not tolerate those who lie, steal, or cheat as a means of success in IT.

**I will conduct my business in a manner that assures the IT profession is considered one of integrity and professionalism.**

1. I will not injure others, their property, reputation, or employment by false or malicious action.
2. I will not use availability and access to information for personal gains through corporate espionage.
3. I distinguish between advocacy and engineering. I will not present analysis and opinion as fact.
4. I will adhere to Industry Best Practices (IBP) for system design, rollout, hardening and testing.
5. I am obligated to report all system vulnerabilities that might result in significant damage.
6. I respect intellectual property and will be careful to give credit for other's work. I will never steal or misuse copyrighted, patented material, trade secrets or any other intangible asset.
7. I will accurately document my setup procedures and any modifications I have done to equipment. This will ensure that others will be informed of procedures and changes I've made.

**I respect privacy and confidentiality.**

1. I respect the privacy of my co-workers' information. I will not peruse or examine their information including data, files, records, or network traffic except as defined by the appointed roles, the organization's acceptable use policy, as approved by Human Resources, and without the permission of the end user.
2. I will obtain permission before probing systems on a network for vulnerabilities.
3. I respect the right to confidentiality with my employers, clients, and users except as dictated by applicable law. I respect human dignity.
4. I treasure and will defend equality, justice and respect for others.
5. I will not participate in any form of discrimination, whether due to race, color, national origin, ancestry, sex, sexual orientation, gender/sexual identity or expression, marital status, creed, religion, age, disability, veteran's status, or political ideology.

## ISSA Code of Ethics[7]

Information Systems Security Association (ISSA) is an international community of cybersecurity leaders and is dedicated to advancing individual growth, managing technology risks and protecting critical information and infrastructure. The ISSA code of ethics covers the following aspects:

1. Perform all professional activities and duties in accordance with all applicable laws and the highest ethical principles;
2. Promote generally accepted information security current best practices and standards;
3. Maintain appropriate confidentiality of proprietary or otherwise sensitive information encountered in the course of professional activities;
4. Discharge professional responsibilities with diligence and honesty;
5. Refrain from any activities which might constitute a conflict of interest or otherwise damage the reputation of or is detrimental to employers, the information security profession, or the Association; and
6. Not intentionally injure or impugn the professional reputation or practice of colleagues, clients, or employers.

# Credentialing

Credentialing is a process that establishes qualifications of licensed professionals. The credentialing process assesses individual and organizational backgrounds to ensure legitimacy. While university-level education in specialized areas is a means to provide credentials to individuals, there is many who either cannot afford university-level education or enter professions through unconventional routes, who may require some level of credentialing. There are several credentialing organizations which ensure that requisite skill sets exist in the workforce. Many organizations depend upon the professional certifications to ascertain the proficiency level of an individual. Currently many organizations and institutions have made it mandatory for new cybersecurity hires to acquire some form of certification.

## (ISC)²

(ISC)2 is an international membership based organization that is best known for the Certified Information Systems Security Professional (CISSP) certification. (ISC)2 provides a host of certifications, some targeting the managerial cadre and others that are more technical. (ISC)2 certifications include the following:

- CISSP—Certified Information Systems Security Professional. This certification recognizes IS security leaders and their experience in designing, developing, and managing the security posture of a firm.
- CCSP—Certified Cloud Security Professional. This certification focuses on best practices in cloud security architecture, design and operations.

---

7 Source: http://www.issa.org/?page=CodeofEthics

- SSCP—Systems Security Certified Practitioner. This certification focuses on hands on technical skills to monitor and administer the IT infrastructure as per the security requirements.
- CAP—Certified Authorization Professional. This certification recognizes managers responsible for authorizing and maintaining information systems.
- CSSLP—Certified Secure Software Lifecycle Professional. This certification targets qualification of developers for building secure applications.
- CCFP—Certified Cyber-Forensics Professional. This certification focuses on techniques and procedures to support investigations.
- HCISPP—Health Care Information Security and Privacy Practitioner. This certification is targeted at health care information security and privacy professionals.
- CISSP concentrations—Including ISSAP (Information Systems Security Architecture Professional), ISSEP (Information Systems Security Engineering Professional), and ISSMP (Information Systems Security Management Professional).

The CISSP exam comprises 250 multiple choice questions and advanced innovative questions. The length of the exam is six hours. The exam tests applicants' abilities in eight domains:

1. Security and risk management
2. Asset security
3. Security engineering
4. Communications and network security
5. Identity and access management
6. Security assessment and testing
7. Security operations
8. Software development security

To appear for the exam, the applicant must have a minimum of 5 years of full-time experience in at least two of the eight domains. CISSP certification, along with the successful completion of the exam, requires the endorsement by a qualified third party (the applicant's employer, another CISSP, or another commissioned, licensed, certified professional) to ensure that the candidate meets the experience requirement. To retain the certification, the CISSP certification holders must earn a specified number of ongoing education credits every three years.

For individuals in an operational role, the SSCP certification is more useful. SSCP certification indicates proficiency of a practitioner to implement, monitor, and administer IT infrastructure according to the information security policies and procedures that establish confidentiality, integrity, and availability of information. It also confirms the technical ability of a practitioner to deal with security testing, intrusion detection/prevention, cryptography, incident response and recovery plan, malicious code countermeasures, incident response and recovery, and more.

## ISACA Certifications

The Information Systems Audit and Control Association (ISACA) offers five different certifications, each focusing on a specific domain:

1. CISA—Certified Information Systems Auditor, for those who audit, control, monitor, and assess an organization's information technology and business systems
2. CISM—Certified Information Security Manager, for those who design, build, and manage enterprise information security programs
3. CGEIT—Certified in the Governance of Enterprise IT, for those engaged in critical issues around governance and strategic alignment of IT to business needs
4. CRISC—Certified in Risk and Information Systems Control, for those who are involved in the management of enterprise risks
5. CSX and CSX-P—Cybersecurity Nexus certification, for those who want to demonstrate their skills and knowledge regarding standards

The flagship certification for the association is the CISA, which covers five domains:

1. IS auditing
2. IT management and governance
3. IS acquisition, development, and implementation
4. IS operations, maintenance, and service management.
5. Protection of information assets

The CISM certification is also popular and covers four practice areas:

1. **Information security governance.** Initiate and maintain a security governance framework and the processes to establish that the information security strategy and organizational goals are aligned with each other.
2. **Information risk management.** Ascertain and manage the risks to the information systems to a sustainable level based on the risk appetite (i.e., to achieve the organizational goals and objectives).
3. **Information security program development and management.** Initiate and maintain information security program to identify, manage, and protect the organization's assets, while aligning business goals and strategy and information security strategy.
4. **Information security incident management.** Project, indicate, and manage the ability to detect, analyze, react, and recover from information security incidents.

## GIAC

Global Information Assurance Certification (GIAC), founded in 1999, is another certification that authenticates the expertise of information security professionals. The objective of GIAC is to impart the guarantee that the certification holder has the knowledge and skills imperative for practitioners in significant areas of information and software security. GIAC certification addresses the variety of skill sets involved with broad-based security and entry-level information security essentials, along with higher level subject areas such as audit, hacker techniques, secure software and application coding, firewall and perimeter protection, intrusion detection, incident handling, forensics, and Windows and Unix operating system security.

GIAC certifications are distinct since they assess certain skills and knowledge areas, as opposed to general InfoSec knowledge. Regardless of the many different entry-level certifications are available, GIAC provides exclusive certifications covering advanced technical subject areas. GIAC certifications are effective for four years. To remain certified, the students must revise the recent course information and take the test every four years to remain certified. There are several different kinds of certifications, which are typically linked to the various SANS Institute programs. The available certifications are as follows:

1. GIAC Security Essentials (GSEC)
2. GIAC Certified Incident Handler (GCIH)
3. GIAC Certified Intrusion Analyst (GCIA)
4. GIAC Certified Forensic Analyst (GCFA)
5. GIAC Penetration Tester (GPEN)
6. GIAC Security Leadership (GSLC)
7. GIAC Web Application Penetration Tester (GWAPT)
8. GIAC Reverse Engineering Malware (GREM)
9. GIAC Systems and Network Auditor (GSNA)
10. GIAC Information Security Fundamentals (GISF)
11. GIAC Certified Windows Security Administrator (GCWN)
12. GIAC Exploit Researcher and Advanced Penetration Tester (GXPN)
13. GIAC Assessing and Auditing Wireless Networks (GAWN)
14. GIAC Certified UNIX Security Administrator (GCUX)
15. GIAC Mobile Device Security Analyst (GMOB)
16. GIAC Security Expert (GSE)
17. GIAC Python Coder (GPYC)
18. GIAC Advanced Smartphone Forensics (GASF)
19. GIAC Certified Project Manager (GCPM)
20. GIAC Law of Data Security and Investigations (GLEG)
21. GIAC Certified Web Application Defender (GWEB)

## Security+

Security+ is a CompTIA certification. This certification focuses on key principles for risk management and network security, making it a significant stepping stone for anyone seeking entry into an IT security career. Security+ demonstrates an individual's ability to protect computer networks. The CompTIA Security+ certification is approved by the US Department of Defense to accomplish the directive 8570.01-M prerequisite. It also meets the ISO 17024 standard.

The exam content originates from the contribution of subject matter experts and industry-wide survey feedback. The topics covered include operations security, network security, compliance, cryptography, identity management, access control and data, and application and host security. There are a maximum of 90 multiple choice and performance-based questions.

## In Brief

- It is important to inculcate ethical value systems in information systems security professionals.
- It is more important now than ever before for good ethics training. This is because of
  - Increased unemployment
  - Increased technological reliance
- There are different kinds of attacks that have become problematic. These are largely because of teleworking and the virtual nature of work. Some of these attacks include
  - Replay attacks
  - Message dropping
  - Delay and drop attacks
  - Message spoofing
- Ethical behavior can be inculcated through
  - Monitoring and preventing deviant behavior
  - Communicating appropriate behavior and attitudes
  - Making the employees believe in the mission of the organization
- Adhering to various codes of conduct is an important first step.
- Credentialing such as CISSP and others assist in certifying information systems security knowledge.

# Questions and Exercises

**Discussion questions.** These are based on a few topics from the chapter and are intentionally designed for a difference of opinion. These questions can best be used in a classroom or a seminar setting.

1. What are the differences between ethical and unethical hackers? Should ethical hacking be allowed?

2. If an employee of an organization is disgruntled, will increased technological surveillance help, or should some motivational techniques be used? Discuss the pros and cons of either of the approaches.

3. "CyberEthics is not definable, is not implementable, because it is not conscious; it involves not only our thinking, but also our feeling." Discuss.

**Exercise.** Imagine that you have recently graduated with a BS in information systems. While you did take a few cybersecurity courses in your undergraduate education, you do not have formal training in the subject matter. Sketch out a path for yourself such that you will be a successful security professional. What extra courses should you take? What certifications should you focus on? What path do you see your career taking over the next five to seven years?

See Appendix A (after the Index) for a set of Short Questions for this chapter.

# Case Study: The DoubleClick Case

Headquartered in New York City with more than 30 offices around the world, DoubleClick went public in February of 1998, offering 3.5 million shares of stock at $17.00 per share. Since that time, DoubleClick has embraced a strategy of acquisition and IT development that has led to leaps of commercial successes mired by significant controversy.

By 1999, DoubleClick had been taking aggressive moves to build their company and increase their ability to provide accurate and increasingly detailed information about online users. Essentially they wanted to increase their already strong competitive advantage. DoubleClick felt that they had the IT ability to continue this competitive edge. To implement this strategy, DoubleClick merged with Abacus Direct, the nation's largest cataloging company, which tracks the purchasing habits of catalog shoppers. Abacus's databases included more than 2 billion consumer catalog transactions with information on more than 88 million American households, essentially 90% of the United States.

It soon became evident that DoubleClick's intention was to merge the Abacus database, which specifically contained the name and address information on purchasers, and merge that information with DoubleClick's user databases to create a tremendous database of online user information, destroying the anonymity aspect of the Internet. It appears that DoubleClick had reversed their initial strategy of protection and anonymity of the user when DoubleClick acquired Abacus.

Before the acquisition of Abacus, DoubleClick made its money by targeting banner ads in less direct ways, such as checking the Internet addresses of people who visit participating sites. This way, people in their homes or offices would see different ads as they browsed from different sites, whether it was at work in a factory for General Motors, a machine shop in Ohio, or at school in Nevada. Every time someone clicked on a DoubleClick client ad, DoubleClick would add that fact to its databases, which were built from the cookies that it placed on users' hard drives. DoubleClick utilized the cookie technology to help them target the ads more precisely by understanding the habits of the user and compiling them into a user profile with a unique identification code.

However, now DoubleClick had undertaken a pursuit of a serious strategy to merge their coded habit tracking information with the identity information of Abacus's databases. DoubleClick's advertising network would now correlate the names and addresses of net shoppers of the Abacus Alliance database, made up of more than two billion consumer catalog transactions, with DoubleClick's consumer buying habits database, thus providing information about web customers that would allow marketers to send media ads to precisely the right targeted potential customer more efficiently. The idea was that DoubleClick would now be able to offer to its clients more directed and detailed information regarding the web surfers that hit their sites. That may not seem like that big of a deal. But once the pervasive nature of each of the companies is understood, the ramifications quickly come into view, and a massive and accurate database can visibly be developed.

The strategic decision resulted in a serious public backlash and quickly revealed that DoubleClick's actions were not performed ethically. DoubleClick had always viewed this new database information as simply an additional product that it would be able to offer its clients. Quickly they realized that the public saw it as much more. Several lawsuits were filed by private citizens against DoubleClick for the invasion of privacy, among other allegations. DoubleClick had been forced to defend themselves in these highly publicized cases. In addition, several independent privacy protection groups filed complaints with federal administrative organizations, such as the Federal Trade Commission (FTC), complaining of violations of federal privacy statutes.

Today DoubleClick is part of Google. It continues to provide digital marketing services. While the privacy landscape has evolved ever since DoubleClick came into being, the problem of consumer privacy and access to personal data continues.

1. How can a balance be established between consumer needs for privacy and company (business) needs to know their customers?
2. Is it ethical to collect user data and track their movement on the Internet?
3. What kind of codes of conduct should security managers at DoubleClick (and Google) should be familiar with?

# References

1. Ed, D, et al. 1999. Subjective well-being: Three decades of progress. *Psychological Bulletin*, 125(2): 276–302.
2. Kaufman, C., R. Perlman, and M. Speciner. 2002. *Network security: Private communication in a public world*. Chicago: Prentice Hall Press.
3. Baskerville, R. 1988. *Designing information systems security*. New York: John Wiley & Sons.
4. Moores, T., and G. Dhillon. 2000. Software piracy: A view from Hong Kong. *Communications of the ACM*, 43(12): 88–93.
5. Schneider, F., and U. Windischbauer. 2008. Money laundering: Some facts. *European Journal of Law and Economics*, 26(3): 387–404.
6. Agarwal, J., and A. Agarwal. 2004. International money laundering in the banking sector. *Finance India*, 18(2): 767.
7. Schneider, F., K. Raczkowski, and B. Mróz. 2015. Shadow economy and tax evasion in the EU *Journal of Money Laundering Control*, 18(1): 34–51.
8. Lam, B. 2015. How dirty money gets into banks. *The Atlantic*, February 23. https://www.theatlantic.com/business/archive/2015/02/how-dirty-money-gets-into-banks/385774/. Accessed September 24, 2017.

# PART IV
# REGULATORY ASPECTS OF INFORMATION SYSTEMS SECURITY

# CHAPTER 12

# Legal Aspects of Information System Security

*The prestige of government has undoubtedly been lowered considerably by the Prohibition law. For nothing is more destructive of respect for the government and the law of the land than passing laws which cannot be enforced. It is an open secret that the dangerous increase of crime in this country is closely connected with this.*

—Albert Einstein,
"My First Impression of the U.S.A.," 1921

One rather interesting challenge confronted Joe Dawson: What would happen if there were a security breach and his company had to resort to legal action? Were there any laws concerning this? Since his company had a global reach, how would the international aspect of prosecution work? Although Joe was familiar with the range of *cyberlaws* that had been enacted, he was really unsure of their reach and relevance. Popular press, for example, had reported that the state of Virginia was among the first states to enact an antispamming law. He also remembered reading that someone had actually been convicted as a consequence. What was the efficacy level, though? And how could theft of data be handled? What would happen if someone defaced his company website, or manipulated the data such that the consequent decisions were flawed? What would happen if authorities wanted to search someone's phone? *Clearly,* it seemed to Joe, *these were serious consequences.*

While the obvious solution that came to mind was that of training—and there is no doubt that training helps overcome many of the cybersecurity compliance issues—there was something more to managing security. One required a good understanding of the law. Joe was reminded of the San Bernardino shootings and the fallout between Apple and the FBI. The FBI has spent nearly 900,000 to break into the iPhone and at one time had insisted that Apple provide a backdoor. As Joe thought about the issues, he was flipping through the pages of an old issue of *Business Horizons*. What caught his eye was an article titled "Protecting corporate intellectual property: Legal and technical approaches." As Joe flipped through the pages, he read that the arguments between Apple and the FBI were serious. There were many concerns about security and privacy. Apple's Tim Cook was quoted as saying:

> The government is asking Apple to hack our own users and undermine decades of security advancements that protect our customers…the same engineers who built strong encryption into the iPhone to protect our users would, ironically, be ordered to weaken those protections and make our users less safe.
>
> While we believe the FBI's intentions are good, it would be wrong for the government to force us to build a backdoor into our products…and ultimately we fear that this demand would undermine the very freedoms and liberty our government is meant to protect.[1]

*This is a legal nightmare,* Joe thought. *And it requires some in-depth study.* What was also perplexing to Joe was the large number of laws. How did all of these come together? What aspects of security did each of the laws address? Joe was also a subscriber to *Information Security* magazine. As he scanned through his earlier issues, he found an old issue of the magazine. It identified seven legislations related to cybersecurity: the Computer Fraud and Abuse Act (1985; amended 1994, 1996, and 2001); the Computer Security Act (1987); the Health Insurance Portability and Accountability Act (1985); the Financial Services Modernization Act (a.k.a. GLBA) 1999; the USA Patriot Act (2001); the Sarbanes Oxley Act (2001); and the Federal Information Security Management Act (FISMA; 2002). Joe wondered if these were still relevant. At face value, he needed legal counsel to help him wade through all these acts and their implications.

---

Today we have widespread use of individual and networked computers in nearly every segment of our society. Examples of this widespread penetration of computers include the federal government, health care organizations/hospitals, and business entities ranging from small "mom and pop" shops to giant multinational corporations. All of these entities use computers (and servers) to store and maintain information with varying degrees of computer security and storage of confidential personal and business data. Many of these computers and servers are now accessible via the Internet or Local-Area-Networks by their users. Much of the data are also stored in the cloud.

Computers are vulnerable without the proper safeguards—software and hardware—and without the proper training of personnel to minimize the risk of improper disclosure of that data, not to mention theft of said data, for ill-gotten financial gain. The fact that many computers and servers can be accessed via the Internet increases the risk of theft and misuse of data by anyone with sufficient skills in accessing and bypassing security safeguards.

In the United States, Congress has mandated several pieces of legislation to help safeguard computers in order to combat the ever present security threat. The legislation is meant to provide safeguards and penalties for improper and/or illegal use of data stored within computers and servers. The International Court of Justice passed a ruling that one country's

---

1 https://www.apple.com/customer-letter/

territory cannot be used to carry out acts that harm another country. While this ruling was in the context of the *Corfu Channel* case, it did make a call for "cybersecurity due diligence norms" such that nations and companies establish a framework to deal with cybersecurity issues within their jurisdiction (see [4]). Three fundamental considerations form the basis of any cybersecurity legislation:

1. States should maintain an ability to protect consumers. Nations and companies should formulate a strategy to protect consumers by developing a method to notify if and when a breach occurs. In addition, states should maintain adequate standards and have an ability to enforce civil penalties in case there is a violation.
2. Effective sharing of information with states. All federal efforts encourage the private sector to share information on cyberthreats with the federal government. There should also be increased sharing of information among state and local authorities.
3. Support of the National Guard in cybersecurity. Efforts of the Army and Air National Guard to develop cybermission forces should be encouraged. The National Guard, in particular, is uniquely positioned to leverage private sector skills.

In the United States, there are three main federal cybersecurity legislations:

1. The 1996 Health Insurance Portability and Accountability Act (HIPAA)
2. The 1999 Gramm-Leach-Bliley Act
3. The 2002 Homeland Security Act, which included the Federal Information Security Management Act (FISMA)

The three laws mandate that health care, financial, and federal institutions adequately protect data and information. In recent years a number of other federal laws have been enacted. These include:

1. **Cybersecurity Information Sharing Act (CISA)**. The objective of this act is to enhance cybersecurity by sharing information. In particular, the law allows sharing Internet traffic information between the government and manufacturing companies.
2. **Cybersecurity Enhancement Act of 2014**. This act stresses public-private partnerships and helps enhance workforce development, education, public awareness, and preparedness.
3. **Federal Exchange Data Breach Notification Act of 2015**. This act requires a health insurance exchange to notify individuals if and when their personal information has been compromised.
4. **National Cybersecurity Protection Advancement Act of 2015**. This act amends the original Homeland Security Act of 2002. It now includes tribal governments, information sharing and analysis centers, and private entities as its non-federal representatives.

In the following sections a select number of cybersecurity-related laws and regulations are discussed.

## The Computer Fraud and Abuse Act (CFAA)

The first version of the CFAA was passed in 1984. This law was meant to protect classified information stored within federal government computers, as well as to protect financial records and credit information stored on government and financial institution computers from fraud. It originally protected computers used by government or in national defense. Congress broadened the CFAA in 1986 by making amendments that extended protection to "federal interest computers." Then, as computer technology and the Internet evolved, the CFAA was amended again in 1996, with the phrase "protected computer" replacing the previous concept of "federal interest computer." This increased the reach of the CFAA to include all computers involved in interstate and international commerce, whether or not the US government had a vested interest in a given computer or storage device.

The key elements of the CFAA are to provide protections and penalties for violating the law. The criminal penalties for violating the CFAA can range from 1 to 20 years in prison, and fines. The civil penalties also are severe.

According to the CFAA, the legal elements of computer fraud[2] consist of

1. Knowingly and with intent to defraud
2. Accessing a protected computer without authorization, or exceeding authorization
3. Thereby furthering fraud and obtaining anything of value (other than minimal computer time)

The first part means that the offender is aware of the natural consequences of his action (i.e., that someone will be defrauded). The second part refers to the act of the offender accessing a computer without authorization, or in a manner that exceeds what he is normally allowed/authorized to do. Finally, the third part refers to the purpose of the fraud (i.e., to take information that can be used for financial gain).

The implementation of the CFAA made it easier to prosecute complaints of theft of sensitive information (financial, military, legal, etc.) or passwords that allowed one access to sensitive information and commit fraud in the private sector, not just in the federal government. It also allowed plaintiffs to pursue actions against defendants in federal court, not just in state courts. In effect, this allowed a double-whammy against the defendant, and allowed the plaintiff to attempt to recover more in damages.

In an illustration of the impact of the CFAA, the case *Shurgard Storage Centers v. Safeguard Self Storage* demonstrated the increased scope of the "protected computer" concept to include the private sector (see [2]). In *Shurgard*, certain managers of a self-storage business (Shurgard Storage) left to work for a competitor (Safeguard). Prior to leaving and without

---

[2] Fraud, as defined in Gilbert's Law Dictionary, is "An act using deceit such as intentional distortion of the truth of misrepresentation or concealment of a material fact to gain an unfair advantage over another in order to secure something of value or deprive another of a right. Fraud is grounds for setting aside a transaction at the option of the party prejudiced by it or for recovery of damages."

notifying the former employer, these employees allegedly used the plaintiff's computers to send trade secrets to the defendant via email.

The defendants (the former managers of Shurguard and their new employer, Safeguard Self Storage) argued that they did not access computers "without authorization," since they were employees of the plaintiff at the time they allegedly emailed trade secrets. However, the court said that the "authorization" presumed to have been held by the employees of Shurguard ended as soon as they began sending the information to their new company. In effect, the moment they emailed proprietary information to Safeguard, they acted as an agent of Safeguard. Next, the defendants argued that they had not committed "fraud" because the plaintiff had been unable to show the traditional elements of common law fraud. However, the court held that allegations such as those described here stated a claim for redress under the CFAA. Adopting a very broad definition of fraud, *Shurgard* held that "wrongdoing" or "dishonest methods" qualified as fraud under the CFAA; proof of the elements of common law fraud is not required.

Further, *Shurgard* held that when an employee accesses a computer to provide trade secret information to her prospective employer, the employee is unauthorized within the meaning of the CFAA. The court found that "the authority of the plaintiff's former employees ended when they allegedly became agents of the defendant." As a result, the disloyal employee was in effect treated as a hacker—from and after the time he started acting as an agent for Safeguard.

Finally, under the statute, "damage" is defined as any "impairment to the integrity" of the computer data or information. In other words, the sanctity of the data stored on Shurgard's computer was violated—which is impairment to its integrity. *Shurgard* held that the employee's alleged unauthorized use of the employer's email to send trade secret information, in that case confidential business plans, qualified as "damage" under the statute.

The court found that as soon as the former managers accessed and sent the proprietary information—the impairment to its integrity—the act was an implicit revocation of the employee's authorization to access that information. In all likelihood, the odds of any employee's negotiating a contrary agreement into his or her employment agreement would seem very slight, as that would mean giving away proprietary information that was needed to ensure that the business functioned well not to mention survived in a competitive environment.

The court concluded that the extensive language in the legislative history demonstrated the broad meaning and intended scope of the terms "protected computer" and "without authorization." By giving broad interpretations to these phrases, the court in effect created an additional cause of action in favor of employers who may suffer the loss of trade secret information at the hands of disloyal employees who act in the interest of a competitor and future employer.

## The Computer Security Act (CSA)

After several years of hearings and debate, Congress passed the Computer Security Act of 1987. Motivation for the CSA was sparked by the escalating use of computer systems by the federal government and the requirement to ensure the security and privacy of unclassified, sensitive information in those systems. A broad range of federal agencies had assumed responsibility for various facets of computer security and privacy, prompting concerns that federal computer security policy lacked focus, unity, and consistency, and contributed to a duplication of effort. The purpose of the CSA was to standardize and tighten security controls on computers in use throughout the federal government, and those in use by federal contractors, as well as to train its workforce in maintaining appropriate security levels.

There were several issues that shaped debate over the Computer Security Act:

1. The role of the National Security Agency (NSA) versus the National Institute of Standards and Technology (NIST) in developing technical standards and guidelines for federal computer privacy and security. Congress balanced the influence of the NSA upon the federal government's security systems by giving NIST responsibility for developing standards and guidelines for civilian federal computer systems, drawing upon the technical advice and assistance from NSA.
2. The need for greater training of personnel involved in federal computer security.
3. The scope of the legislation in terms of defining a "federal computer system." The CSA defines a federal computer system not only as a "computer system operated by a federal agency," but also "operated by a contractor of a federal agency or other organization processing information (using a computer system) on behalf of the federal government to accomplish a federal function," such as state governments disbursing federal funds.[3]

The CSA requires the following:

1. The identification of systems that contain sensitive information, and the establishment of security plans by all operators of federal computer systems that contain sensitive information
2. Mandatory periodic training in computer security awareness and accepted computer security practices for all persons involved in management, use, or operation of federal computer systems that contain sensitive information
3. The National Institute of Standards and Technology (NIST) must establish a Computer Standards Program. The primary purpose of the program is to develop standards and guidelines to control loss and unauthorized modification or disclosure of sensitive information in systems and to prevent computer-related fraud and misuse.
4. The establishment of a Computer System Security and Privacy Advisory Board within the Department of Commerce. The duties of the board are
    a. To identify emerging managerial, technical, administrative, and physical safeguard issues relative to computer systems security and privacy
    b. To advise NIST and the Secretary of Commerce on security and privacy issues pertaining to federal computer systems

---

3 https://csrc.nist.gov/csrc/media/projects/ispab/documents/csa_87.txt

c. To report its findings to the Secretary of Commerce, the director of the Office of Management and Budget, the director of the National Security Agency, and the appropriate committees of Congress.

# Health Insurance Portability and Accountability Act (HIPPA)

In today's health care environment, whether it be patient, provider, broker, or third-party payer, personal health information can be accessed from multiple locations at any time from any of these integrated stakeholders. The spirit of HIPAA is to promote a better health care delivery system by broad and sweeping legislative measures. One way this can be accomplished is by the adoption of lower cost Internet and information technology. It is clear the Internet will probably be the platform of choice in the near future for processing health transactions and communicating information and data. Therefore IS security is of paramount importance to the future of any health care program.

Whether you are a large health care provider/insurance company or a small rural physician practice or benefits consulting firm, you will have to consider a security strategy for personal history information (PHI) to be in compliance with HIPAA. Otherwise your operation could be subjected to hefty fines and potential lawsuits.

## Requirements

In 1996, the Health Insurance Portability and Accountability Act (HIPAA PL 104-191) was passed with provisions subtitled "Administrative Simplification." The primary purpose of this act was to improve Medicare under title XVIII and XIX of the Social Security Act, as well as the efficiency and effectiveness of the health care system through the development of a health information system with established standards and requirements for the electronic transmission of health information. HIPAA is the first national regulation on medical privacy and is the most far-reaching federal legislation involving health information management affecting the use, release, and transmission of private medical data [1]. When health records are not protected, there can be several negative consequences:

- There can be unauthorized secondary use of medical records.
- Inaccuracies in medical records may not get corrected.
- Hackers can discover and disclose medical records.
- Employers could use medical records for personnel decisions.
- Revelation of medical records by insurance company employees, who may not be adequately trained.

HIPAA has important implications for all health care providers, payers, patients, and other stakeholders. Although the Administrative Simplification standards are lengthy and complex, the focus of this discussion will examine the following areas regarding PHI privacy and security:

- Standardization of electronic patient administrative and financial data
- Unique identifiers for providers, health plans, and employers
- Changers to most health care transaction and administrative information systems
- Privacy regulation and the confidentiality of patient information.
- Technical practices and procedures to insure data integrity, security, and availability of health care information.

HIPAA mandates a set of rules to be implemented by health providers, payers, and government benefit authorities, as well as pharmacy benefit managers, claims processors, and other transaction clearinghouses. It is important to note that HIPAA security and privacy requirements may be separate standards but they are closely linked. *Privacy* concerns what information is covered, and *security* is the mechanism to protect it. The privacy and proposed security standards of HIPAA can apply to any individual health information, whether it is oral or recorded in any form or medium. The information identifies the individual or can be used to identify the individual. This is a significant departure from the previous draft rules that covered only electronic information. As a much broader definition of the law, it will require a significant change in the way health information is handled, disseminated, communicated, and accessed.

## Compliance and Recommended Protection

The first place to consider is to examine PHI vulnerabilities and exposure by completing a business impact analysis and a risk assessment to determine compliance with HIPAA. This should include the following:

1. **Baseline assessment.** The baseline assessment inventories an organization's current security environment with respect to policies, processes, and technology. This should include a thorough assessment of information systems that store, transact, or process patient data.
2. **Gap analysis.** The goal of the gap analysis is to compare the current environment with the proposed regulatory one in terms of the level of readiness and to determine whether "gaps" exist and, if so, how large they are.
3. **Risk assessment.** The risk assessment should address the areas identified in the gap analysis requiring remediation. A risk assessment should provide an analysis of both likely and unlikely scenarios in terms of probability of occurrence and their impact on the organization.

HIPAA does provide a "common sense" approach to implementing recommended and required security procedures. The list of tools and techniques to protect web applications include authentication, encryption, smart cards or secure identification cards, and digital signatures. Furthermore, HIPAA mandates that security standards must be applied to preserve health information confidentiality and privacy in four main areas:

- Administrative procedures (personnel procedures, etc.)
- Physical safeguards (e.g., locks, etc.)
- Technical security services (to protect data at rest)
- Technical security mechanisms (to protect data in transit)

The security standard mandates safeguards for physical storage and maintenance, transmission, and access to individual health information. The standard also requires safeguards such as encryption for Internet use, as well as security mechanisms to guard against unauthorized access to data transmitted over a network. An incident at the University of Washington Medical Center highlights the sensitivity as well as the vulnerability of health care data systems connected to the Internet to outside threats. A hacker called "Kane" managed to download admission records for 4,000 heart patients in June/July 2000. The hospital would have faced stiff penalties if HIPAA had been enforced. As one can imagine, the risks to a health care provider of inadequate computer security could include harm to a patient, liability of leaked information, loss of reputation and market share, and fostering of public mistrust of the technology. As a result of this breach of security, the University of Washington Medical Center recommended several precautionary steps to protect and secure PHI [3]:

1. **Risk Analysis:** Acknowledge potential vulnerabilities associated with both the internal or external process of storing, transmitting, handling, disseminating, communicating, and accessing PHI. Therefore each business unit should access potential vulnerabilities by:
   a. Identifying and documenting all electronic PHI repositories
   b. Periodically re-inventorying electronic PHI repositories
   c. Identifying the potential vulnerabilities to each repository
   d. Assigning a level of risk to each electronic PHI repository

   All repositories of electronic PHI will be identified and logged into a common catalogue, in the appropriate medium form used, with the appropriate level of file, system, and owner information.

   Some of the user/owner identifiers should include: Repository name, custodian name, custodian contact information, number of users that access the repository, number of records, system name, system IP address, system location, system manager and contact information. Further, each Business Unit should update its electronic PHI inventory at least annually to ensure that the electronic PHI catalogue is up to date and accurate.

2. **Risk Management:** Each Business Unit must implement security measures and safeguards for each electronic PHI repository sufficient to reduce risks and vulnerabilities to a reasonable and appropriate level. The level, complexity and cost of such security measures and safeguards must be commensurate with the risk classification of each such electronic PHI repository. For example, low risk electronic PHI repositories may be appropriately safeguarded by normal best-practice security measures in place such as user accounts, passwords and perimeter firewalls. Medium and high- risk EPHI repositories must be secured in accordance with HIPAA Security Policies #1–17.

3. **Sanctions for Noncompliance:** Unfortunately, WU experienced a serious breach in the security of electronic PHI repositories and had to adopt sanctions for noncompliance to prevent both lawsuits and fines.
4. **Information System Activity Review:** It is imperative that internal audit procedures must be implemented to regularly review records of information system activity, such as audit logs, access reports, and security incident tracking reports. This is to ensure that system activity for all systems classified as medium and high risk is appropriately monitored and reviewed. Each Business Unit should implement an internal audit procedure to regularly review records of system activity (examine audit logs, activity reports, or other mechanisms to document and manage system activity) every 90 days or less.
5. **HIPAA Compliance / Risk Management Officer:** Finally, all health care organizations should have a HIPAA compliance/Risk Management officer with proper training and credentials, such as Information Systems Security Professional (CISSP) and/or the Certified Information Systems Auditor (CISA). This person should works closely with the Information Systems personnel and management to ensure compliance, continuity, conformity and consistency with the protection of PHI privacy and security.

Most important is to develop a corporate culture that communicates with all levels of the organization's workforce. This involves writing periodic reminders, providing in-services, and orienting new hires to the intent of the policies. All of these training activities must be conducted using easily understood terms and examples. Secondly, all of these standards should be a part of an organization's overall security strategy and are critical from a risk mitigation standpoint. Finally, PHI security needs to have the full support and cooperation from the executive level of the organization.

## HIPAA: Is It a Help or Hindrance?

Increasingly sophisticated technology presents opportunities in advancing integrated health care. Clearly, automation and technology help improve the access and quality of care, while reducing administrative costs. Unfortunately, when PHI information is shared both internally and externally by multiple users, a health care organization must put safeguards in place to prevent a compromise to the security of PHI by a disgruntled employee or outside "hacker."

### Positive Aspects of HIPAA

There are many positive aspects that have come from the legislation of the act—the first being a standardization of identifiers that makes it possible to communicate effectively, efficiently, and consistently with regard to PHI. Whether it is the pharmacist, doctor, hospital, or insurance company, the standardization of electronic PHI data helps in the accessibility and dissemination of data needed to process claims and deliver health care effectively. Thus efficiencies have been gained, in this respect, as a result of HIPAA compliance.

A second benefit is that it has made the health care provider/insurance–related industry more cognizant of associated risks related to the storage, access, and retrieval of sensitive PHI. Doctor's offices, hospitals, and ancillary providers have had a primary focus on treating the patient at hand. Organization of medical data that included sensitive PHI is a necessary by-product of the paperwork that it generates. Prior to HIPAA, physical patient files were stored on the walls, halls, or periphery of the practice without much thought to exposure of sensitive PHI. Electronically stored PHI was handled in a similar haphazard manner. The mandatory HIPAA compliance for safe storage, retrieval, and transmission of physical and electronic PHI has led to a "best practice" standard for the responsibilities associated with PHI. Further, this increased awareness promotes a secure feeling for the patient that the provider/insurance company is making a conscious effort to protect the privacy of such sensitive PHI.

A third benefit is accountability through the use of monitoring and updating the security aspect of PHI. HIPAA demands an ongoing effort to make sure PHI privacy and security is maintained and protected. This can ensure that sensitive PHI will have a lesser chance of being compromised.

The final benefit is that of disaster planning. The tragedy of September 11, 2001, in conjunction with HIPAA mandates, have made all health care providers and associated industries acutely aware of business continuity in the event of disaster. A patient may need to be seen suddenly at a hospital in a disaster zone that requires specific PHI and patient data that has been stored on an electronic file. Having backup/recovery systems can help ensure the continuity and quality of health care delivery for any patient.

## Negative Aspects of HIPAA

HIPAA has some serious residual negative challenges as health care providers and insurance related industries become compliant to the Act. The first is cost. Since April 14, 2003, when the privacy rule of the Health Insurance Portability and Accountability Act took effect at the end of 2004, health care organizations have spent years and well over $17 billion dollars in an effort to comply with HIPAA. The additional cost of a security compliance officer in larger organizations, the cost related to training all employees, the cost of security, the cost of the physical facilities, and the cost of maintaining integrity of IT systems create a drain on cash flow and help decrease profitability.

Complications of interpretation and compliance are another negative aspect that the act imposes on the health care industry. Clearly, meeting HIPAA mandates is a complex and arduous task. The security standard was developed with the intent of remaining technologically neutral in order to facilitate adoption of the latest and most promising developments in evolving technology and to meet the needs of health care entities of different sizes and complexity. As previously stated, the Health Insurance Portability and Accountability Act was passed with provisions subtitled Administrative Simplification. It appears to be anything but simple. Instead, the standard is a compendium of security requirements that must be satisfied. The problem is how the law is applied from provider to provider in a compliant manner. Regardless of the difficulty, each provider must meet the basic requirements. A concern expressed by health care providers and administrators, besides cost, is how to address all or some of the standards, especially when compliance requirements are vague.

Fines and penalties are another still negative by-product of the act, experienced by those who do not comply. Entities that come under the purview of HIPAA include

- Any health care provider (hospitals, clinics, regional health services, individual medical practitioners) that uses an electronic means for their transactions
- All health care clearinghouses
- All health plans—including insurers, health maintenance organizations, Medicaid, Medicare prescription drug card sponsors, flexible spending accounts, public health authority (in addition to employers, schools, or universities that collect, store, or transmit electronic protected health information [ePHI]), or enroll employees or students in health plans
- All business associates for the previous entities (including private sector vendors and third-party administrators)

The cost of instituting a security program in health organizations is certainly an issue, but the fact that the regulation has forced various institutions to ensure security awareness training, risk assessment, and disaster recovery planning is more than beneficial.

Some have argued that HIPAA has resulted in loss of productivity. Insurance companies, prior to HIPAA, were for the most part fairly compliant, as proprietary safeguards under physical constraints protected much of the PHI kept in repositories. Doctors, dentists, and hospitals, on the other hand, had loose policies and procedures regarding the protection and security of PHI. This is primarily because these frontline health care providers are more concerned with treating the patient than the details and business of record-keeping. Many of these frontline health care providers are now spending more time and resources in the area of lost productivity regarding patient care delivery by trying to be HIPAA compliant.

In the era of managed care and thin financial margins, the competitiveness of providers will depend on the use of IT to streamline clinical and other business operations. Much of this will require the transmission of PHI through various communication mediums. Therefore it is crucial how this PHI is handled, disseminated, communicated, and accessed by health care organizations. The increased computerization of medical information requires surveillance of policies and procedures to protect the confidentiality of private medical data. Failure to develop, implement, audit, and document information security procedures could result in serious consequences, such as penalties and loss of reputations, market share, and patient trust. Some of the steps that health care organizations can take include

- Computers storing or displaying sensitive PHI records should automatically log off or lock up after use to prevent any unauthorized access.
- Organizations should establish policies for shredding documents and locking file cabinets.
- Health care providers should ensure practices that prevent eavesdropping.
- Health care organizations should have a designated privacy officer, who ensures that staff understand and follow HIPAA guidelines.

Finally, as to whether or not HIPAA is helpful or a hindrance, it comes as a mixed blessing. No matter how much effort we put into PHI security and protection, bad guys still break in. The only thing we can do is take precautionary measures that make compromising the security of PHI difficult. Developing a common protective culture and awareness goes a long way to ensure compliance and vigilance. Unfortunately, those involved with patient care delivery and related services must recognize that there will be additional costs, redirection of resources, and the loss of productivity in the protection and security of PHI. It is the cost of doing business.

# Sarbanes-Oxley Act (SOX)

The Public Company Accounting Reform and Investor Protection Act, which was sponsored in Congress by US Senator Paul Sarbanes and US Representative Michael Oxley, was signed into law July 30, 2002. The law is more commonly referred to as the Sarbanes-Oxley Act, or SOX for short. The law is aimed at strengthening the corporate governance of enterprise financial practices.

In late 2001, a rash of financial scandals at prominent US companies like Enron and Arthur Andersen, ImClone, Global Crossing, and others began to come to light. Although there were investigations, no official laws were passed by Congress. Then, during the summer of 2002, another wave of financial improprieties surfaced at other major companies like WorldCom and Adelphia. After this second blow, Congress rallied together to form a defense. Soon after, SOX was passed in response to these corporate scandals.

Most of the provisions of Sarbanes-Oxley apply only to United States domestic publicly traded corporations, nonpublic companies whose debt instruments are publicly traded, and foreign companies registered to do business in the United States. The Securities and Exchange Commission (SEC) administered act is composed of 11 titles that are designed to cover the areas discussed in the following sections.

## External Auditor Oversight and Standards

The act calls for the establishment of the Public Company Accounting Oversight Board (PCAOB) under the SEC. The board will oversee (i.e., investigate, discipline) accounting firms that audit public companies.

Auditing standards will also be set by the PCAOB that accounting firms must follow. One area that the standards must address is regulation of the capacity that accounting firms can serve a company. This area speaks to the fact that an accounting firm may provide other services (i.e., consulting, legal) to a company, which could potentially conflict with their interests in auditing the company.

## Internal Audit Committee Responsibility

The act establishes new standards that impacted companies' audit committees must adhere to. The new standards provide more responsibility to and regulations for the internal auditors. Among other things, the audit committees are responsible for approving auditing services, establishing audit policies and procedures, and working with external auditors.

Some standards are also designed to address the separation of interests of the audit committees from the board of directors. In other words, the act hopes to put in place measures that will ensure that the audit committees are not controlled by the top management of their organizations. Another area of concern that the audit committee must address is the strengthening of whistleblower protection through defined and executed policy and procedures. The increased protection is extended to any report of fraudulent actions and is not just limited to securities violations.

## Executive Management Accountability

The act establishes standards that require corporate management to certify the accuracy of company financial reports. Knowingly false certifications carry stiff criminal penalties for the executives. In addition, under the act, the SEC can prohibit executives from receiving bonus and/or benefits compensation if it deems that financial misconduct has occurred.

## Financial Disclosure Strengthening

The act increases the requirements that organizations must adhere to for financial disclosure. There must be full disclosure in financial reports of

- Off-balance sheet transactions and special purpose entities
- Financial results, if generally accepted accounting principle would have been used

Disclosure requirements for other areas are also provisioned, like legal insider trading and financial ethics code and adherence.

## Criminal Penalty

The act establishes (or reinforces) federal crimes for obstruction of justice and securities fraud. The penalties for violations carry high penalties (up to 20 or 25 years, depending on the category of a crime). In addition, maximum fines for some securities infractions increased up to 10 times. For some violations, the maximum fine can be as high as $25 million. Also, under SOX, criminal penalty can be pursued for management that persecutes employees that reported misconduct under the whistleblower protection.

It should be noted that the areas addressed by SOX are not to mandate business practices and policies. Instead, the act provides rules, regulations, and standards that businesses must comply with and which result in disclosure, documentation, and the storage of corporate documentation.

## IT-Specific Issues

Although Sarbanes-Oxley establishes rules and regulations for the financial domain of corporations, it inadvertently impacts the IT domain. IT can be greatly leveraged by an organization to comply with the requirements of the law

The titles and sections of the law will need to be scrutinized to determine what is important to the organization and, furthermore, how IT will enable the organization to be compliant with the specific sections. This scrutiny of the law will need to be translated into requirements for the IT domain of the organization.

Overall, some of the main themes of requirements that IT will be presented with are as follows:

- Analyze and potentially implement/integrate software packages on the market that assists with SOX compliance.

- Provide authentication of data through the use of data integrity controls.
- Capture and document detailed logging of data access and modifications.
- Secure data by means such as firewalls.
- Document and remediate IT application control structures and processes.
- Provide storage capacity for the retention of corporate data assets related to the law (i.e., email, audits, financial statements, internal investigations documentation).
- Provide recoverability of the archive.

Organizations had to be in compliance with Sarbanes-Oxley by November 15, 2004. The compliance date carried a major milestone, which is Section 404 compliance. This section requires that companies report the adequacy and effectiveness of their internal control structure and procedures in their annual reports. In order to meet this, IT really is under pressure to fulfill the requirements to document and remediate, if necessary, application controls, their risks, and deficiencies.

In terms of the impact to IT, companies will have to decide how best to work with the IT domain to accomplish implementation. While some of the changes that must occur within the organization to be compliant with Sarbanes-Oxley can be accomplished through process and procedure, it is almost impossible to believe that an avenue can be pursued that doesn't involve the IT domain.

# Federal Information Security Management Act (FISMA)

The Federal Information Security Management Act (FISMA) was passed in late 2002 as a requisite of the Department of Homeland Security. Among other things, the act mandates that federal organizations establish a framework that facilitates the effective management of security controls in their IT domain. FISMA applies to all federal agencies, as well as other organizations (i.e., contractors, governments) that utilize or have access to federal information systems.

Some have suggested that FISMA came about due to increased awareness of the need for protection of United States information and controls to ensure that the information is secure. Others suggest that the impetus for the act stemmed from the realization that the key to effective security of information assets does not come from purely technical means, but rather from effective management processes that focus on security throughout all stages of decision-making—from strategic planning to project implementation.

There are several facets to FISMA. A brief overview of some of the components is provided as follows.

## Security Program

The security program requires the chief information officer (CIO) of each federal agency to define and implement an information security program. Some of the aspects that the security program should include are

- A structure for detecting and reporting incidents
- A business continuity plan

- Defined and published security policies and procedures
- A risk assessment plan

## Reporting

At regular intervals, each impacted agency has to report its compliance to the requirements mandated by the law. This report has to include any security risks and deficiencies in the security policy, procedures, controls, and so on. Additionally, the agency must report a remediation plan that the agency plans to follow to overcome any high risks and deficiencies.

## Accountability Structure

The FISMA holds IT executives accountable for the management of a security policy. With the act, an accountability structure is defined. Some players in the structure are

- CIO—Responsible for establishing and managing the security program
- Inspector general—An independent auditor responsible for performing the required annual security assessments of agencies

## National Institute of Standards and Technology (NIST)

Under the act, the Office of Management and Budget (OMB) is responsible for the creation of policies, standards, and guidelines that each agency must adhere to in their information security program. The OMB selected the National Institute of Standards and Technology (NIST) to develop the key standards and guidelines that agencies must utilize. Some topics that NIST must cover are

- Standards that agencies use to categorize information and information systems
- Security requirements, by category, that agencies should implement or address in their security program
- Standards for incident management
- A methodology for assessing the current state of security policies and procedures

## Categorization of Federal Information and Information Systems

One of the main standards of NIST is that of categorization of federal systems. NIST's Federal Information Processing Standards (FIPS) Publication 199 is a key product in the implementation of the act because it establishes the standards used to categorize information. The appropriate level of information security controls that an agency must address in its security program are driven by the organization's categorization. Additionally, the categorization is used by inter/intra agencies to determine the level of security and sensitivity that should be applied when sharing information.

Overall, the mandates of the FISMA will help ensure that information security is ingrained into the overall practices (culture) of an agency. It also recognizes that federal agencies must address security in a cost-effective manner, so it tries to mandate the level of risk and security controls that should be established based on the classification of the information that the agency is responsible for.

FISMA implementation is mandatory for federal agencies and public/private sector organizations that handle federal information assets. But the concepts that the act hopes to instill are applicable for any organization, public or private, regardless of whether or not they handle federal information assets. The management of security in all processes of the information domain—from strategy to actual implementation of projects and products—is prudent.

## Concluding Remarks

In conclusion, there are always security threats and risks facing the information systems of organizations. Some organizations are proactive in their establishment of cybersecurity protections. However, a lot of the cybersecurity concerns that are being instituted in organizations are due to the mandates of the government through legislative acts. Hopefully, through compliance to the cybersecurity laws, organizations will have information system security controls and measures ingrained throughout their organizations.

Legislative controls come into being when the nation state feels that there is a need to protect citizens from potential harm, or when there is lack of self-regulation. Clearly any enacted law imposes a set of controls, which in many cases might be a hindrance to the daily workings of people. However, legal controls are mandatory and have to be complied with. It is prudent, therefore, to be aware of them and their reach. In this chapter we have largely focused on US-based laws, but note that many other countries have laws that are rather similar in intent.

> **In Brief**
> - In the United States there are various laws that govern the protection of information.
> - Besides various smaller pieces of legislation, the following have a major security orientation:
>   - The Computer Fraud and Abuse Act (CFAA)
>   - The Computer Security Act (CSA)
>   - The Health Insurance Portability and Accountability Act (HIPPA)
>   - The Sarbanes-Oxley Act (SOX)
>   - The Federal Information Security Management Act (FISMA)
> - These laws are not all inclusive in terms of ensuring protection of information. Commonsense prevails in ensuring IS security.

## Questions and Exercises

**Discussion questions.** These are based on a few topics from the chapter and are intentionally designed for a difference of opinion. These questions can best be used in a classroom or a seminar setting.

1. A February 13, 2017, issue of the *USA Today* reported the following: "An estimated 15.4 million consumers were hit with some kind of ID theft last year. 2016 will be remembered as a banner year for fraudsters, as numerous measures of identity fraud reached new heights. Fraud losses totaled $16 billion. About 1 in every 16 U.S. adults were victims of ID theft last year (6.15%)—and the incidence rate jumped some 16% year over this year. This despite 2016 being the first full year that brick-and-mortar retailers were forced to accept more secure EMV chip cards or face liability consequences."[4] Comment on the efficacy of prevalent laws in ensuring confidentiality of personal information.
2. Consider HIPPA and SOX as two cases in point. Consider aspects of each law and comment on the extent to which the laws demand extraordinary measure as opposed to regular good management.

**Exercise.** Given the discussion of various North American laws in this chapter, identify corresponding laws in the European Union. To what extent do these differ in terms of nature and scope. Discuss.

See Appendix A (after the Index) for a set of Short Questions for this chapter.

# Case Study: FTC versus Wyndham Worldwide Corporation

Let's consider the case of *Federal Trade Commission v. Wyndham Worldwide Corporation*, a civil suit brought in the District of Arizona by the Federal Trade Commission (FTC). The case relates to a cybersecurity breach at Wyndham. The FTC sued the hospitality company and three of its subsidiaries because of data breaches where millions of dollars of fraudulent charges on consumer credit and debit cards were incurred. To understand why the case matters quite a bit, we need to step back and understand the role of FTC.

The FTC has two grounds on which it can bring a civil lawsuit. One is an allegation of deception—in other words an argument that some consumer service organization (like, say, Wyndham Hotels) had made representations to the consuming public that were false. As you may imagine, allegations of that sort are often very fact-specific and tied to particular circumstances.

The second ground for FTC enforcement is a broader one—that a company has engaged in "unfair" business practices—in other words, that a company "caused or [is] likely to cause substantial injury to consumers that consumers cannot reasonably avoid themselves and that is not outweighed by countervailing benefits to consumers or competition."

The FTC suit against Wyndham is tied to a breach of Wydham's computer systems by a Russian criminal organization that allegedly resulted in more than $10 million in fraud losses. It seeks a permanent injunction, directing Wyndham to fix its cybersystems so that they are more secure and unspecified damages.

The suit asserts two grounds for FTC jurisdiction. It first alleges that Wyndham's privacy policy about how they will maintain the security of information about their customers is

---

4 https://www.usatoday.com/story/money/personalfinance/2017/02/06/identity-theft-hit-all-time-high-2016/97398548/

deceptive—in other words that Wyndham made cybersecurity promises it couldn't keep. The suit also alleges that systematically Wyndham's failure to provide adequate cybersecurity for the personally identifiable information of its customers is an unfair business practice.

This type of lawsuit by the FTC is not unusual. These legal theories have been the foundation, for example, of the FTC's investigation of Google, Twitter, and HTC, and its investigation of data breaches at large consumer companies like Heartland. In almost all of these cases, the FTC deploys some combination of the argument that a company has misled the public about the nature of its cybersecurity ("deception") or that it has failed to invest adequately in cybersecurity measures ("unfair practices"). Until now, all of these actions have resulted in out-of-court settlements, leaving the validity of the FTC's legal theories untested.

FTC's efforts are the only effective aspect of a federal program to compel the business community to adopt more stringent cybersecurity measures. While opinions are divided as to if the effects of FTC efforts are good or bad, it is indisputable that the outcome where companies are paying credence to the possibility of a lawsuit have increased. Since cybersecurity legislation is still to come in the future, and the administration's executive order remains in development. The FTC is the only effective game in town.

But now—in the Wyndham case—the FTC's authority is being questioned. As the *Wall Street Journal* reported, Wyndham is challenging the basic premise of the FTC's suit, arguing that consumer protection statutes cannot be stretched to cover cybersecurity issues. Wyndham has argued that the lawsuit exceeds the FTC's enforcement authority—a position supported by the Chamber of Commerce.

The principal evidence that the FTC may be acting beyond its authority is its own report from 2000, in which it asked Congress to expand its legal authority to consider security breaches as consumer-protection issues. Congress has never acted on that request, but the FTC has decided to proceed anyway. Indeed, as Wyndham notes, there are a host of more specific data-security laws already on the books (HIPAA; COPPA; Graham-Leach-Bliley; Fair Credit Reporting), suggesting that the FTC is acting beyond its remit as a regulatory authority.

Now, we can see why this is a significant matter. In the absence of comprehensive cybersecurity legislation and while we are waiting for the cybersecurity standards of the executive order to be developed, the only effective method for cybersecurity regulation by the government is to use the FTC's enforcement authority. If, in the end, it turns out that the FTC lacks the authority it has been asserting, then the government will be without any real authority to compel cybersecurity improvements. Some will see that as a victory, and others will see that as a defeat, but either way it will be quite important. (Note: The Third Circuit eventually decided the case in favor of the FTC.)

1. Comment on the authority and responsibility aspects of different legislations. What is the best way to give cybersecurity responsibility to an agency and yet have the authority to execute?
2. In situations like that of the FTC, what kind of regulations should be developed so as to oversee follow-through in cybersecurity cases?
3. As technology evolves, what should be done for the organizations to comply with the legislations?

Source: Drawn in part from https://www.lawfareblog.com/most-important-cybersecurity-case-youve-never-heard. Reproduced with permission from the author, Paul Rosenzweig. Accessed May 1, 2017.

## References

1. Baumer, D., J.B. Earp, and F.C. Payton. 2015. Privacy of medical records: IT implications of HIPAA, in *Ethics, computing and genomics: Moral controversies in computational genomics*, edited by H. Tavani, 137–152. Sudburt, MA: Jones & Bartlett Learning.
2. Jakopchek, K. 2014. Obtaining the right result: A novel interpretation of the Computer Fraud and Abuse Act that provides liability for insider theft without overbreadth. *Journal of Criminal Law & Criminology*, 104: 605.
3. Annonymous. 2004. *HIPAA Security Policy #2*. St. Louis: Washington University School of Medicine.
4. Shackelford, S.J, S. Russell, and A. Kuehn. 2017. Defining cybersecurity due diligence under international law: Lessons from the private sector, in *Ethics and policies for cyber operations*, edited by M. Taddeo and L. Glorioso, 115–137. Switzerland: Springer.

# CHAPTER 13

# Computer Forensics*

*Now a deduction is an argument in which, certain things being laid down, something other than these necessarily comes about through them....It is a dialectical deduction, it reasons from reputable opinions....Those opinions are reputable which are accepted by everyone or by the majority, or by the most notable and reputable of them. Again, deduction is contentious if it starts from opinions that seem to be reputable, but are not really such...for not every opinion that seems to be reputable actually is reputable. For none of the opinions which we call reputable show their character entirely on the surface.*

—Aristotle,
*Topics*

Joe Dawson was at a stage where SureSteel was doing well. The company had matured and so had its various offices in Asia and Eastern Europe. In his career as an entrepreneur, Joe had learnt how to avoid legal hassles. This did not mean that he would give in to anyone who would file a lawsuit against him, but he wanted everything as per the procedure so that in case things went wrong, he had the process clarity to deal with it. For instance, Joe had never deleted a single email that came into his mailbox. Of course, the emails piled up regularly. Not deleting emails gave Joe a sense of confidence that nobody could deny anything they had written to him about.

Now with the networked environment at SureSteel and the increased dependence of the company on IT, Joe was a little uneasy with the detail as to how things would transpire if someone penetrated the networks and stole some data. He knew that they had an intrusion detection system in place, but what would it do in terms of providing evidence to law enforcement officials?

While speaking with his friends and staff, one response he got from most people was that the state of affairs is in a "state of mess." To some extent Joe understood the reasons for this mess. The laws were evolving and there was very little in terms of precedence. Joe knew for sure that this area was problematic. He remembered once reading an article by Friedman and Bissinger, "Infojacking: Crimes on the Information Superhighway" [1]. The article had stated:

---

* This chapter was written by Currie Carter under the supervision of Professor Gurpreet Dhillon.

The first federal computer crime statute was the Computer Fraud and Abuse Act of 1984 (CFAA), 18 U.S.C. § 1030 (1994)....Only one indictment was ever made under the C.F.A.A. before it was amended in 1986....Under the C.F.A.A. today, it is a crime to knowingly access a federal interest computer without authorization to obtain certain defence, foreign relations, financial information, or atomic secrets. It is also a criminal offence to use a computer to commit fraud, to "trespass" on a computer, and to traffic in unauthorized passwords....In 1986, Congress also passed the Electronic Communications Privacy Act of 1986, 18 U.S.C.§§2510-20, §§2710-20 (1992), (ECPA). This updated the Federal Wiretap Act to apply to the illegal interception of electronic (i.e., computer) communications or the intentional, unauthorized access of electronically stored data....On October 25, 1994, Congress amended the ECPA by enacting the Communications Assistance for Law Enforcement Act (13) (CALEA). Other federal criminal statutes used to prosecute computer crimes include criminal copyright infringement, wire fraud statutes, the mail fraud statute, and the National Stolen Property Act.

Although the article was a few years old and some of the emergent issues had been dealt with, by and large the state of the law was not any better. There was no doubt in Joe's mind that he had to learn more. Joe searched for books that could help him learn more. Most of articles and chapters he found seemed to deal with technical "how to" issues. As someone heading SureSteel, his interests were more generic.

Finally, Joe got hold of a book on computer forensics: *Incident Response: Computer Forensics Toolkit* by Douglas Schweitzer (John Wiley, 2003). Joe set himself the task of reading the book.

## The Basics

This chapter is about "computer forensics," which is a new and evolving discipline in the arena of forensic sciences. The advent of this discipline has been produced by the popularization of computing within our culture and society.

The computer's uses are as varied as the individuals and the motivations of those individuals. One's computing activity mirrors one's relationship with society. Thus it is not surprising that person's who are prone to live within the laws of society use computers for events that are sanctioned by and benefit that society. Nor is it surprising that those persons that tend to evade or flout society's norms and laws perform computer-based activities that can be classed as antisocial and/or detrimental to society's fabric. They are detrimental because the behavior and its results tend to run roughshod over other individuals' values and rights. The long-term effect is that they erode the basis of society's existence, which is the ability of its members to trust one another.

One of society's basic rights and responsibilities is to protect itself, its fabric, and its members from egregious acts of others that threaten the foundation and stability of society as well as the benefits that society promises to its members. The subject of this chapter grows out of the tension that exists between those two groups of people and societies' responsibility both to itself and those two groups of people. Of necessity, computer forensics concerns itself more with the actions and deeds of that group that society describes as antisocial and whose actions are deemed by society as posing a threat to society and its law-abiding citizens.

Computer forensics is society's attempt to find and produce evidence of these antisocial computer-related behaviors in such a way that the suspect's rights are not trampled by the process of evidence collection and examination, and the suspect's standing within society is not damaged by the presentation of evidence that does not accurately represent the incident that it is purported to describe. Thus, and finally, computer forensics is a sociological attempt to balance society's need to protect itself and the rights of the individuals that are perceived as threatening society's survival and prosperity.

## Types and Scope of Crimes

US law defines two types of law: civil and criminal. Civil crimes are those committed against private individuals, be they persons or corporations. Examples of civil crime include stealing data from an individual's computer, or intrusion with the intent to damage a computer's functionality or to steal private data. Criminal acts are those committed against the state. Examples of criminal acts are acts such as treason, espionage, or terrorism. Criminal perpetrators commit both types of crime.

The virtual nature of electronic data storage and exchange provides a rich, vast arena in which those that are criminally inclined can function. This is primarily because electronic data storage and transmission no longer requires that the user to be physically present to the data they are accessing or observable to the gatekeeper of that data. These two factors permit a functional anonymity that makes identifying criminals very difficult. This virtual quality also makes it difficult to apprehend and bring the criminal to trial, since the perpetrator may live in one jurisdiction and the crime may be committed in another, without the criminal ever being physically present in the jurisdiction in which the crime was committed. Since physical apprehension is the only way to limit those persons that won't voluntarily limit themselves, thorough, competent computer forensics becomes a "must" within the chain of events that leads to apprehension, trial, and punishment.

The anonymity factor and the increasingly large amounts of information that are stored electronically, coupled with the fact that a larger and larger percentage of the total amount of data that is created is stored electronically, present a vast rich arena in which misdeeds can be perpetrated and in which finding the miscreants becomes infinitely more difficult.

Such losses are hard to measure because much cybercrime goes unreported. However, as reported by Stambaugh et al.: "The statistics and losses remain staggering....A recent report on cybercrime by the Center for Strategic and International Studies (CSIS) says, 'almost all Fortune 500 corporations have been penetrated electronically by cybercriminals. The Federal Bureau of Investigation (FBI) estimates that electronic crimes are running about $10 billion a year but only 17 percent of the companies victimized report these losses to law enforcement agencies'" [2].

Indeed, data are sketchy at best because the instruments for reporting and tracking cybercrime remain underdeveloped. Clearly there is a general porosity of our electronic infrastructure, and our current inability to understand the magnitude of the problem is great.

In general, this porosity mirrors the open quality of our society. Even when we attempt to secure or harden that infrastructure against intrusion and loss, the infrastructure remains fragile. The proof of this fragility lies in the successful number of mass attacks against the infrastructure. The success of viruses such as Melissa, Nimda, Blaster, and SoBig are vital reminders that much remains to be done to make the infrastructure resistant to attack. And these not infrequent occurrences corroborate the words of the 1997 report of the President's Commission on Critical Infrastructure Protection, which sums up the urgency of the situation: "We are convinced that our vulnerabilities are increasing steadily, that the means to exploit those weaknesses are readily available and that the costs associated with an effective attack continue to drop."[1]

It can be assumed from this information that the estimated $10 billion annual loss that is reported is a guesstimate and, as such, the figure could be much higher. The lack of accurate figures prevents one from quantifying the amount of loss. However, this anecdotal reporting coupled with the relatively frequent and debilitating viruses can only lead to the conclusion that the effects of cybercrime are greater than what can be documented.

## Lack of Uniform Law

The nature of law is such that it is primarily reactive rather than proactive. We tend not to pass laws until there is evidence that some activity needs to be restricted. In fact, it is almost impossible to enact effective law until the scope of the problem is revealed through real-life, everyday experience. Even when one has the experience in hand to inform law, almost always we find that statutory law is inadequate to produce as true a form of justice as society prefers and needs. Hence, even after statute law is enacted, we find we tend to modify that statute law through the process of case law. By definition, case law can only be decided based upon events that have occurred. Therefore most law and almost all effective law is reactionary. As such the attempt to reign in cybercriminals (as well as most other crimes committed within the realm of "new" fields that contain as yet undefined and undescribed potentialities, such as computing) is a process of catch-up after the fact.

The attempt to define, respond to, and discourage cybercrime; the attempt to restrain cybercriminals; and the attempts to develop the needed tools to do so is still developing, because it is the law that permits those attempts to stop crime and criminals and defines what tools must be used to produce the evidence needed to succeed in those attempts.

One of the great lacks induced by the lag-time described here is the lack of uniform state, national, and international law. The lack of passage of uniform state, national, and international law results as much from the lack of common definitions as it does from a lack of political will. Before such laws can be passed, all parties have to agree to (1) a description of the individual acts that compose cybercrime, (2) appropriate punishments for those acts, and (3) other such indirect though complex, knotty problems, such as questions of extradition. Beyond those procedural issues lies the even thornier problem of political will. It is hard to convince a third-world nation that has a minimal computing infrastructure and a maximal hunger problem to spend much time dealing with cybercrime when it is so much less a pressing problem than feeding its people. Thus we find ourselves in the middle

---

1 http://www.sei.cmu.edu/reports/97sr003.pdf

of world not yet ready for prime time confrontation and containment of the cybercrime that we find rampant in that world. It is as if the computer revolution has set free forces we neither foresaw nor with which we were ready to cope. But that is, of course, the very definition of a revolution.

It is in that maelstrom that computer forensics finds itself being born. And it is within that maelstrom that computer forensics seeks to both define itself and its procedures, so that it can better afford society the promised benefits of beneficent computing while enabling society to avoid falling prey to malevolent computing. It is not unlike finding oneself in the midst of the proverbial "perfect storm" while simultaneously having to both build and bail out the lifeboat.

## What Is "Computer Forensics"?

According to the Oxford English Dictionary, "forensics" entered the language by 1659 as an adjective. The OED defines "forensics" as that of "pertaining to, connected with, or used in courts of law, suitable or analogous to pleadings in court." The Merriam-Webster Online Dictionary adds this: "relating to or dealing with the application of scientific knowledge to legal problems." Examples that are given are "<*forensic* medicine> <*forensic* science> <*forensic* pathologist> <*forensic* experts>." So, then, *computer* forensics is the application of scientific knowledge about *computers* to legal problems. The term itself was coined in 1991 at the first training session held by the International Association of Computer Specialists.[2]

In real-world terms, there are at least two time frames in which computer forensics can be said to occur. There is real-time computer forensics, which might occur during the use of a sniffer on a network to watch the actual, contemporaneous transmission of data. There is also reconstructive or post facto computer forensics in which data or processes that have occurred in a past time are recreated or revealed via the tracing or extraction of data from within a given system. For our purposes, this chapter is concerned with the latter type of computer forensics.

Like any other forensic science, computer forensics deals with the application of a science to a question of law. This question of law has two parts: (1) "Did a crime occur?" And (2) "If so, what occurred?" Thus, computer forensics is concerned with the gathering and preserving of computer evidence, as well as the use of this evidence in legal proceedings. Computer forensics deals with the preservation, discovery, identification, extraction, and documentation of computer evidence.

The forensic process applies computer science to the examination of computer related evidence. The methodology used is referred to as "scientific," because the results that are obtained meet the required ends of science: The results provide a reliable, consistent, and nonarbitrary understanding of the data in question. This, in turn, gives credibility to both the investigation and the evidence so that the evidence can be said to be both authentic and accurate. This work is done through managing the investigation of the crime scene and its evidentiary aspects through a thorough, efficient, secure, and documented investigation. It involves the use of sophisticated technological tools and detailed procedures that must be

---

2 See "Computer Forensics Defined," an article published on the New Technologies website, at http://www.forensics-intl.com/def4.html. Accessed September 9, 2003.

exactly followed to produce evidence that will stand up in court. To provide that level of proof the evidence must pass courts tests both for authenticity and continuity.

"Authenticity" is the test that proves the evidence is a true and faithful copy of that which was present at the crime scene. "Continuity," often referred to as "chain of custody," is proof that the evidence itself and those persons handling and examining the evidence are accounted for since the evidence was seized. This guarantees that the evidence hasn't been tampered with or contaminated.

The tools and techniques used to examine the evidence consist of hardware devices and software applications. They have been developed through our knowledge of the means and manner in which computers work. And their algorithms have been provided reliable through repeated testing, use, and observation. Computer forensic specialists guarantee accuracy of evidence processing results through the use of time tested evidence-processing procedures and through the use of multiple software tools, developed by separate and independent developers.

For our purposes, such evidence is both "physical" and "logical." These terms are used in the traditional sense in which computer science understands these terms. "Physical" refers to evidence that can be touched, such as a hard drive. "Logical" refers to data that, in its native state (bits), cannot be understood by a person. Such data is said to be "latent" in the same way in which a fingerprint is often latent. To make latent evidence comprehensible, some technique must be applied. In the case of data, this might mean using an application to translate the bits into a visual representation of words on a computer monitor. Without the application of some interpretive vehicle, the latent evidence is unintelligible. The physical side of computer forensics involves taking physical possession of the evidence of interest, such as what occurs during the process of search and seizure. The logical side of computer forensics deals with the extraction of raw data from any relevant information resource. This information discovery usually involves, for example, combing through log files, searching the Internet, retrieving data from a database, and so on.

Computer forensics, like other forensic disciplines, has subdisciplines, such as computer media analysis, imagery enhancement, video enhancement, audio enhancement, and database visualization. These are subdisciplines because the tools, techniques, and skills required to conduct these various functions in a forensically sterile environment are different. There is one major difference between computer forensics and other forensic sciences. While in every other forensic science, the analysis of evidence involves the destruction of part of the evidence, computer forensics does not destroy or alter (through the analytical process) the evidence that was seized at the crime scene. Computer forensics makes an exact bit-for-bit copy of the evidence, and analysis is performed on the mirror image copy while the original evidence is held in a secure environment.

## Gathering Evidence Forensically

Since gathering forensic data imposes additional constraints beyond those required for simply copying data, it is important to point out the difference in requirements and to address how those requirements are met. For the purposes of this discussion, there are two types of data: "forensically identical data" and "functionally identical data." They differ not in terms of the functionality of the data, but in terms of the bit-for-bit construction of the data.

"Functionally identical data" is a copy of the subject data that when used will perform exactly as the original data. For instance, if a word-processing file is emailed to a colleague, the file that is received is functionally identical to the one retained by the sender. When the receiver and sender open the two copies of the file, the two copies contain the same text and formatting. Thus what has been transmitted is functionally identical.

However, merely copying a file so that it is a functionally identical copy isn't satisfactory for forensic purposes. For forensic purposes, the data that is obtained must be identically bit-for-bit the same. Thus "forensically identical data" is a mirror image of the original data down to the smallest details (the bits!) and the way those bits translate into humanly understandable information, such as time and date stamps. This identity is such that an MD5 Hash of each copy would be identical.

The mandate for this degree of exactness derives from the courts' requirement that the investigator be able to present incontrovertible proof that the evidentiary information procured during forensic examination is 100% accurate.

The process of making a forensically acceptable copy of data for examination involves a whole host of steps to ensure that the data copy is forensically identical to the data source. Some of these steps apply to the source drive, some apply to the destination drive, and some apply to both. To demonstrate the complexity of the process and the delicateness of the task, let us assume a simple case: We have a hard drive that has data of interest, and we need to make a forensically acceptable copy of the source drive in order to examine the data thereon.

In short, the process involves making a copy of the data in question and then establishing through a series of documented tests that (1) the source data was not changed during the copying process (i.e., the source data after the copying process is the same as it was prior to the copying process) and (2) that the copy of the data is identical to the source data.

The testing that proves that both the source hasn't changed and the copy is identical is done through creating an MD5 or a CRC-32 calculation first on the source drive prior to and after the copy operation and then on the destination drive. If all is well, then the calculations will equal each other in all three instances. The key to having the first two calculations equate (those performed on the source drive prior to and after the copy operation) lies in ensuring that no bit is changed on the source drive. One solution to the "write" problem is to interpose a device (hardware or software) in the command circuitry that controls the source drive so that write commands are blocked and the operations that they dictate are not written to the source drive, thus ensuring that the source drive is not changed.

Examples of these devices are Computer Forensic Solutions' CoreRESTORE™ ASIC microchip (which intercepts and controls data signals from the hard drive and motherboard at the ATA command level), Paraben Forensic Tools's software Forensic Replicator that mounts all hard drives in a "read only" mode, and Digital Intelligence Inc.'s FRED line of hardware/software devices.

The key to having the third calculation (performed on the destination drive) match the first two calculations lies in having the destination drive "forensically sterile" prior to the data being copied to it. "Forensically sterile" means that the destination drive is absolutely clean. That is, there is no data whatsoever on the drive prior to copying the data from the source drive. To accomplish this, one must employ an algorithm or process that (1) wipes all existing data from the media surface of the destination drive and then (2) produces a record to document that sterility. Again, there are numerous devices that accomplish this task.

Ideally the forensic tool that one employs will have several features. Among these features one would first look for a tool that is the equivalent to the type of integrated development environment (IDE) that a software developer would use. By IDE we mean that the software environment handles a lot of the necessary details transparently behind the scenes. The ideal tool automatically invokes a series of processes that accounts for the various proofs required by the law in order that the evidence be admissible and convincing. This feature can be thought of as the "automation" function that aids the examiner by helping the examiner not to make technical mistakes that cause the evidence to be discounted. This feature takes care of preventing writes to the source drive, and as part of the process ensures that a forensic wipe of the destination disk is done prior to copying the data from source to destination. Secondly, the tool should have a history of wide usage and acceptance in the forensic community. This can be thought of as the "accuracy" feature. Wide usage signifies to the court that the tool is accepted as accurate by the scientific community and is not some home-brewed, off-the-wall process that has little standing. "Acceptance" prevents the court from disqualifying the evidence as inaccurate on the basis that the processes invoked by the tool are either unproven and/or produce inaccurate results. Third, the tool will be adaptable to different environments. This is the "functionality" feature. For example, can the tool be used not just to gather data from a hard drive but from a number of types of storage media? Is it adaptable to a wide range of needs and environments? Often this is accomplished through a type of hardware modularity that permits (through the use of adapters) the tool to be used, for example, not only with a standard 3.5-inch hard drive but also with a hard drive from a notebook.

## Formal Procedure for Gathering Data

The Fourth Amendment to the United States Constitution and the subsequent stature and case law that derive from that amendment impose the necessity of a formal procedure for forensics work upon the investigator. In brief those laws exist to ensure that the subject of the investigation does not have their rights arbitrarily abridged because of a subjective interpretation and/or behavior of the state's representative. The legal doctrine that derives from the Fourth Amendment, and the law theoretically defines the balance that must be struck between the rights of the individual and the rights of the state. The state and its representative then must define practices that first accord with the legal doctrine and then second do not transgress the legal doctrine. It is these processes that form the formal procedure that must be observed during the search, seizure, examination, and documentation of evidence.

Law enforcement officials have two responses when asked about how forensic evidence should be gathered. The first springs forth from their experience in the technical forensics field, experience with the legal process, and experience with technically proficient individuals and corporations that have IT departments that believe they are capable of either lending a helping hand to the enforcement officials and/or capable of gathering evidence themselves.

Uniformly, law enforcement officials' first response is a resounding "Don't do it!" This response springs forth from the complicated procedure imposed by the law that must be followed in precise detail in order to avoid (1) contaminating the evidence, (2) voiding the chain of custody, (3) and/or infringing on the rights of the suspect. If any event(s) occurs that violates these principles, then either the evidence won't be admissible and/or won't stand up in court. While any technically adept person can master that knowledge and technique, creating that copy within the context of the burdens imposed by the law adds an extraordinarily complex set of parameters that require the investigator to be a legal expert as well as technically proficient. The former is a specialty in itself and one for which an individual who is forensically technically proficient, but legally a novice should have a healthy respect.

For the same reason that network administrators don't allow power users to have administrative rights to the network, those network administrators should not presume they have enough knowledge and/or experience to successfully complete the search, seizure, examination, and documentation processes necessitated by the twin requirements of the law and forensic science.

Law enforcement's second response is "Don't touch anything; call us!" This response springs forth from their commitments to helping those in need, to upholding the law, and to putting their expertise to constructive use.

In general, the first two goals are to ensure that the evidence is neither inadvertently destroyed nor contaminated and then to summon appropriate, competent help. However, in the heat of an event, trying to remember what to do and then doing it is far easier and far more likely to be successful if the individual or corporation has prepared ahead of time. There are several reasons for this:

1. The integrity of the evidence must be maintained in order to seek redress in court.
2. It is of immense help to have foreknowledge of what events might occur once public law enforcement becomes involved. This ensures that you as the potential victim or representative of the victim can make sure that your rights are not infringed and/or that you or your corporation aren't put in legal jeopardy by an inadvertent, well-intended, but misinformed action by public law enforcement.

The formal procedure for gathering data can be summarized in the following three steps.

## The Political Step

First, get to know your local law enforcement officials. Build relationships with the powers that be and the line investigators before you have critical, time-sensitive interactions that are exacerbated by the high emotions that occur in the midst of crises. The operative word in any relationship is "trust." If trust is established ahead of time, subsequent interactions will go much more smoothly.

## The Policy Step

Second, put in place a clear, written policy for event response. The policy development needs to be a coordinated effort of (1) senior management; (2) the legal department; (3) the IT department; (4) an outside firm that specializes in security, intrusions, and recovery from those events; (5) law enforcement officials; and (6) any other stakeholders within the company.

The purpose of involving all the different parties is twofold: (1) so each party can bring to its perspective and experience to the discussion and thus the full range of corporate considerations can be addressed, and (2) political, so that all stakeholder interests are incorporated and any possible resistance is managed adequately. Any policy is much more likely to succeed if all who are affected by are involved in its development. Thus management brings permission, authority, champions, and political expertise. Legal addresses the issue of legal requirements and permissibility. IT brings the knowledge of technical feasibility. Outside aid brings several things to the discussion. First, it brings a breadth of security expertise that enables a corporate policy to embrace a larger set of possibilities than that which might spring from the limited view of those within the company. Second, it brings an objective freedom that helps defeat both "sacred cow" syndrome and enables objective criticism of the proposed policy. Law enforcement brings its experience of merging the requirements of legal doctrines with the practical processes that are necessary to affect adequate computer forensics. Others bring viewpoints that may be particular to the individual circumstances that surround the company for which the policy is being prepared.

Broadly speaking, the policy should address

1. What security procedures are needed
2. How they will be implemented
3. How they will be maintained
4. How the implantation and maintenance will be documented
5. How and how often the procedures for the policy will be reviewed and revised
6. The detailed procedures that will be followed when an event occurs

The following is a stepwise detailed procedure to be followed when an event occurs:

1. The person discovering the event should do the following:
   a. Notify the appropriate response team manager that will take charge of managing the event
   b. See that the physical event scene is secured and undisturbed until the appropriate response team arrives to take charge so that any evidence is not disturbed;
   c. Make detailed written and dated documentation of how, when, and where the event was discovered, including an explanation of what drew discovers' attention to the event and who was present at the time
   d. Standby until relieved by the response team manager
2. The response team manager should notify the appropriate response team members that will supervise the various event aspects such as
   a. Technical
   b. Legal
   c. Public relations
   d. Internal corporate security
   e. Necessary public law enforcement

3. First responders should be summoned with the rest of the team being put on standby alert. The specialties of the first responders will depend on the situation
4. Event analysis begins
    a. A first attempt to understand and define the rough outlines of the event should begin. The controlling rubrics are
        i. Evidence must be preserved
        ii. The civil rights of all concerned must be observed
    b. Subsequent to management's satisfaction that all evidence and all civil rights are protected; then a more granular analysis is undertaken.
5. Follow-up activities begin based upon
    a. What the evidence analysis reveals
    b. What steps management decides to take regarding perpetrators that can be identified
        i. These steps are driven by the needs of the corporation
        ii. The practical possibilities that can be pursued
        iii. The economic costs to the corporation
6. Debriefing—After the facts are in and have been digested, the response team should prepare a final critique of the event and the performance of the response team. This report should address matters and details such as
    i. What happened
    ii. How various members respond to the event
        a. Were helpful/not helpful
        b. What communications were effective/not effective
    iii. What steps might be taken to ensure against similar events in the future
7. Incorporation—Finally, the results of the study should be incorporated into the existing policy to further refine that policy.

Lastly, since both statute and case law change constantly, the written policy should be reviewed annually to determine whether any changes in the law merit updating the policy and its derivative procedures.

## The Training Step

Third, make sure that all persons that will play a part in carrying out the policy and the procedures: (1) know what the policy and procedures are and then (2) practice, rehearse, drill, and run through those policies until people can carry out those procedures without mistakes.

While training or the lack thereof obviously has ramifications for effective, efficient handling of events, it may also have unintended legal consequences. There is a whole class of civil lawsuits (the end of which is monetary damages) that are brought against individuals whose actions cause harm to others. It is not at all unusual to see an individual's employer named in these suits on the basis that the individual was either inadequately trained or supervised. A principle means of seeking the corporations release from the lawsuit is to show that the individual knew what they were supposed to do, but for whatever reasons did not do it. The corporation also needs to show that due diligence was exercised.

# Law Dictating Formal Procedure

In this section, we seek to address the law that governs the formal procedure as if that procedure was made up of three distinct activities. In this scenario, the three activities follow each other sequentially. The first activity is "search and seizure." This is the activity that procures the computer hardware and software so that forensic analysis can be done. The second activity is the actual forensic analysis that discovers the information pertinent to the investigation. The third activity is the presentation in court of that information discovered during the second activity, forensic analysis.

One of the problems with adopting the three-stage approach is that the totality of the law doesn't neatly divide into three parts. While a particular law (such as the Electronic Communications Privacy Act) might deal with the rights of an individual to protect their electronic communications and the explanation(s) that the state must provide in order to abridge those rights, no individual law deals only with one of the three activities. The three activities are actually a unity that the totality of the law addresses. To further complicate this division, the nuances and interpretations of a given statute law have a way of being written and interpreted in light of other pertinent laws. Therefore this division is somewhat artificial.

The nuances of the law and the brevity of this treatment don't allow a full discussion of either the law or the procedures dictated by the law. Thus the treatment herein of these topics is superficial and introductory. For detailed information, IS managers should seek competent legal counsel from legal practitioners experienced in the Fourth Amendment, the doctrine that both drives and is derived from the Fourth Amendment, the Statute and Case Law that is in force as a result of that doctrine, and the way those laws and doctrines apply to the field of computers.

# Law Governing Seizure of Evidence

The laws that we will consider forthwith are as follows:

1. The Fourth Amendment of the United States Constitution and the limitations that the amendment places on both warranted and warrantless searches.
2. The law that governs search and seizure pursuant to search warrants and the manner in which those laws are modified by the Privacy Protection Act, 42 U.S.C. Section 2000aa and Rule 41 of the Federal Rules of Criminal Procedure
3. The Electronic Communications Privacy Act, 18 U.S.C., Sections 2701–12, hereafter referred to as the "ECPA" and modifications imposed upon the ECPA by the Patriot Act (USA PATRIOT Act of 2001, Pub. L. No 107-56, 115 Stat. 272 [2001]). This law governs how stored account records and contents are obtained from network service providers.
4. Statutory laws regarding privacy hereafter known as "Title III" as modified by
    a. The ECPA
    b. The Pen Register and Trap and Trace Devices statute, 18 U.S.C., Sections 3121–27, as modified by the Patriot Act (2001)

These laws govern real-time electronic surveillance of communications networks.[3] These laws will be considered from the particular standpoint of their application to search and seizure of computer hardware and software. Most of the material presented is drawn from the US Department of Justice publication on "Searching and Seizing Computers and Obtaining Electronic Evidence in Criminal Investigations."

## The Fourth Amendment to the United States Constitution
The amendment states:

> The right of the people to be secure in their persons, houses, papers, and effects, against unreasonable searches and seizures, shall not be violated, and no Warrants shall issue, but upon probable cause, supported by Oath or affirmation, and particularly describing the place to be searched, and the persons or things to be seized.

Broadly, the amendment provides that individuals may not be forced to endure searches by governmental authority without a government official having testified as to the reason for the search and that for which is being searched. The only sufficient reason is that the government reasonably expects to find evidence of crime. That evidence may be something used to commit a crime or the fruits of crime, or to bear witness to the crime that is believed to have been committed.

### Exceptions to Search Limitations Imposed by the Fourth Amendment
There are exceptions to the Fourth Amendment that allow investigators to perform warrantless searches. Warrantless searches are permissible if one of two conditions occurs. First, the search does not violate an individual's "reasonable" or "legitimate" expectation of privacy. Second, the search falls within an established exception to the warrant requirement (even if reasonable expectation is violated).

Generally one has a reasonable expectation of privacy if the information isn't in plain sight. Thus the analogy applied to determining reasonable expectation is that of viewing the computer as a "closed container" such as a briefcase, file cabinet, or the trunk of one's car. If there are no established exceptions to perform the search in question without a warrant, then a warrant is required. The courts are still debating the fine points of just what constitutes a closed container. Thus if there is any chance that the case is going to go to trial, the investigator should take a conservative view and obtain a warrant beforehand.

The Fourth Amendment doesn't cover searches by private individuals that are not government agents and aren't acting with the participation or knowledge of the government. Bear in mind that the Fourth Amendment protects individuals only from unwarranted government intrusion. Investigators coming into possession of information provided by a private individual can reenact the scope of the search that produced the information without a warrant. They may not exceed the scope of the first search.

The right of a reasonable expectation of privacy ceases to exist if

---

3 See "Searching and Seizing Computers and Obtaining Electronic Evidence in Criminal Investigations," published by the Computer Crime and Intellectual Property Section, Criminal Division, United States Department of Justice, July 2002, pp. 7–8.

1. Information is in plain sight, such as displayed on a monitor.
2. The subject voluntarily conveys the information to a third party, regardless of whether the subject expects the third party not to convey that information to others. So if a subject sends information via email to a third party and that party conveys it to an investigator, then the investigator doesn't have to get a warrant to use that information.

## Specific Exceptions That Apply to Computer-Related Cases

A warrantless search that violates reasonable expectation of privacy is allowed when it is preceded by

1. Consent
2. Implied consent
3. Exigent circumstances
4. Plain view
5. Searches incident to a lawful arrest
6. Inventory searches
7. Border searches

### Consent

Warrantless searches may be conducted if a person possessing authority gives the inspector permission. Two challenges to this type of search may arise in court. The first has to do with the scope of the search. The second concerns the authority to grant consent.

The legal limit of the search is limited by the scope of permission given. And the "scope of consent depends on the facts of the case." If the legality of the search is challenged, the court's interpretation becomes binding. Thus the ambiguity of the situation makes this type of search fraught with the possibility that the evidence will be barred because the search failed to account for the defendant's Fourth Amendment rights. With respect to granting authority, generally a person (other than the suspect) is considered to have authority if that person has joint access or control of the subjects' computer.

Most spouses and domestic partners are considered to fall into this category. Parents can give consent if the child is under 18; if the child is over 18, a parent may be able to consent, depending on the facts of the case.

A system administrators' ability to give consent generally falls into two categories. First, if the suspect is an employee of the company providing the system, the administrator may voluntarily and legally consent to a search. However, if the suspect is a customer, such as is the case when a suspect is using an ISP, the situation becomes much more nuanced and much more difficult to determine. In this situation, the ECPA is called into play. To the extent that the administrator's consent complies with requirements for the ECPA, the search will then be legal. We comment on the ECPA later in this section.

### Implied Consent

If there exists a requirement that individuals consent to searches as a condition of their use of the computer and the individual has rendered that consent, then typically a warrantless search is not considered a Fourth Amendment violation.

### Exigent Circumstances

The "exigent circumstances" exception applies when it can be shown that the relevant evidence was in danger of being destroyed. The volatile nature of magnetically stored information lends itself well to this interpretation. The courts will determine whether exigent circumstances existed based on the facts of the case.

### Plain View

If the plain view exception noted previously is to prevail, the agent must be in a lawful position to see the evidence in question. For instance, if an agent conducts a valid search of a hard drive and comes across evidence of an unrelated crime while conducting the search, the agent may seize the evidence under the plain view doctrine. However, the plain view exception cannot violate an individual's Fourth Amendment right. It merely allows the agent to seize material that can be seized without a warrant under the Fourth Amendment exception.

### Search Incident to a Lawful Arrest

In situations in which a lawful arrest is occurring, an agent may fully search the arrested person and perform a limited search of the surrounding area. The current state of this doctrine allows agents to search pagers, but it has not yet been determined whether this doctrine covers other types of portable personal electronic devices such as personal digital assistants. Decisions about devices that can be legally searched seem to hang on the question of whether the search is reasonable.

### Inventory Searches

An inventory search is one that is conducted routinely to inventory items that are seized during the performance of other official duties. For instance, when a suspect is jailed, his material effects are inventoried to protect the right of the individual to have his property returned. For this type of seizure to be valid, two conditions must be met: The search must not be for investigative purposes and the search must follow standardized procedures. In other words, the reason the search occurred must be for a purpose other than accruing evidence, and it must be able to be shown that the search procedure that occurred would have occurred in the same way in similarly circumstanced cases. These conditions don't lend themselves to a search through computer files.

### Border Searches

There exists a special class of searches that occur at borders of the United States. These searches are permitted to protect the country's ability to monitor and search for property that may be entering or leaving the US illegally. Warrantless searches performed under this exception don't violate Fourth Amendment protection, don't require probable cause, or even reasonable suspicion that contraband will be discovered.

Obviously, when the evidence is in a foreign jurisdiction, US laws do not apply. Obtaining evidence in this scenario depends upon securing cooperation with the police power of the foreign jurisdiction. However, such cooperation depends at least as much, if not more, on political factors than on legal ones.

## *Workplace Searches*
There are two basic types of workplace searches: private and public. Generally speaking, the legality of warrantless searches in these environments hinges on subtle distinctions such as whether the workplace is private or public, whether the employee has relinquished their privacy rights by prior agreement to policies that permit warrantless searches, and whether the search is work-related or other.

In a private setting, a warrantless search is permitted if permission is obtained from the appropriate company official. Public workplace searches are permissible if policy and/or practice establish that reasonable privacy cannot be expected within the workplace in question. But be forewarned—while Fourth Amendment rights can be abridged by these exceptions, there may be statutory privacy requirements that often are applicable and therefore complicate an otherwise seemly straightforward search.

**Private Sector Workplaces**
Rules for private sector warrantless searches are generally the same for other private locales. That is, a worker usually retains a right to a reasonable expectation of privacy in the workplace. However, because the worker does not own the workplace, the reasonable expectation of privacy can be trumped if a person having common authority over the workplace consents to permit the search, because private sector employers and supervisors generally have broad authority to consent to searches. As such, warrantless searches rarely violate the reasonable expectation of privacy because of this authority.

**Public Sector Workplace Searches**
However, warrantless searches in the public-sector workplace are a far different matter. Such searches are controlled by a host of case law interpretations that make these searches much more difficult to effect in such a way that the search doesn't become grounds for the violation of Fourth Amendment rights. Generally, warrantless searches succeed best when they are performed under the aegis of written office policy and/or banners on computer screens that authorize access to an employee's workspace. Another exception to reasonable expectation occurs when the search is deemed both "reasonable" and "work-related."

As usual, the devil is in the details. The details involved herein are those factors that determine the definitions of both "reasonable" and "work-related." The definition of these terms is extensive and detailed. However, the principal difference lies here: a public-sector employer may not consent to a search of an employee's workspace. The reasoning behind this doctrine is that since both employer and investigator represent the same agent (the government), the government cannot give itself permission to search.

## *General Principles for Search and Seizure of Computers with a Warrant*
In general, courts of competent jurisdiction grant investigators the right to invade an individual's privacy when an investigator, under oath, persuades the court that there is reason or "probable cause" to suspect criminal activity. The presentation must include a description of the person and/or place to be searched, and the items that are being searched for and that will be seized if they are found.

Because of (1) the detail that the legal system requires in the aforementioned oath and (2) the complexity and portability of data storage, warrants for computer and computer-related data searches can be difficult both to obtain and to execute properly. Bear in mind that in order to do a legally admissible search and seizure, the investigator must not only obtain the warrant, but the warrant must properly describe the scope of the search and the evidence being sought. In addition, failure to execute the search within the confines of the permission granted within the warrant can cause the court to disqualify any evidence that is found during the search.

As a result of these complexities, the work that must go into the warrant application and the subsequent search and seizure has to be informed by

1. That which is permissible under the law
2. A technical knowledge of what is possible in a search of computer and computer-related evidence
3. That which is needful during the search in order to recover the desired evidence

As a result, the Department of Justice's *Searching and Seizing Computers and Obtaining Electronic Evidence in Criminal Investigations* publication (2015) recommends a four-step process to ensure that warrants obtained are adequate for the necessary search and seizure. These four steps are

1. Step 1: Assemble, in advance, a team to write the warrant application that consists of
    a. The investigator, who organizes and leads the search
    b. The prosecutor, who reviews the warrant application for compliance with applicable law
    c. The technical (forensic) expert, who creates (and may execute) the plan for the search of computer-related evidence
2. Step 2: Develop as much knowledge as possible about the systems to be investigated. This is because (a) its impossible to specify the scope of search until one knows what it is that one will be searching and (b) failure to do so can create a situation in which the search is declared illegal and inadmissible because of an inadvertent violation of another piece of law
3. Step 3: Develop a plan for the search. This plan should include a backup plan in case the initial plan is foiled or becomes impossible to execute.
4. Step 4: Draft the warrant request. The draft should describe
    a. The object of the search
    b. Accurately and particularly the property to be seized
    c. The possible search strategies and the legal and practical considerations that inform the proposed search strategy

Part of the planning procedure has to take into account the various types of search warrants and the factors that determine whether those types of warrants can be obtained and then executed within the limits of those warrants.

Some of the other types of warrants are

1. Multiple warrants for one search
2. "No-knock" warrants
3. "Sneak-and-peek" warrants

In addition, there are other factors that must be considered when planning the application for a warrant. These include

1. The possibility and effect of uncovering privileged information
2. The liabilities that can occur if the search runs afoul of limitations imposed by other acts of law

### Considerations Affecting the Search Strategy
Given that the role of search/seizure is to obtain evidence, the gathering of evidence can take four principle forms:

1. Search the computer and print out a hard copy of particular files at the time of the on-site search.
2. Search the computer and make an electronic copy of particular files at the time of the on-site search.
3. Create a duplicate electronic copy of the entire storage device during the on-site search, and then later recreate a working copy of the storage device off-site for review.
4. Seize the equipment, remove it from the premises, and review its contents off-site.

The role of the device (read "CPU" and/or other "peripheral devices") in the crime is a principle consideration that informs the decision to either seize the device or perform an on-site search. If the device is contraband, evidence, and/or the instrumentality and/or fruit of the crime, then the computer is likely to be seized, barring other mitigating factors. If the device is merely a storage container for evidence of the crime, the computer will likely not be seized, again barring other mitigating factors.

In the first instance (device as instrumentality, fruit, and so on of the crime), the device might not be seized if such seizure will post a hardship for an innocent third party. For instance, if the suspect is an employee that has been using part of an integrated network to commit the crime, then seizure of that device may bring the network down and harm the employer. In such an instance, seizure may be foregone. On the other hand, if the device is merely a storage vehicle for evidence, the device might not be seized if on-site data recovery is feasible. However, the extremely large size of hard drives may prohibit thorough searching on-site. In such a situation, the device may be seized to enable a thorough examination at a later date. The general rule is to make the search/seizure "the least intrusive and most direct" that is consistent with obtaining the evidence in a sound manner. As one might expect, there are a whole host of other gradations.

## The Privacy Protection Act (PPA) 42 U.S.C. § 2000aa
The purpose of this act is to protect persons involved in First Amendment activities (those activities that have to do with the practice of free speech), such as newspaper publishers, who are not themselves suspected of criminal activities for which the materials that they possess are being sought. The act gives a very wide scope to what can constitute First Amendment activities.

However, at the time of the act's passage (1980), publishing was primarily a "hard copy" event, and electronic publishing wasn't an understood practicality. Therefore the act makes no distinction between the two forms. The practical result is that many electronic files that are of a criminal nature are protected by this act and are therefore not subject to the usual criminal search and seizure process. Thus because almost anybody can now "publish" on the Internet, the "publication" may fall within the protection of this act.

While violation of the act doesn't produce suppression of the evidence, it does open the authorities to civil penalties. Further, the act as it stands and the surrounding case law leave many points ambiguous, in part because PPA issues aren't often tried by the courts. When PPA considerations bar the use of a search warrant, using processes or subpoenas that are permitted by the Electronic Communications Act can often circumvent these considerations.

## The Electronic Communications Privacy Act 18 U.S.C. §§ 2701–2712

This act creates privacy rights for those persons that subscribe to network services and are customers of network service providers. The act is designed to provide privacy rights in the virtual arena that are akin to those that individuals enjoy in a physical arena. In a physical arena, an investigator must go into an individual's property to obtain information about that individual. Ingress to that property is protected because individuals have a right to a reasonable expectation of privacy. However, in a virtual arena, information about one is stored with a third party such as a network services provider. Thus an investigator could approach the third party about desired information, were it not for the protections of the ECPA.

As we have seen previously, the Fourth Amendment guarantees privacy rights within one's physical domain. But that amendment doesn't address privacy rights within a virtual domain. It is this arena that the ECPA seeks to address. The act offers various levels of protection, depending upon the importance of various privacy interests that an individual is perceived to have. In seeking to comply with the act, the investigator must address the various classifications of privacy rights to the facts of the particular case in question in order to successfully obtain the desired evidence.

In its attempt to fairly address the various competing needs, the act makes two key distinctions. The first distinction has to do with the type(s) of service offered by the provider to the customer. The second distinction has to do with the type(s) of information possessed by the provider as a result of the business relationship that the provider enjoys with the customer.

### Types of Service Provided

Under the act, providers are classed as either a provider of "electronic communication service" or a provider of "remote computing service." A communication service is any service that makes it possible for a user to send or receive communications across a wire or in an otherwise electronic form (e.g., telephone companies and email providers). This designation is limited to those who provide the equipment that conveys the information. It does not apply to those who use someone else's equipment to sell such services to a third party. For instance, in the scenario where Business "A" hosts websites on Business "B's" equipment, Business "A" is not considered as providing electronic communication services. Therefore

Business "B" is not covered under the ECPA. Data stored in a manner that is seen to be held in "electronic storage" hangs upon whether the data in question is determined for another or final destination that is different from the place in which it is stored. A "remote computing service" is defined as a provision of storage and computing services by means of an "electronic communications system." An "electronic communications system" is defined as any vehicle that moves data or stores data electronically.

## Types of Information That May Be Held by Service Providers

The ECPA classifies information that providers hold as follows:

1. Basic subscriber information that reveals the customer name and the nature of the relationship that the customer has with the provider
2. Other information or records that reveal information about the customer. This category includes everything that is neither basic information nor content of data such as the text of email and so on. It can include items such as numbers that the customer called, logs about call usage, and so on.
3. Content, including the actual files stored by the provider on behalf of the customer. Included in this category are things such as the actual text of emails, word processing files, spreadsheets, and so on.

## Means of Obtaining Information Protected by the ECPA

Information protected under the ECPA may be obtained by subpoena, court order, or search warrant. The ECPA divides these devices into five categories, each of which requires a different level of proof in order for authorities to obtain the desired device. From lowest to highest burden of proof, these devices are

1. Subpoena
2. Subpoena with prior notice to the customer
3. Court order issued under Section 2703(d) of the act
4. Court order issued under Section 2703(d) with prior notice given to the customer
5. Search warrant

The lowest order of proof is required for a subpoena; the highest order for a search warrant. The privacy value of the information determines which of the processes is required to force disclosure of information from the service provider: the higher the privacy value of the information, the greater the level of proof required in order to obtain the document that compels the provider to release that information. Generally speaking, a level 2 process can be used to obtain both level 1 and level 2 information; a level 3 process can be used to obtain level 3, level 2, and level 1 information, and so forth.

A subpoena is adequate to force disclosure of customer information as described in the previous section. A subpoena with prior notice to the customer can force disclosure of opened email or documents stored for a period of time greater than 180 days. The force of a Section 2703(d) order is the same as a subpoena without notice. The standard for obtaining this order is higher than that of a subpoena, but the range of effectiveness is greater.

The proof required is that the government must offer specific and articulated facts showing that there is sufficient reason to believe that they are relevant and materially useful to the ongoing criminal investigation. Such facts must be included in the application for the order. Though the standard is higher than that of a subpoena, it is lower than that required for a probable cause search warrant. The purpose of raising the standard is to prevent the government from going on "fishing expeditions" for transactional data.

A Section 2703(d) order accompanied by notification to the customer can compel the types of information mentioned previously, plus any transactional data, except for unopened email or voicemail stored for less than 180 days. This type of order permits the government to obtain everything that the provider has concerning the customer, except the unopened email or voicemail less than 180 days old. A search warrant does not require prior notification to the customer and can compel disclosure of everything in a customer's account, including unopened email and voicemail. Thus a search warrant may be used to compel disclosure of both informational and transactional data.

## Voluntary Disclosure and the ECPA

Any provider that doesn't offer service "to the public" is not bound by the ECPA and may voluntarily disclose any data, informational or transactional, without violating the act. In addition, there are provisions within the ECPA for voluntary disclosure of both informational and transactional data under the following justifications.

**Voluntary Disclosure of Informational Data**

Informational data may be disclosed when

1. The disclosure is incidental to rendering the service or protecting the rights of the provider.
2. The provider reasonably believes that a situation exists in which if the information is not disclosed, a person is in immediate danger of death or injury.
3. The disclosure is made with permission of the person who is described by the information.

**Voluntary Disclosure of Content Data**

Transactional data may be voluntarily disclosed by the provider when

1. The disclosure is incidental to rendering the service or protecting the rights of the provider.
2. The disclosure is made to a law enforcement official if the contents were inadvertently obtained by the provider and it appears that the data pertains to the commission of a crime.
3. The provider reasonably believes that a situation exists in which if the information is not disclosed, a person is in immediate danger of death or injury.
4. The Child Protection and Sexual Predator Punishment Act of 1998 requires such disclosure.
5. The disclosure is made
    a. To the intended recipient of the data
    b. With the consent of the sender or intended recipient

c. To a forwarding address
   d. Pursuant to a legal process

## ECPA Violation Remedies

Violations of ECPA do not lead to suppression of evidence. However, violations do lead to civil actions that can result in civil and punitive damages and disciplinary actions against the investigator that violate the act.

## Real-Time Monitoring of Communications Networks

This section of law dictates how and electronic surveillance, commonly called "wiretaps," can be used. There are two pertinent sections of the law that need to be considered. Legally, the first is described as 18 U.S.C. Sections 2510–2522, otherwise known as Title III of the Omnibus Crime Control and Safe Streets Act of 1968 and informally referred to as "Title III." This is the law that governs wiretaps. The second is the Pen Registers and Trap and Trace Devices chapter of Title 18, 18 U.S.C. Sections 3121–3127 and informally known as the "Pen/Trap Statue." This law governs pen registers and trap and trace devices. Each of these statutes addresses and regulates the collecting of different types of information. Title III pertains to obtaining the content of communications; Pen/Trap pertains to collecting noncontent data such as address headers.

### The Pen/Trap Statute

Pen/Trap covers both hardware and software applications—that is, anything that collects information in any way to be used to identify the senders or recipients falls under the regulation of Pen/Trap.

To obtain a Pen/Trap order, the applicant must

1. Identify themselves.
2. Identify the law enforcement agency under whose aegis the investigation occurs.
3. Certify that the information likely to be gained is pertinent to an ongoing criminal investigation.

In addition, the court granting the order must have jurisdiction over the suspected offence. A Pen/Trap order may be granted so that it has force beyond the jurisdiction in which the court is located. Thus an official may get an order to trace to a particular IP address, regardless of whether the IP address terminates at a computer within the geographical jurisdiction of the court that issues the order. The orders are granted for 60 days, with one 60-day renewal period.

A nongovernmental entity may use Pen/Traps on their own network without having to obtain a court order subject to the following conditions. Such use must be for

1. The operation, maintenance, and testing of the hardware
2. To protect the rights or property of the provider
3. For the protection of the users
4. To protect the provider from charges of malfeasance from another provider
5. Where the consent of the user has been obtained

## The Wiretap Statute (Otherwise Known as Title III)

This statute prohibits a third party from eavesdropping on the content of a conversation between two or more other parties in which the third party is not a participating party.

In deciding the legitimacy of a surveillance action, agents need to consider three questions:

1. Is the monitored communication considered "oral" or "electronic" and thereby protected under the act?
2. Will the surveillance lead to an interception of communications that are protected under the act?
3. If so then does a statutory exception exist that obviates the need for a court order?

The answer to the first two questions hangs on the act's definition of

1. Interception
2. Wire communications
3. Oral communications
4. Electronic communications

A "wire communication" is one that travels across a wire and has a voice component crossing the wire, and the "wire" must be provided by a public provider. However, stored wire communications are covered under the ECPA, not under Title III.

The term "electronic communication" covers most Internet communications. An "electronic communication" is just about any transmission of data via any means that isn't a "wire or oral communication." Exceptions to this definition include tone-only paging devices, tracking devices, such as transponders, or a communication that involves electronic storage or transfer of funds.

An "intercept" refers to the acquisition of communications in "real time." An intercept does not occur when data is acquired from a storage medium, even if that storage medium is an intermediary stop along the path from sender to recipient.

### Obtaining a Court Order

To obtain an order, the applicant must show

1. Probable cause that a felony offence will be found
2. That normal procedures to discover the evidence have failed
3. Probable cause that the communication is being used for illicit purposes

### *Exceptions to the Act*

Since Title III is so broad, the legality of a non-court-ordered surveillance hangs upon the statutory exceptions of the court order discussed previously. The exceptions and their interpretations follow:

1. Consent is deemed to be given if
   a. The interceptor is a party to the conversation.
   b. One of the parties to the conversation has given consent.

Such consent can be implied or explicit. Implied consent exists in situations where one or more of the parties is aware of the "monitoring." A banner that advertises that the system is monitored and that appears at the security dialog is deemed to be consent. The best banner is one that gives the broadest definition to what monitoring is and what is being monitored. While it is relatively easy to determine who is a party to a voice conversation, the nature of electronic network communications adds complexities that make the successful application of this exception much less likely. Therefore one should use it as a fallback position if no other exception can be fitted to the circumstances.
2. There exists an exception for a provider. A provider may monitor and disclose to law enforcement without violating the law so long as the monitoring is to protect the provider's rights or property. Again, this exception has limitations, and investigators are advised to use it cautiously. Also, the provider exception tends to be easier to apply when the provider is a private entity. It is much more difficult to meet the tests for this exception when the provider is a governmental entity.
3. The "computer trespasser" exception exists to allow victims of hacking to authorize investigators to monitor the trespasser. It is applicable when a person accesses a computer without authorization and has no existing contractual relationship with the owner. When investigators monitor under this exception, they should obtain written permission of the owner or a person in a position of authority. This exception also requires that (a) the monitoring be solely for the purpose of a lawful investigation, (b) the investigator reasonably believes that the contents of the intercepted data will be relevant to the investigation, and (c) the intercepts must be limited to those coming from the trespasser.
4. The "extension telephone" exception works a bit like the banner consent exception in that it allows a person who works for a company and uses a telephone extension provided by the company to be monitored. The presumption is that if the person is using something (in this case the extension) and it is clear that the device can be monitored, then the person gives consent to monitoring. However, case law interpretations of this exception are varied, and therefore investigators are encouraged to avoid this exception if a more widely acceptable exception is applicable.
5. "Inadvertently obtained criminal evidence" can be divulged to law enforcement officials.
6. The "accessible to the public" exception states that any person may intercept a communication if that data is readily accessible to the public. However, this exception has no case law behind it. Thus it is actually an undefined exception for which no safe harbor exists.

In the event of violation by the practitioner, Title III provides statutory suppression of oral or wire data, but not electronic data; Pen/Trap does not provide statutory suppression. Law enforcement officials are generally not liable for Title III violations when such occur as the result of reasonable, good faith decisions.

# Law Governing Analysis and Presentation of Evidence

The body of law that governs both the analysis of evidence and the presentation of that evidence in court is the Federal Rules of Evidence.[4] In addition, each state may have its own rules of evidence. There is also a piece of case law that goes to the issue of competence of the evidence, *Daubert v. Merrill Dow Pharmaceuticals*. Although it is not possible to undertake an extensive review of the relevant legal issues, there are concepts that the rules try to embody that we can examine with respect to both the analysis and presentation of evidence in court. The concepts of custody and admissibility are discussed as follows. Chain of custody and the matters that are substantiated by a proper chain of custody precede admission of evidence at trial. To this end, one can view the establishment of chain of custody as the procedural requirement that guides the work of evidence detection and ensures the veracity of the evidence, whereas admissibility is the determinant of whether a piece of evidence can be used at trial.

Up to this point, we have discussed the law as if it breaks cleanly into two sections. For the sake of discussion, we have said that the first section of law (which we discussed previously) has to do with "search and seizure" and the second (which we will now discuss) has to do with analysis and presentation of the evidence that is discovered. This is true. But it is helpful to make a subtler distinction at this point: the rules that must be followed to ensure the veracity and admissibility of evidence come into effect at the time the first bit of evidence is seized, not when the evidence reaches the site where the analysis will be performed. Enforcement officials must start documenting the chain of custody as soon as anything that can yield or be evidence is seized at the site of the search. Thus the first part of this section (which dealt with abridging constitutional rights) more properly deals with obtaining the right to search. While the application for a search warrant must specify what evidence is likely to be found and seized, the documentation of that seizure falls more properly within this section.

## The Chain of Custody

The thrust of this section revolves around maintaining what is known as "the chain of evidence" or "the chain of custody." The doctrine that the Federal Rules of Evidence (hereafter "FRE") seeks to embody and which mandates the practice of establishing the chain of evidence is one that says that evidence must be proven true in order to be presented at trial. So chain of custody procedures are designed to guarantee the veracity of the evidence so that impure or tainted evidence may not be used to produce an inaccurate portrayal of the events in question and thus lead to the conviction of innocent defendants. Thus FRE imposes requirements to guarantee the veracity. Such requirements include documentation of the location(s) of the evidence and any person(s) interaction(s) with the evidence. This proof takes the form of verifiable (read: "written") documentation that demonstrates that the state or quality of the evidence has not been changed intentionally or inadvertently.

Clearly, if the location of the evidence can't be documented from the point of seizure to the point of admission at trial, then there can be no guarantee that the evidence has not

---

4 http://www.law.cornell.edu/rules/fre/overview.html.

been altered or even that the evidence produced is the same piece that was taken at the point of seizure. Thus the whereabouts of the evidence must be documented from start to finish. But it is not enough to prove that the evidence hasn't been altered. The prosecution must also be able to prove that the evidence produced at trial is the same and identical to that which was seized. The distinction between these two burdens is subtle but different. One seeks to prove no alteration could have occurred. The other seeks to prove that the evidence produced at trial is identical to that which was seized. It is the latter proof that gives rise to the need to document that each and every person that handled the evidence has maintained the purity of that evidence.

Thus each interaction must be documented. Such interactions include documenting the seizure; the transportation from the point of seizure to the point of storage; that the storage is secure; the names of each and every person that has contact with and/or custody of the evidence; the date and time and type of contact; the time, place, and nature of bit-stream copies of the original evidence; the production of an MD5 hash on the original prior to and after the bit-stream backup; the production of an MD5 hash of the bit-stream copy postproduction prior to any analysis; and the production of an MD5 hash after each investigation of the copy.

Establishing and maintaining the chain of custody meets these requirements. While any form of written notation is acceptable, an ad hoc format puts a burden on the examiner. The examiner must remember to notate all the details, because failure to record a transaction can invalidate the chain of custody. Here an established, proven case management tool has two benefits. First, it relieves the examiner of the burden of remembering what details need to be documented. And second, it allows the examiner to focus on the work of extracting the data.

## Admissibility

The FRE define the factors that determine admissibility of evidence. There are general rules of admissibility. There are also specific rules of admissibility that apply to evidence, depending upon which of the four categories of evidence a particular item falls into.

### General Rules of Admissibility

In order for evidence to be admitted, it must be shown to be true, relevant, material, competent, and authenticated. We have covered the issue of veracity previously, which, as can be seen, relates to the issue of competence. Evidence is deemed relevant if it shows the fact that it is offered to prove or disprove an event. Evidence is material if it substantiates a fact that is at issue in the case. Evidence is competent if the proof being offered meets the requirements of reliability. There are four types of evidence: real, testimonial, demonstrative, and documentary. For our purposes, we are primarily interested in real and testimonial evidence, since the bulk of forensic cases involve the production and admission of real evidence and testimony by a forensics expert.

### Real Evidence

Real evidence relates to the actual existence of the evidence. For instance, data found on a hard drive is real evidence. To be admissible, real evidence must meet the tests of relevance, materiality, and competency. It is argued that relevance and materiality are usually self-evident; however, competence must be proven.

Establishing competence is done through the process of authenticating that the evidence is what it purports to be. Authentication is proven either through identifying that the item is unique, or that the item was made unique through some sort of identifying process, or establishing the chain of custody. It is fairly hard to establish that one of a manufacturer's hard drives is different from another. However, a hard drive can be made unique through affixing a serial number on that hard drive. Another way to make a hard drive unique is by having the seizing officer mark the hard drive with some distinguishing mark at the point of seizure (i.e., evidence tags that bear the initials of the seizing officer) or establishing the chain of custody. In this instance, chain of custody is used to show that the drive is what it is claimed to be—the drive that was seized. Chain of custody establishes the fact that no other drive could have been the drive that was seized, since the drive that was seized has been in protected and documented possession since the point of seizure. The last fact that must be authenticated is that the seized item remains unchanged since its seizure.

**Testimonial Evidence**

Testimonial evidence relates to the actual occurrences by a witness in court. As such, it doesn't require substantiation by another form of evidence. It is admissible in its own right. Two types of witnesses may give testimony: lay and expert. A lay witness may not give testimony that is in itself a conclusion. Expert witnesses can draw conclusions based upon the expertise that they bring to the field of inquiry and the scientific correctness of the process that they used to produce the data about which they are testifying. Obviously, a forensic witness is more often than not an expert witness.

The strength of expert testimony lies in not only the experienced view that the expert brings to the stand, but in the manner he derives the data offered as evidence. *Daubert v. Merrill Dow Pharmaceuticals* is the case that defines the manner that must be followed for testimony to be received as expert.[5] There is a lot of controversy about the exact requirements that *Daubert* makes on expert testimony, but given the fact that it is the most stringent interpretation of what constitutes expert testimony, it is wise to keep this in mind when one intends to introduce evidence that will have expert testimony as its underlying foundation.

The US Supreme Court's opinion in *Daubert* defines expert testimony as being based on the scientific method of knowing and is considered expert when four tests are met:

1. **Hypothesis testing.** Hypothesis testing involves the process of deriving a proposition about an observable group of events from accepted scientific principles, and then investigating the truthfulness of the proposition. Hypothesis testing is that which distinguishes the scientific method of inquiry from nonscientific methods.
2. **Known or potential error factor.** This is the likelihood of being wrong that the scientist associates with the assertion that an alleged cause has a particular effect.
3. **Peer review.** Peer review asks the question of whether the theory and/or methodology has been vindicated by a review by one's scientific peers. That is, does the process and conclusion stand up to the scrutiny of those who have the knowledge to affirm that the process is valid and the conclusion correct?

---

5 Stephen Mahle, "The Impact of *Daubert v. Merrell Dow Pharmaceuticals, Inc.*, on Expert Testimony," http://www.daubertexpert.com/FloridaBarJournalArticle.htm. Accessed May 6, 2004.

4. **General acceptance.** This is the extent to which the findings can be generalized and hence qualify as scientific knowledge.

As one can see, to get forensic evidence admitted and the witness qualified as expert, the examiner must be able to show that a scientific procedure has been followed, which has been generally accepted by the forensic community. The expert can handle this most easily by using proven case management tools based on the scientific method. Again, this argues for not doing "freehand" examination, but rather using a methodology that is demonstrably the same time after time.

In summary, FRE and *Daubert* are intended to provide a safe harbor for both parties to the suit. They form a rule and an agreed upon standard for what constitutes a fair presentation of the evidence under consideration. The larger goal is to produce a fair hearing of the evidence; the more immediate goal is to prevent the defense from crying "procedural foul" and thus muddying the waters of the trial and/or having the court prohibit the admission of the evidence.

# Emergent Issues

In broad terms, the issues that need to be discussed evolve from within two arenas: the international arena and the national arena. In the former area, the issues revolve around the politics of international cooperation as affected by both political good will and the legal compatibility between two or more sovereign states' governing laws. Within the second area, the issues tend to center on the more traditional logistical issues such as skill sets, information dissemination, forensic equipment, and the like.

## International Arena

The virtual nature of computing makes it possible for a perpetrator to live in one country and to commit a crime in another county. This possibility leads to two scenarios: Perhaps the country in which the perpetrator lives (the "residential country") discovers that the individual has committed a crime and needs evidence from the country in which the crime was committed (the "crime-scene country"). Or the reverse occurs: the crime scene discovers the crime and seeks to have the residential country bring the perpetrator to justice.

In international law, the cooperation of two or more sovereign countries to resolve a common concern hinges upon at least five factors: a common interest, a base of power, political good will, the existence of a legal convention that both countries agree to observe (though such observance is necessarily voluntary, hence the need for political good will), and the particulars within that convention such as some means of trying the individual that honors the needs of both countries' systems of laws. The absence of any of these factors can derail the attempt of either or both countries to bring the perpetrator to effective justice—that is, balanced action(s) imposed by the concerned states upon the individual which prevent the individual from being able to continue in his criminal conduct while protecting that individual's other rights.

## Common Interest and a Base of Power

We discuss these two factors in conjunction with each other—for they, together, make up the basis of effective negotiation. For two sovereign states to reach an agreement or legal convention about any matter, both countries must have a desire to achieve the same outcome from the circumstances that lead to negotiation. Otherwise, there will be no negotiation. Also necessary is the ability to negotiate from a stance of power. That is, Country 1 has something Country 2 wants but also the ability to prevent Country 2 from attaining that "something." If either of these power differentials is missing, then no fruitful negotiation can occur, because either the powerful country will have no incentive to negotiate with the weaker country or neither country has anything to bring to the table with which to entice the other country.

## Political Good Will

Once a convention is effected, the observance of that convention, though agreed to legally by both countries, is functionally voluntary. Since both countries are sovereign there exists no legal means for either country to force the observance by the other country. Thus the practice of observance by a particular country is always hedged around by the way that countries' policy makers feel toward both the other country and the other countries' policy makers. So it behooves both countries and their leaders to work diligently toward fostering continual political good will toward each country and among the respective leaders. Enough significant failures of good will leads to failure of the convention.

## The Convention and Its Particulars

Once the political factors are present, the convention itself can be constructed. While the crafting isn't easy, it certainly is possible. Witness The Hague Convention, which defines jurisdiction and such matters as the practice of legal service by one sovereign state within the jurisdiction of another sovereign state. Look also as the European Union and the convention that surrounds and undergirds such things as a common currency.

In order for the convention to be acceptable to all countries, the convention must prescribe practices that are acceptable to all participants' existing legal systems and their underlying senses of both justice and preservation of national sovereignty. With respect to trying computer crimes, the convention must provide the particulars that form the basis for mutually satisfactory means for the seizure and analysis of the evidence; the procedure for the trial; the sanctions that may be imposed; a manner of publishing the verdict of guilt or innocence and the resulting sanctions or remunerations; and the treatment of those found guilty during the imposition of any punishment imposed. The determination of and agreement to these particulars only prove that the "devil is in the details."

Here are just a few examples: It seems to be hard to imagine that the United States would be a signatory to any convention that allowed a foreign power to try a US citizen wherein the gathering of evidence violated Fourth Amendment guarantees. Again, how would the United States try a person who was a resident non-US citizen if that individual's Fourth Amendment rights were violated by the crime-scene country's gathering of evidence, even though the evidence might have been gathered using the due legal process of the country that gathers the evidence? Would that evidence be admissible? Does the Fourth Amendment promise that the government won't violate the Fourth Amendment rights of a noncitizen or just that the United States won't violate the Fourth Amendment rights of a noncitizen in the gathering of evidence by US authorities within the United States? In other words, what is the legal admissibility of evidence into US courts that has been gathered both beyond the confines of the United States by non-US authorities and in accordance with the gathering countries' due process? As one can see, the issues imposed upon the crafting of multinational conventions by the contradictions inherent in each state's legal due process are daunting. It is the substance of true statesmanship.

While the particulars of the convention revolve around physical aspects such as how much and what type of punishment is fair and how the convicted person shall be treated during the punishment phase, much of the work of crafting such an agreement and then the asking by one country that another country honor that agreement in a particular case revolves around the intangibles of the political process between the two states. In contrast, within the US national arena, the bulk of the work falls in the area of determining the practical needs of a cooperative arrangement between subjudicatories within the United States.

## National Arena

The principle difference, described previously, is due primarily to the overarching umbrella produced by a federal form of government and a fairly uniform cultural sense of justice. That is, any political differences that impede the process between subjudicatories, such as individual states, can finally be trumped, if necessary, by federal law that takes precedence over conflicting laws resulting from two competing states' interests. Thus a whole difficult and very different layer of work that must be resolved in the international arena is made moot in the US national arena. Therefore the discussion can move toward defining the particular needs that must be met for the effective prevention and prosecution of computer crime and the particulars of our topic: What must occur for forensics to produce convincing, legally admissible evidence?

There are 10 areas that define the critical issues at a national level. These are:

1. **Public awareness.** Public awareness is needed because such awareness helps the public stay alert to the problem. It puts more eyes on the ground. Better awareness means better data in order to better understand the scope of the problem and respond thereto.
2. **Data and reporting.** More reporting in more detail is needed to better understand and combat cybercrime
3. **Uniform training and certification courses.** Officers and forensic scientists need continual training to familiarize themselves with technological innovations and the manner in which they are employed to commit crimes. Within this area, there needs to be established standard forensic training certifications that when applied are adaptable to local circumstances.
4. **Management assistance for local electronic crime task forces.** Due to the often cross-jurisdictional nature of cybercrime, the high cost of fielding an effective forensic team, and the high level of training need for effective forensic work,

regional task forces are perceived as a way to achieve economies of scale and effectiveness

5. **Updated laws.** As the events surrounding 9/11 unfolded, one of the striking ironies was that by law the FBI and the CIA were prevented from sharing information with each other. This is a wonderful, though poignant, example of a state of law that is inadequate to the demands of the times. The law must arm enforcement officers with the wherewithal to "support early detection and successful prosecutions."

6. **Cooperation with the high-tech industry.** High-tech industry management must be both encouraged and compelled to aid law enforcement. We say "compelled" because often industry and corporate interests (such as profit-motivation) run counter to the public good. For instance, the hue-and-cry raised by the industry over the government's former refusal to allow exportation of greater-than-40-bit encryption software led directly to the repeal of that law

7. **Comprehensive centralized clearing house of resources.** Time and opportunities are lost if the resources that are needed in a particular case exceed the abilities of the individual assigned to that case and that individual is unable to find additional resources. Examples of print resources include technical works on decryption and forensically sterile imaging of disks and data, to name just two.

8. **Management and awareness support.** Law enforcement management need to better back the forces on the ground. Often senior management in public service are answerable to the electorate and encumbered with a bureaucracy that resists change. Hence our first need, public awareness, becomes a vehicle for promoting change. As the public becomes more aware of the effects of cybercrime on the good fabric of culture, they bring change to the public agenda and thus encourage management to place a higher priority on stemming the tide of cybercrime.

9. **Investigative and forensic tools.** Here the sense of urgency may be even greater—because here is the tactical spot where the lack or richness of proper tools shows up first. This is the spot in which the battle is fought hand-to-hand by the opposing forces. It is here that officials first know the abilities they possess. If the essential tools, software, and technology aren't available, the crime goes unsolved and/or untried and the perpetrator goes unstopped. The cost of forensic tools and the rate of change of technology are the two biggest drivers of this need.

10. **Structuring a computer crime unit.** Both the organization of the computer crime unit as well as its placement within the existing enforcement infrastructure is a concern. Local and regional law enforcement officials are hopeful that the US Department of Justice will bring together the best of the thinking on these issues so that local and state officials can benefit from that critical thought as they organize their respective units.

# Concluding Remarks

The confluence of the constraints that we have mentioned—time, education, rate of change, and cost—makes possible several observations about and at least one implication for the forensics:

- The field is in great flux.
- The field is in great need.
- The resources that can be brought to bear in forensics are expensive.
- These resources have a short lifetime.
- Constant continuing education is the order of the day.
- The practice of forensics lags behind the development of the technology about which the forensic practice is supposed to produce knowledge and evidence.
- The law that governs the fruit of forensics needs to be updated to support new demands on forensics.
- The only way to keep forensics viable is through a centralized coordinated approach to technology, tools, and experts.

In this chapter we have attempted to address the milieu from which the necessity of computer forensics springs—that of a society trying to do justice, which is an attempt to balance the competing needs of its members and those members with society's needs. Computer forensics is the tool we use to try to gain a fair and full accounting of what occurred when there is a dispute between society and one or more of its citizens with respect to whether a citizen's action is just. Forensics seeks to answer this question: "Does the action meet the legal test of being responsible to the society in which that action occurs?" Forensics' technical work is guided by a formal procedure, which when followed is both lawful and accurate and therefore promotes the cause of justice. This procedure is guided and dictated by society's doctrine and body of law. The body of law is society's attempt to give particulars to its desire to do justice. We have discussed this body of law as it influences the aspects of seizure, analysis, and presentation of evidence.

### In Brief

- Computer forensics is the application of scientific knowledge about computers to legal problems.
- Forensic process is an attempt to discover and present evidence in a true and accurate manner about what transpired in a computer-related event so that society can determine the guilt or innocence of the accused.
- Because of the nascent state of the law governing computer crimes and forensics, there is a lack of uniformity among jurisdictions that makes the apprehension and trial of criminals difficult in cases that cross jurisdictional lines.
- Procedural quality is paramount in collecting, preserving, examining, or transferring evidence.
- Acquisition of digital evidence begins with the collection of information or physical items collected and stored for examination.
- Collection of evidence needs to follow the rules of evidence collection as prescribed by the law.
- Data objects are the core of the evidence. These may take different forms, without necessarily altering the original information.

## Questions and Exercises

**Discussion questions.** These are based on a few topics from the chapter and are intentionally designed for a difference of opinion. These questions can best be used in a classroom or a seminar setting.

1. In a US Supreme Court case of *Illinois v. Andreas*, 463 US. 765 (1983), the court held that a search warrant is not needed if the target does not have a "reasonable expectation of privacy" with respect to the area searched. In *U.S. v. Barth*, 26 F. Supp. 2d 929 (1998), a US. District Court held that the computer owner has a reasonable expectation of privacy, especially with respect to data stored on the computer. However, if the ownership of the computer is transferred to a third party (e.g., for repair), the expectation for privacy can be lost. How can this issue be addressed from the perspective of the Fourth Amendment?

2. "Information provided in an Intrusion Detection System is useful in dealing with computer crimes." Comment on the legal admissibility of such information.

3. Today security executives perform the difficult task of balancing the art and science of security. While the art relates to aspects of diplomacy, persuasion, and understanding different mind-sets, the science deals with establishing measures, forensics, and intrusion detection. Given that security is indeed an art and a science, comment on the role of computer forensics in the overall security of the enterprise.

**Exercise.** Scan the popular press for computer crime cases. Collect information on the kinds of evidence that were collected, their usefulness in apprehending the culprit, emergent problems, and concerns, if any. Draw lessons and present best practices for computer forensics.

See Appendix A (after the Index) for a set of Short Questions for this chapter.

## Case Study: The Deathserv Case

In June 2002, a young hacker later identified as a then 16-year-old Joseph McElroy of the UK was able to break into servers at a secret weapons lab in the United States. Upon realizing that a breach of security had occurred, part of the labs network was shut down for three days, fearing that it was related to a terrorist attack. It was later learned that the teen had merely wanted to download music and films and was using the lab's servers as a location to store the copyrighted loot. He had developed a software program to enable him to download the copyrighted materials and named it Deathserv. He may not have been caught as quickly if he had not told friends about his scheme, and it was the increased traffic on the server by him and his friends that alerted authorities. The investigation required the use of forensic science to track the hackers back to the UK, but the teen received a light sentence, in part due to his age. The judge, Andrew Goymer at London's Southwark Crown Court, sentenced McElroy to 200 hours of community service after pleading guilty to unauthorized modification to the contents of a computer. The US Department of Energy had sought 21,000 pounds in compensation due to the seriousness of the offence and to recover damages due to the network disruption and investigation.

1. What kind of message does this light sentence send to future offenders?
2. Would an international set of sentencing guidelines help in such situations, given the borderless nature of the Internet?
3. What investigative tools might have been used to track the hacker and learn of his identity?

Source: Compiled from *The Register*, ZDNet UK (April 14, 2005), and *The Financial Times* (February 18, 2004).

# Case Study: Trade-Offs in Eavesdropping

Australian police have greater powers to use electronic means of eavesdropping since the passing of the Surveillance Devices Act. The act authorizes federal and state police to use Trojans or other key-logging software on a suspect's computer after a warrant is obtained. Police can obtain a warrant to investigate offences with a maximum sentence of three years or more. As with similar legislation passed in the United States, there are concerns that the act gives police too much power.

The US laws prohibit electronic eavesdropping on telephone conversations, face-to-face conversations, or computer and other forms of electronic communications in most instances. They do, however, give authorities a narrowly defined process for electronic surveillance to be used as a last resort in serious criminal cases. When approved by senior justice department officials, law enforcement officers may seek a court order, authorizing them to secretly capture conversations concerning any of a statutory list of offences. The US laws generally do not allow law enforcement to plant Trojans or keystroke devices, but can intercept communication passing through ISPs, which could provide considerable information regarding a suspect's computer usage.

Some laws permit pen register and trap and trace orders for electronic communications (e.g., email); authorize nationwide execution of court orders for pen registers, trap and trace devices, and access to stored e-mail or communication records; treat stored voicemail like stored email (rather than like telephone conversations); and permit authorities to intercept communications to and from a trespasser (hacker) within a computer system (with the permission of the system's owner). A warrant could then be obtained to seize computer equipment during a criminal investigation and additional evidence could be determined forensically. The surveillance of US residents is strongly resisted. Lawmakers have suggested that surveillance of any kind could violate the Fourth Amendment to the Constitution. But most experts believe that Fourth Amendment protections do not apply to foreigners.

In Australia, however, no warrant is necessary for organizations to seek access to meta data of private parties. In practical terms, this means that there is no access to content, but there is complete access to the form of the communication (i.e., who called who, when, and where). In the UK a warrant signed by the Secretary of State is necessary in order to access the content of the communication.

1. Given that police can obtain a warrant to seize computer equipment, and use pen traps to trace the source or destination addresses of communications, is there a need to allow police to use key-logger programs to monitor a criminal's activities?

2. If government officials do not follow proper security guidelines when storing data they have legally obtained, is there a danger that information regarding innocent citizens might be misused or fall into the hands of criminals?
3. What is the trade-off for security versus privacy, and how can society tell when police are restricted from apprehending known criminals?

Source: ZdNet Australia.

## References
1. Friedman, M.S., and K. Bissinger. 1995. Infojacking: Crimes on the information superhighway. *New Jersey Law Journal*, May 22.
2. Stambaugh, H., et al. 2001. *Electronic crime needs assessment for state and local law enforcement*. Washington, DC: National Institute of Justice.

# CHAPTER 14

# Summary Principles for IS Security

*You can have anything you want, if you want it badly enough. You can be anything you want to be, do anything you set out to accomplish if you hold to that desire with singleness of purpose.*

—Abraham Lincoln

As Joe Dawson sat and reflected on all that he had learnt about security over the past several months, he felt happy. At least he could understand the various complexities involved in managing IS security. He was also fairly comfortable in engaging in an intelligent discussion with his IT staff. Joe was amazed at the wealth of security knowledge that existed out there and how little he knew. Certainly an exploration into various facets of security had sparked Joe's interest. He really liked the subject matter and thought that it might indeed not be a bad idea to seek formal training. He had seen programs advertised in *The Economist* that focused on issues related to IS security. *Well …that is an interesting idea. I need to sort out my priorities*, thought Joe.

From his readings Joe knew that IS security was an ever-changing and evolving subject. He certainly had to stay abreast. Perhaps the best way would be to get personal subscriptions to some of the leading journals. There was the *Information Security Magazine* (http://informationsecurity.techtarget.com/) that Joe was aware of, but that was a magazine. *Where there any academic journals?* he thought. Who else could he ask but his friend Randy from MITRE. Joe called Randy up and shared with him his adventures of learning more about IS security. Following the usual chit-chat, Randy recommended that Joe read the following three journals on a regular basis:

1. *Computers and Security* (http://www.elsevier.com/). Randy told Joe that this was a good journal that had been around for more than two decades. It was a rather expensive journal to subscribe to, but it did make a useful contribution in addressing a range of IS security issues.

2. *Journal of Information System Security* (http://www.jissec.org). This was a relatively new journal, publishing rigorous studies in IS security. As an academic journal, it strived to further knowledge in IS security.
3. *Journal of Computer Security (http://www.iospress.nl/)*. This journal was a little technical in orientation and dealt with computer security issues with a more computer science orientation. It was a good solid technical publication nevertheless.

"Thank you. I will take a look at these," said Joe. There was a wealth of information out there that Joe had to acquire. After all, a spark had been kindled in Joe's mind.

---

In this book we adopted a conceptual model to organize various IS security issues and challenges. The conceptual model, as presented in Chapter 1, also formed the basis for organizing the material in this book. A large number of tools, frameworks, and technical and organizational concepts have been presented. This chapter synthesizes the principles for managing IS security. The principles are classified into three categories—in keeping with the conceptual framework in Chapter 1. It is important to adopt such a unified frame of reference for IS security, since management of IS security permeates various aspects of personal and business life. It may be simply buying a book from an online store, or engaging in Internet banking—there is no straightforward answer to the protection of personal and corporate data. While it may be prudent to focus on the ease of use and functionality in some cases, in others, maintaining confidentiality of private data may be the foremost objective. Clearly no individual or company wants private and confidential data to get in the hands of people they do not want to see it, yet the violation of safeguards (if there are any) by organizational personnel or access to information by covert means are things we hear about on a rather regular basis.

Although complete security is a goal that most organizations aspire for, it is often not possible to have complete security. Nevertheless, if companies consider certain basic principles for managing IS security, surely the environment for protecting information resources will improve. In this chapter we synthesize and summarize the key principles for managing IS security. The principles are presented as three classes:

- Principles for technical aspects of information system security
- Principles for formal aspects of information system security
- Principles for informal aspects of information system security

## Principles for Technical Aspects of IS Security

As has been discussed and presented in previous chapters, the success of technical aspects of security is a function of the efficacy of the formal and informal organizational arrangement. And clearly regulatory aspects have a role to play. This is an important message, since it suggests that exclusive reliance on technical controls is not going to result in adequate security. Traditionally organizations have been viewed as purposeful systems and security has for the most part not been considered part of the "useful system" designed for the purposeful activities (see arguments proposed by Longley [1], page 707). Rather, security has always been considered as a form of guarantee that the useful activities of an organization will continue to be performed and any untoward incidents prevented. This mind-set has been challenged by a lot of researchers [2, 3], and there have been calls for developing security visions and strategies, where IS security management is considered a key enabler in the smooth running of an enterprise.

Fundamental principles that need to be considered for managing the technical aspects of IS security are as follows:

**Principle 1: In securing the technical systems, the changing local conditions should be proactively considered.** In managing the security of technical systems, a rationally planned grandiose strategy will fall short of achieving the purpose. There are many organizations that spend a lot of time formulating security policies and then hope that through some magical way, the organization is going to become more secure. Clearly policies are an important ingredient to the overall security of the organization. However, exclusive emphasis on policy and designing it in a top-down manner is counterproductive. There are two reasons for this. First, as noted earlier in the book, a rationally planned strategy does not necessarily consider the ground realities. There is enough evidence in the literature that emphasizes the importance of emergent strategies and policies (see, for example, the research done by Mintzberg [4], Osborn [5], among others). Second, the constantly changing and dynamic nature of the field makes it rather difficult to formulate grandiose strategies and wait for them to play out. In an era where organizations remained stable, where reporting and accountability issues were clearly defined, it made sense to develop policies and practices at a given point in time and then hope that organizations adapt to these. Herein is also the criticism of US security standards, which are created at a certain point in time with the hope that systems and organizations will adopt them. However, changes in technology and the dynamic nature of business makes most standards obsolete even before they get published.

**Principle 2: Formal models for maintaining the confidentiality, integrity, and availability (CIA) of information are important. However, the nature and scope of CIA needs to be clearly understood. Micromanagement for achieving CIA is the way forward.** There is no doubt that *Confidentiality*, *Integrity*, and *Availability* of data are key attributes of IS security. Technical security can only be achieved if CIS aspects have been understood. *Confidentiality* refers mostly to restricting data access to those who are authorized. The technology is pulling very hard in the opposite direction with developments aimed at making data accessible to the many, not the few. Trends in organizational structure equally tug against this idea—less authoritarian structures, more informality, fewer rules, empowerment.

*Integrity* refers to maintaining the values of the data stored and manipulated (i.e., maintaining the correct signs and symbols). But what about how those figures are interpreted for use? Businesses need employees who can interpret the symbols processed and stored. We refer not only to the numerical and language skills, but also to the ability to use the data in a way that accords with the prevailing norms of the organization. A typical example is checking the creditworthiness of a prospective loan applicant. We need both data on the applicant and correct interpretation according to company rules. A secure organization not only needs to secure the data but also its interpretation.

*Availability* refers to the fact that the systems used by an organization remain available when they are needed. System failure is an organizational security issue. This issue, although not trivial, is perhaps less controversial for organizations than the previous two principles.

An important point to note, however, is to assess how requirements for IS security are derived. Clearly any formal model is an abstraction of reality. The preciseness of a model is judged on the basis of the extent to which it represents a given subset of the reality. Most information security models were developed for the military domain and to a large extent are successful in mapping that reality. It is possible to do so since the stated security policy is precise and strictly adhered to. In that sense, a security model is a representation of the security policy rather than the actual reality. This suffices as far as the organization works according to the stated policy. In recent years, however, with the widespread reliance on the Internet to conduct business, problems arise at two levels. First, the organizational reality is not the same for all enterprises. This means that the stated security policy for one organization is bound to be different from that of the other. Second, a model developed for information security within a military organization may not necessarily be valid and true for a commercial enterprise. It follows therefore that any attempt to use models based on the military are bound to be inadequate for the commercial organizations. Rather, their application in a commercial setting is going to generate a false sense of security. This assertion is based on a definition of security that goes beyond simple access control methods. This book has addressed the issues of various IS security attributes, but a more succinct definition can be found in [2]. The way forward is to create newer models for particular aspects of the business for which information security needs to be designed. This would mean that micro-strategies should be created for unit or functional levels.

## Principles for Formal Aspects of IS Security

As discussed in Chapter 1, the emergence of formal organizational structures usually gets formed as a result of increased complexity. This is very elegantly presented by Mintzberg [6] by narrating Ms. Raku's story of her pottery business and how she organized her work as her business evolved from a basement shop to Ceramics Inc. (see page 1). Computers are subsequently used to automate many of the formal activities in a business. And there is always a challenge in deciding as to which aspects one should computerize and which should be left alone (for a detailed description, see Liebenau & Backhouse [7], pages 62–63). It is thus important to understand the nature and scope of formal rule-based systems and evaluate as to how IS security could be adequately designed into an organization.

The following paragraphs present principles that need to be considered in instituting adequate control measures:

**Principle 3: Establishing a boundary between what can be formalized and what should be norm based is the basis for establishing appropriate control measures.** Clearly security problems arise as a consequence of "overformalization" and managerial inability to balance the rule- and norm-based aspects of work. Establishing a right balance

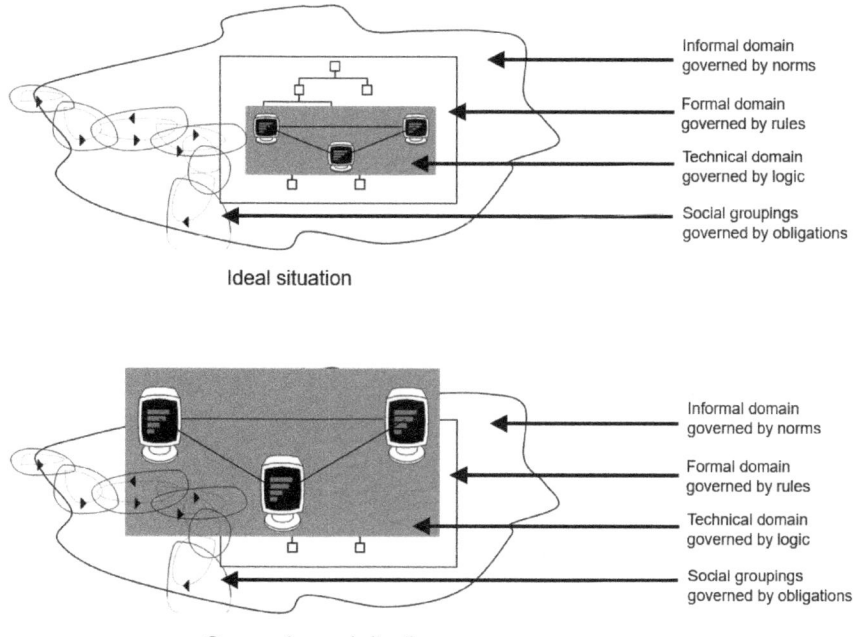

**Figure 14.1.** Ideal and overengineered solutions

is important. Problems of overformalization are usually a consequence of the "big is beautiful" syndrome. In many cases project teams tend to feel that if the system is big and interconnected with other systems in the organization, it suggests that a "good" solution has been designed. Ideally a computer-based system automates only a small part of the rule-based formal system of an organization, and commensurate with this, relevant technical controls are implemented (Figure 14.1).

At a formal level an organization needs structures that support the technical infrastructure. Therefore formal rules and procedures need to be established that support the IT systems. This would prevent the misinterpretation of data and misapplication of rules, thus avoiding potential information security problems. In practice, however, controls have dysfunctional effects. This is primarily because isolated solutions (i.e., controls) are proposed for specific problems. These "solutions" tend to ignore other existing controls and their contexts.

**Principle 4: Rules for managing information security have little relevance unless they are contextualized.** Following on from the previous principle, exclusive reliance on either the rules or norms falls short of providing adequate protection. Clearly an inability to appreciate the context while applying rules for managing information security is perhaps detrimental to the security of an enterprise. It is therefore important that a through review of technical, formal, and informal interventions is conducted. Many times, a security policy is used as a vehicle to create a shared vision to assess how the various controls will be used and how data and information will be protected in an organization. Typically a security policy is formulated based on sound business judgment, value ascribed to the data, and related risks associated with the data. Since each organization is different, the choice of various elements in a security policy is case-specific, and it's hard to draw any generalization.

## Principles for Informal Aspects of IS Security

It goes without saying that a culture of trust, responsibility, and accountability, which has been termed at the "security culture" in this book, goes a long way in ensuring IS security. Various chapters in this book have touched upon a range of informal IS security aspects that are important for security. Central to developing a good security culture is the understanding of context. As research has shown, an *opportunity* to subvert controls is one of the biggest causes of breaches—others being *personal factors* and *work situations* (see Backhouse and Dhillon [8]). It becomes apparent therefore that organizations need to develop a focus on the informal aspects in managing IS security. The various principles that need to be adopted are as follows.

**Principle 5: Education, training and awareness, although important, are not sufficient conditions for managing information security. A focus on developing a security culture goes a long way in developing and sustaining a secure environment.** Research has shown that although education, training, and awareness are important in managing the security of enterprises, unless or until an effort to inculcate a security culture exists, complete organizational integrity will be a far-fetched idea. A mismatch between the needs and goals of the organization could potentially be detrimental to the health of an organization and to the information systems in place. Organizational processes such as communications, decision-making, change, and power are culturally ingrained and failure to comprehend these could lead to problems in the security of information systems. While discussing issues in disaster recovery planning, Adam and Haslam [9] note that although managers are aware of the potential problems related with a disaster, they tend to be rather complacent in taking any proactive steps. Such an attitude could be a consequence of the relative degree of importance placed on revenue generation. As a consequence, while automating business processes and in a quest for optimal solutions, backup and recovery issues are often overlooked.

**Principle 6: Responsibility, integrity, trust, and ethicality are the cornerstones for maintaining a secure environment.** As has been argued in this book and elsewhere, traditional security principles of confidentiality, integrity, and availability are very restricted [10]. In response to the changing organizational contexts, the RITE (responsibility, integrity, trust, and ethicality) principles have been suggested [10]. The RITE principles hark back to an earlier time period when extensive reliance on technology for close supervision and control of dispersed activities was virtually nonexistent. Beniger [11] terms this as the "factorage system of distributed control," where the trade between cotton producers in America and British merchants was to a large extent based on trust (see pages 132–133). The extensive reliance on information technologies today questions the nature and scope of individual responsibilities, and many times challenges the integrity of individuals. Trust is also broken especially when technology is considered as an alternative supervisor.

The RITE principles are as follows:

- **Responsibility**. In a physically diffuse organization, it is evermore important for members to understand what their respective roles are and what their responsibilities should be. Today vertical management structures are disappearing as empowerment gains in stature as a more effective concept for running organizations well. Furthermore, members are expected to be able to develop their own work practices on the basis of a clear understanding of what they are responsible for.
- **Integrity**. Integrity of a person as a member of an organization is very important, especially as information has emerged as the most important asset/resource of organizations. It can be divulged to a third party without necessarily revealing that it has been done. Business sensitive information has great value, and organizations need to consider whom they allow to enter the fraternity. But cases still abound where new employees are given access to sensitive information without their references being properly checked out.
- **Trust**. Modern organizations are emerging to be ones where there is less emphasis on external control and supervision but more on self-control and responsibility. This means that there is a need to have mutual systems of trust. Principles of division of labor suggest that colleagues be trusted to act in accordance with company norms and accepted patterns of behavior. This may, however, not happen in practice. Inculcating a level of trust is therefore important.
- **Ethicality**. There has been a lowering of ethical standards generally in recent years, eliminated in part by the loss of middle management, job tenure perhaps, and this has resulted in an increasing numbers of frauds. No longer is it possible to assume unswerving loyalty to the employer. As a result, elaborate systems of control are implemented, which are more expensive compared with informal secure arrangements.

## Concluding Remarks

The various chapters in this book have essentially focused on four core concepts: the technical, formal, informal, and regulatory aspects for IS security. This chapter synthesizes the core concepts into six principles for managing IS security. IS security has always remained an elusive phenomenon, and it is rather difficult to come to grips with it. No one approach is adequate in managing the security of an enterprise, and clearly a more holistic approach is needed. In this book a range of issues, tools, and techniques for IS security has been presented. It is our hope that these become reference material for ensuring IS security.

### In Brief

The contents of this book can be synthesized into six principles for managing IS security:

- Principle 1: In managing the security of technical systems, a rationally planned grandiose strategy will fall short of achieving the purpose.
- Principle 2: Formal models for maintaining the confidentiality, integrity, and availability (CIA) of information are important. However the nature and scope of CIA needs to be clearly understood. Micromanagement for achieving CIA is the way forward.
- Principle 3: Establishing a boundary between what can be formalized and what should be norm-based is the basis for establishing appropriate control measures.
- Principle 4: Rules for managing information security have little relevance unless they are contextualized.
- Principle 5: Education, training, and awareness, although important, are not sufficient conditions for managing information security. A focus on developing a security culture goes a long way in developing and sustaining a secure environment.
- Principle 6: Responsibility, integrity, trust, and ethicality are the cornerstones for maintaining a secure environment.

See Appendix A (after the Index) for a set of Short Questions for this chapter.

## References

1. Longley, D. 1991. Formal methods of secure systems, in *Information security handbook*, edited by W. Caelli, D. Longley, and M. Shain, pp. 707–798. Macmillan, UK: Basingstoke.
2. Dhillon, G. 1997. *Managing information system security*. London: Macmillan.
3. Baskerville, R. 1988. *Designing information systems security*. New York: John Wiley & Sons.
4. Mintzberg, H. 1987. Crafting strategy. *Harvard Business Review*, July–August: 66–74.
5. Osborn, C.S. 1998. Systems for sustainable organizations: Emergent strategies, interactive controls and semi-formal information. *Journal of Management Studies*, 35(4): 481–509.
6. Mintzberg, H. 1983. *Structures in fives: Designing effective organizations*. Englewood Cliffs, NJ: Prentice-Hall.
7. Liebenau, J., and J. Backhouse. 1990. *Understanding information*. Basingstoke: Macmillan.
8. Backhouse, J., and G. Dhillon. 1995. Managing computer crime: A research outlook. *Computers & Security*, 14(7): 645–651.
9. Adam, F., and J.A. Haslam. 2001. A study of the Irish experience with disaster recovery planning: high levels of awareness may not suffice, in *Information security management: Global challenges in the next millennium*, edited by G. Dhillon, 85–100. Hershey, PA: Idea Group Publishing.
10. Dhillon, G., and J. Backhouse. 2000. Information system security management in the new millennium. *Communications of the ACM*, 43(7): 125–128.
11. Beniger, J.R. 1986. *The control revolution: technological and economic origins of the information society*. Cambridge, MA: Harvard University Press.

# PART V

# CASE STUDIES

# CASE STUDY 1

# The Anthem Data Breach

On January 29, 2015, Anthem announced to the world that it was the victim of a "sophisticated attack" by cybercriminals. The attackers gained unauthorized access to Anthem's IT network and were able to steal from their customer database. It was reported that 78.8 million people had their names, birthdays, social security numbers, medical identification numbers, home addresses, email addresses, and employment data stolen. Anthem was quick to report the incident to the authorities and hired a cybersecurity firm, Mandiant, to begin an investigation of their IT systems and to begin cleaning up the mess that the cybercriminals had left behind. The media is calling it one of the biggest data breaches of all time.

Anthem discovered suspicious activity in early December 2014, almost two months before the public announcement of the data breach. The attacks were persistent for the next several weeks as the cybercriminals looked for vulnerabilities in Anthem's IT systems. Experts theorize that the cybercriminals were active in Anthem's system for some months prior to December 2014. However, Anthem's security measures deflected their initial attempts. Eventually, the cybercriminals were able to obtain network credentials from at least five Anthem associates (employees) who had high-level IT access. The means by which this was done was most likely through a technique called "spear phishing" where the cybercriminals sent targeted emails to these individual Anthem associates to trick them into revealing their network IDs and passwords, or by making the associates unintentionally download software that would allow cybercriminals long-term access to their computers. From there the criminals could take their time and glean all the data they wanted.

Anthem customers were notified individually about the breach through mailed letters and email notifications. The website, anthemfacts.com, was set up to give a chronology of events and to provide information in case a customer's identity was stolen. On the website Anthem posted instructions for customers to check their credit reports and to set up fraud alerts in an attempt to prevent identity theft. AllClear ID was hired for two years to provide identity protection services at no cost to customers and former customers of Anthem. AllClear ID is to monitor the affected customers' credit reports and send them alerts once fraud is detected. The detection will reduce the time it takes to clean up the damage done by fraudsters. It cannot, however, stop fraudsters from making new credit applications in a person's name or keep creditors from pulling your credit report.

The financial consequences of the data breach are estimated to be well beyond $100 million but actual figures have remained classified even a year later. What is known is that

Anthem holds a cybersecurity insurance policy with American International Group which covers losses up to $100 million. That limit has most likely been exhausted due to the costs of notifying customers individually and hiring AllClear ID to provide their services free of charge to stakeholders. Anthem still faces multiple lawsuits, government investigations, and regulation fines that will have to come out of their pocket. Anthem still must patch up the vulnerabilities and beef up security so that they are prepared the next time around. The Data Breach of 2015 shows that cybercrime is going to be the price of online business for organizations and their customers.

## Anthem's History and Background of the Industry

Anthem, Inc., started in its current form back in 2004 when for-profit health care networks WellPoint Health Care Networks, Inc., and Anthem, Inc., merged to form the second largest health insurer in the United States. Situated in Indianapolis, Indiana, Anthem operates under the subsidiary branches Anthem Blue Cross, Anthem Blue Cross Blue Shield, Amerigroup, and Healthlink across 14 states. Anthem has 53,000 employees and has made around $79.24 billion in sales so far (2016). Among the organization values are accountability, innovation, customer care, and trustworthiness.

The health care industry is the largest service industry in the United States—a trillion-dollar industry globally as of 2014. It is comprised of health care providers, such as hospitals and private practices, and insurers who aid their customers in paying for health-related services. For such a large market there are relatively few players in the industry. There are only about 35 private health insurance companies that are open to the American public, not including Medicare providers. Though the pool is small, there is still healthy competition among the companies that allows consumers to find a health insurance plan that fits their needs.

Unfortunately, the health care industry holds the record for the largest amount of data breaches in all of the service industries for this year. Yet the health care industry at large is notoriously lax about their network security profiles. According to the information systems blog DavidMarilyn.wordpress.com, about 90% of health care organizations have lost data in breaches over the past two years. That is a staggering majority in an industry whose business is to hold the private information of its customers. The security intelligence vendor Norse did a study in the first half 2014 and found that in 357 health care-related organizations in the United States, 14,729 unique malicious events occurred, with 567 unique malicious source IP addresses found that should not have been communicating with Norse. The malicious events have compromised these health care organizations [1].

In a lot of cases it is found that devices other than computers are what is being hacked into by cybercriminals. In particular, multi-functional printers seem to draw a lot of attention from those seeking to bypass network security measures. They have become a source point for access due to being shared among many units on a local network and because most owners do not take the time to change the factory default username and password for network configuration. Not updating login credentials makes it very easy for cybercriminals to guess passwords for the printer or to look up common out-of-the box login information for network printers. Once they have the login credentials, they have access to the network and can begin working their way through security measures to find the information that they're looking for.

It isn't just the health care industry that faces this crisis either. Alongside Anthem, large companies such as Sony, Target, JP Morgan Chase, and Home Depot all have had their databases hacked into. In fact, using the term "sophisticated attack" is becoming a norm in the post–data breach world. Cyberattack victims can use this as a conventional defense without providing any further explanation.

## Who Is at Stake?

The data breach affected millions of people. Not only were customer and former customers affected, but non-Anthem customers, Anthem's associates, and other health care organizations were all affected by the security breach. Their information was not only accessed in Anthem's system but was confirmed stolen by law enforcement. Anthem has to think about restitution with all of these parties. All of this is going to cost them more time and money.

Tens of millions of customers and former customers across 14 states had their personal information stolen in the breach. Names, addresses, employment information, social security numbers, all stolen in a flash. Customers now face the uncertainty of how their information will be used against them. Will it be sold in marketing campaigns? Will fraudsters use it to open credit accounts in their names? Will criminals use the information to track their whereabouts? According to DavidMarilyn.wordpress.com, Anthem's stolen data can easily sell for as much as $20 per record on the black market. That is $1.6 billion if the cybercriminals decide to sell the records. On top of those concerns, current customers will likely feel the costs of the breach by way of increased insurance premiums. Anthem will most likely increase the price for their health insurance products as a way to make back some of the money lost in the aftermath of the data breach. There is a silver lining for customers though. It has been confirmed that bank and credit card information was not compromised so there is no reason for customers to change bank account numbers or cancel cards.

Customers of other health insurance organizations aren't safe either. Anthem is contracted out by some independent insurance companies to manage their paperwork. Some 8.8 to 18.8 million people whose health insurance belongs to other insurers in the Blue Cross Blue Shield network were unwittingly victims in the data breach [2]. These people had no idea that Anthem processed their claims or even had a database of their personal details. Now cybercriminals came and stole their information. The outcome plays out much the same as it does with Anthem's own customers. Their personal information stolen, AllClear ID's services are offered at no charge, and no payment information was taken. But Anthem has sparked outrage with this particular group of stakeholders. Why is it necessary for Anthem to maintain a database of their personal details? What are their plans for holding such information?

Anthem's own associates were not safe from the data breach: 53,000 employees had their information stolen. This includes current CEO Joseph Swedish. Alongside all the sensitive information that was stolen about their customers, associates' income information was also taken. Also, Anthem's associates now have to work around the clock to clean up the mess that was made and take flak and criticism from an untrusting community. Associates will most likely need to receive more training on cybersecurity vulnerabilities and to make sure that high-level network access credentials aren't exposed again.

Other health care organizations are on the short list as potential victims for data breaches. Because of the laxity in network security across the industry, cybercriminals will continue to plunder health care databases for easy treasure. On top of that, regulatory agencies are investigating Anthem and will bring heavy sanctions across the industry, pushing for cybersecurity reforms and fining organizations who fail to meet the standards. Worst of all, the health care industry has a tarnished image in the eyes of the public. Customers will have a hard time trusting organizations to keep their information safe and not collect more than they need.

## Lawsuits and Policy Issues

For what it's worth, the Federal Bureau of Investigation praised Anthem's quickness to publicly address the data breach. Anthem also immediately hired the cybersecurity firm Mandiant on retainer to aid authorities in the investigation. The FBI's investigation is ongoing but their chief suspects are hackers affiliated with the Chinese government. Authorities so far haven't found any evidence that the stolen information has been sold, shared, or used fraudulently in any way. It is believed that if the Chinese government was behind the cyberattack, their goal was to see how American health care is organized. Also in a bout of good fortune, none of the patients' medical information was taken so the Health Insurance Portability and Accountability Act (HIPAA) doesn't come into effect for legal investigations.

Even with the good news that the stolen information hasn't been used, it has not saved Anthem from backlash. Many of the details of the aftermath of the breach have been under wraps, most likely having to do with the legal process involved. Anthem is facing multiple class-action lawsuits from customers, former customers, and customers of other Blue Cross Blue Shield insurers. These stakeholders have a great deal of distrust of Anthem since they were collecting unnecessary data and seek compensation for their loss. The data breach has given Anthem's reputation a black eye from those who once trusted the organization.

Anthem and other health care organizations have also come under the scrutiny of state governments and other regulators. Anthem faces multiple investigations from state insurance regulators and state attorneys general that could come with heavy fines for failing to comply with state data privacy laws. The National Association of Insurance Commissioners has also called for a multistate examination of Anthem and all of its affiliates for their inconsistencies in data security. Regulators are calling for Anthem and other insurers to be more responsible with the data they own. Cyberattacks are more and more frequent on the health care industry and lawmakers seek to make stronger sanctions on companies to wake up these organizations and make cybersecurity a major issue.

## Significance of the Problem and What Anthem Is Doing About It

Anthem is just one piece of a much larger problem. Cybersecurity threats are increasing each year. Cybercriminals and hackers poke and prod networks to find easy vulnerabilities so that they can go in and take what they want. Anthem has stated that the cybercriminals used a "very sophisticated external cyber-attack" but many critics challenge that and say it is an excuse so that they and other companies like them cannot be held liable for the theft. Cyber-attack victims use this as a way to get out of explaining the failures of their security measures. Usually, what happens is that cybercriminals use known vulnerabilities in the system to gain access. Michael Hiltzik of the *Los Angeles Times* postulates that this is what happened in the case of Anthem. He writes that the cybercriminals exploited the "human element" and that it points to a flaw in the security culture at Anthem [3]. What he means is that high-level IT professionals at Anthem were duped into giving up their network credentials. It would be very hard for any technological security measure to counter that.

All of this highlights the big problem that Anthem, the health care industry, and all other online businesses are dealing with: What is the protocol for network security in a marketplace that is increasingly moving online? The health care industry as a whole has bad protocols for cybersecurity and it has opened them up to not being able to catch and patch the vulnerabilities in their systems before breaches are made. The culture of security that Michael Hiltzik mentions in his opinion piece on Anthem is a major contributor to the success of companies that have a secured online presence. Logically, companies should prioritize the appropriate use of technology to ensure the security of the network. In Anthem's case, those five IT associates should have been extensively trained that network credentials must be protected at all costs. Another critique (and an often controversial one) of Anthem's security protocols is Anthem's failure to encrypt any of their data. Encryption is the process of making data unreadable except by those authorized to view it. This critique is controversial among experts because it most likely would not have stopped the attack despite what many of Anthem's critics contend. It would, however, have been another layer of security potentially slowing down or throwing another roadblock in front of the cybercriminals to discourage them from continuing.

Cybersecurity is a complex and overwhelming problem. Cybercriminals only need to seek out one vulnerability to breach the system while the organization on the defense has to find them all and come up with a plan to patch or monitor them. But if cybersecurity is not addressed in a serious manner, it can do irreparable damage to the organization. Data breaches damage a brand's image, productivity, revenue, and compliance. Anthem has always been about trustworthiness and care for its customers. Yet trust in the organization was shaken because more wasn't done to prevent the theft. Right after the breach Anthem's productivity and revenue were slowed down or even brought to a halt so that the organization could focus on clean-up, patching vulnerabilities in the network, and paying out of pocket to fix everything. Associates were working around the clock to figure out and fix what the cybercriminals stole. In the wake of the breach, Anthem has faced a slew of lawsuits, regulation investigations, and protocol critiques. It has created quite a legal mess for them, most of which still can't be addressed publicly because of the ongoing court cases. It could have been much worse if medical patient records were stolen, however. Then Anthem would also have had to deal with HIPAA violations and be under intense scrutiny from the federal government.

Despite all of this Anthem has continued to press forward. As of this writing, the breach has not significantly impacted overall profits or customer retention according to their quarterly financial earnings calls. Anthem has taken some steps to bring a resolution to the breach and hopefully to secure its network in the future. Authorities praised Anthem for its quick response time and for openly working with law enforcement. Anthem chose to individually notify every current and former customer that their data had been stolen. They could have just posted an announcement online or through media outlets but instead chose to personalize their admission that there was a problem. They offered free credit monitoring and identity protection through AllClear ID. Many criticize this as a public relations stunt to save face, a tactic that is becoming a normal part of corporate apologies, and that the offer is too little too late. To AllClear ID's credit though, they come highly praised. Security experts agree that AllClear ID is very professional and is very good at identity protection monitoring. Anthem, also, has had its associates working around the clock to do everything they can to secure their data and locate and minimize vulnerabilities to their network.

The cause of this breach was ultimately Anthem's inability to line up their business strategy with their information systems strategy. This has caused them to have a relaxed culture of security and to have the human element exploited because of the lack of attention and care of their associates. It is unfortunate that this is not out of the ordinary for the health care industry. If Anthem is going to overcome this data breach they will need to address these issues and maybe then they can set the new standard for health care network security.

## Strategic Alignment—A Possible Cause of the Breach

The primary failure Anthem has to address is the fact that their business strategy and their information systems strategy did not line up. Anthem takes pride in being trustworthy and wanting to help their customers. They value being accountable, innovative, and caring. It can be argued that they have displayed those values at different points in their history but their very core, their service to their customers, their trustworthiness, accountability, and innovation all took a big hit with the data breach. This happened because the infrastructure set in place to help them in their mission was set up haphazardly. Where they should have been trustworthy and customer oriented with the stored and recorded information they were lazy and didn't have proper security measures set in place. Where they could have set the example with their IT infrastructure they followed the crowd of other health care organizations with lackluster security protocols. No one security measure is going to keep an organization secure but Anthem had/has the opportunity to be ahead of the curve. They could have implemented some methodological review of the policy where the leadership and operations people coordinated on a range of security issues. Questions could have been asked, such as:

1. Why was encryption not implemented?
2. What organizational and technological challenges prevented encryption from being implemented?
3. What kind of a cybersecurity strategy should have been crafted?

Both business and IS leadership could have been coordinating goals and then evaluating their infrastructure to see where they stood against cybercrime. IS leaders in Anthem could have been looking ahead to see what techniques cybercriminals were using and could have been planning with business leaders in Anthem to hedge those attacks. Nothing is foolproof, but active management is a requirement to prevent these disasters. Unfortunately, this isn't a new thing. The health care industry as a whole struggles with this very problem. Anthem has become an example of a much greater problem. The following critiques and recommendations can easily be applied to the rest of the health care industry and to good network security habits in general.

**Table C.1.** Synopsis of network security in health care

| The Good | The Bad | The Ugly |
| --- | --- | --- |
| Anthem is a good example of a quick and timely response | Millions of records were stolen | Breaches of this kind are more common in the health care sector |
| Anthem was extremely open and honest with the concerned authorities | Security practices, including encryption and training, was either not ideal or was non-existent | Health care sector suffers from poor security practices and protocols |
| Anthem had a well-defined plan to help key stakeholders | Business and IS Strategy were not aligned | Health care industry requires a proactive implementation of controls |

The inability to align business and IS strategies has created other problems as well. Most prominently, it has not fostered a culture of security among the people in the Anthem organization. Network security is not a one-off event. Like washing one's hands regularly to get rid of germs, organizations need to continue to monitor their networks, lock down passwords and sensitive information, and constantly improve their security methods. Anthem's failure in maintaining a culture of security is evident in the practices that their associates put in place. The cybercriminals were able to obtain network login credentials of at least five high-ranking Anthem IT associates through common scamming techniques like email phishing. Anthem also chose not to encrypt the data that was being stored. While this isn't a sufficient measure in and of itself it should be a step put into place as another fence that cybercriminals would need to get around. Anthem also wasn't forthright about what happened in the attack. They said that it was a "sophisticated attack" with no further explanation, which drew the ire of many critics. Anthem did well by letting authorities know about the data breach and by individually telling customers, but they gave no solid information on the attack. They gave no hint of admittance of fault. They didn't acknowledge the weaknesses that their network had. There was a faint glimmer of transparency but Anthem fell back on being a victim and not admitting that the attack was something that could have been prevented. A healthy culture of security would have had Anthem training its associates on known cybercriminal schemes such as spear phishing. A culture of security would have taught associates the importance of securing passwords and checking the credentials of websites and emails before they divulged sensitive information. There should have been regularly updated training that kept associates abreast of the latest attempts by cybercriminals. And lastly, a healthy culture of security would have made Anthem acknowledge that the attack wasn't so sophisticated and instead that mistakes were made on their part. Playing the victim doesn't help anyone in this situation. Instead, Anthem should have shown where the weakness was and what their plan was to fix it. A sweeping change needs to take place in all of the health care industry. A culture of security needs to be the norm for not just Anthem but all health insurers, health care providers, and their affiliates.

Another issue that has arisen due to Anthem not being able to align their business and IS strategies is that their human element was exploited. What that means is that their employees were taken advantage of by cybercriminals. The human element is both one of IS's greatest weaknesses and one of its greatest strengths. It can be one of its greatest weaknesses because in the Anthem example, it is people who allowed the data breach to happen. It wasn't a failure of security software, and it wasn't a collapse of IT infrastructure that allowed cybercriminals in. It was the people that work at Anthem. The gatekeepers of the network compromised their credentials due to being tricked. They were taken advantage of. No software is going to be able to rectify that. The only way to keep that from happening is to train associates on security threats and proper protocol for login credentials. On the other hand, the human element can also be an organization's greatest asset. In the case of Anthem, it was people who found the breach, not security software. It was Anthem IT associates scanning their network that detected the breach. There were no alarms going off or automated alerts set in place. It was the people that work there realizing that something was not right. Humans can be wild variables, either hurting or helping depending on the situation. Luckily, where some failed others succeeded. People were able to use intuition and logic and not just software programming and red flagging patterns to catch on to what had been happening. If it had been left up to the security software itself the breach may never have been detected since the cybercriminals had obtained the credentials to be on the network.

## Strategic Choices

There are several takeaways for Anthem and the health care industry in general. The breach taught Anthem a lesson. If the understanding from the breach is put into practice it could change the whole industry and Anthem could lead the way. They also can start working towards a culture of security in their organization. They can better inform their associates of the need for security and continually search for ways to keep their network safe against the latest threats. They can also use the human element to their advantage. It is everyone's job to safeguard against cyberthreats. The industry has hired skilled people; they just need to train them up and empower them to be on the lookout for suspicious activity (see Figure C1).

During the data breach of 2015, Anthem's business strategy and IS strategy did not meet up. Often this is the very problem that creates so many issues within the industry. But it doesn't have to stay that way. Because of the breach, Anthem knows that its IT infrastructure needs to be bolstered in order to catch up with the values they hold as a company. Innovation and trustworthiness can be pushed to the forefront of their company if they learn from past mistakes. Samanage.com posted six lessons learned from the Anthem security breach that would help Anthem to get back on track and provide tips to other cybersecurity professionals. The first is to be tougher to crack than others. The idea is the same as running from a bear. You don't have to be the fastest, you just have to be faster than the person behind you.

**Figure C.1.** Different cybersecurity postures

Cybercriminals are looking for easy prey. Putting more barriers up will dissuade a lot of criminals from messing with you. The second point is to not depend on a single security solution. If that solution is bypassed then there is nothing else standing between a cybercriminal and his quarry. IT associates need to regularly monitor their network and continually evaluate and add new solutions to their technical portfolio. Third, don't rely on encryption alone. Encryption is a great step in securing data. It keeps unauthorized users out. However, a lot of breach attempts are from individuals who got into the system using authorized users' information and would therefore be able to bypass the encryption process. Fourth, don't collect and keep info that isn't needed. It is becoming more and more common for organizations to store information that they don't need. The less information you have in your database the less the cybercriminal can steal from you. Fifth, have the IT help desk enact better rules for password changes. Have passwords changed frequently. Make them tough to figure out. Doing this will help keep intruders from guessing. The sixth and last point: make customers and employees aware of the potential for phishing. Health care organizations need to make their associates and customers aware of what official communications look like, by contrast with something bogus that a cybercriminal developed to get information.

Health care organizations still have the ability to create a culture of security. To do so they need some best practices for cybersecurity. The first step to do that is training. Train associates on better password management, from making sure not to give out passwords in emails to making sure passwords are strong and changed regularly. Also, train employees on how to monitor the network to catch cybercriminals before they can do wrong. The next step is to minimize security issues. Have employees update their software, encrypt data, erase old passwords—all will help in adding layers of security to the organization. Third, secure the devices that are capable of being online. Organizations need to know what is connected to their network and where it is physically located. In the Anthem example, the website Darkreading.com recommends context-aware access control for Anthem's network. Context-aware access control is a security process for examining where the login attempt was made from, what platform was being used, and the time of day the login attempt was made. The security technique can potentially lock out outsiders even if they have authorized credentials if certain criteria aren't met such as not being within Anthem's offices, logging in after business hours, making a login attempt on a mobile device or outside of Anthem's ISP, etc. Similarly, the National Security Agency/Central Security Service (NSA/CSS; 2015; pp. 1–3) informs that the best practices to protect information systems and networks from a destructive malware attack include:

- Segregate network systems in such a way that an attacker who accesses one enclave is restricted from accessing areas of the network.
- Protect and restrict administrative privileges, especially for high-level administrator accounts, from discovery and use by the adversary to gain control over the entire network.
- Deploy, configure, and monitor application whitelisting to prevent unauthorized or malicious software from executing.
- Limit workstation-to-workstation communications to reduce the attack surface that an adversary can use to spread and hide within a network.
- Implement robust network boundary defense capabilities such as perimeter firewalls, application firewalls, forward proxies, sandboxing, and/or dynamic analysis filters to catch malware as it enters the network.
- Maintain and actively monitor centralized host and network logging solutions after ensuring that all devices have logging enabled and their logs are being aggregated to those centralized solutions to detect anomalous or malicious activity as soon as possible, enabling containment and response actions before significant damage is done.
- Implement Pass-the-Hash (PtH) mitigations to reduce the risk of credential theft and reuse.
- Deploy Microsoft Enhanced Mitigation Experience Toolkit (EMET) or other anti-exploitation capability (for non-Windows operating systems) to prevent numerous initial exploits from being successful.
- In addition to anti-virus services, employ anti-virus file reputation services to gain full benefit from industry knowledge of known malware sooner than normal anti-virus signatures can be deployed.
- Implement and tune Host Intrusion Prevention Systems (HIPS) to detect and prevent numerous attack behaviors.
- Update and patch software in a timely manner so known vulnerabilities cannot be exploited.

To keep themselves secure in the future organizations can take advantage of the human element. Their greatest asset is really their people. If they take the time to train their associates and train them on how to be safe online and how to monitor the network then they'll be better off than if they just had the best network security software that money can buy. It was through Anthem associates that the data breach was both started and discovered. A group of associates forfeited their passwords and another group discovered that the network had been tampered with. IT associates should be trained to look at user activity and compare it to the user's history to see if any red flags pop up. There are software systems that can aid in this analysis too. Real Time Security Intelligence (RTSI) systems can look for patterns in user activities and if a red flag is found can temporarily close down a user's access until the associate can assess the situation, much like when a person's debit card is used in a location that the person doesn't normally frequent, an automated system can lock access to the card and a customer service representative can get in contact with the person to verify their usage.

Anthem has the capability of learning from the breach and becoming a leader in network security for the health care industry. They just need to make sure that all their departments are on the same page and share the same goals. They need to be more conscientious about cybersecurity and make it more difficult for intruders to cause harm. They also need to empower employees to be cybersecurity watchdogs through training and organizational culture change. With these actions Anthem can take the lead and make the health care industry go from being one of the most hacked industries to being the most secure.

## Questions

1. What proactive steps should Anthem take in order to prevent such data braches from occurring again?

2. There is a lot of controversy surrounding the lack of encryption, HIPAA requirements, and choices made by Anthem. Comment on reasons why encryption was not undertaken and how policy compliance can be brought about.

3. The health care industry has had some issues with cybersecurity. Why is it that the health care sector is so vulnerable? What should hospitals and insurance companies do to ensure protection of data?

## References

1. Munro, D. 2014. Just how secure are IT networks in healthcare? *Forbes*, August 3.
2. Humer, C. 2015. Anthem says at least 8.8 million non-customers could be victims in data hack. *Reuters*, February 24.
3. Hiltzik, M. 2015. Anthem is warning consumers about its huge data breach. Here's a translation. *LA Times*, March 6.
4. NSA/CSS. 2015. Defensive best practices for destructive malware. https://www.iad.gov. Accessed September 15, 2017.

# CASE STUDY 2

# Process and Data Integrity Concerns in a Scheduling System*

Since 2014 the Department of Veterans Affairs has undergone ongoing scrutiny for a scandal involving patient wait times increasing and being falsified at hospitals and clinics across the country. Injured veterans have reported not being able to receive critical care for months due to over- or mis-scheduling. Under scrutiny, the VA is currently debating whether to implement a home-grown software system or procure an off-the-shelf solution supplied by the Epic Systems Corporation. This case provides a great illustration of how breakdowns in the circumstances surrounding information technology systems such as organizational structure, business policy, and project planning, coupled with increasing customer demand, equate to negligence for an agency that is trusted to take care of individuals. The case also speaks to issues related to social responsibility, ethics of technology use, and the resultant security and integrity challenges.

## Case Description

In 1930, President Herbert Hoover established the Department of Veterans Affairs (VA) to facilitate services for United States military veterans and take on this mission. The VA is a United States federal agency providing health care and programming to active and retired military veterans all around the country. To facilitate their care, the VA operates in 144 hospitals spread throughout the country and utilizes an additional 1,221 VA outpatient sites. "To care for him who shall have borne the battle, and for his widow, and his orphan" is part of the VA's mission statement; a large aspect of that mission is timely appointments for care. To realize this mission, in 1995 the VA established a goal of fulfilling any primary or specialty appointments within 30 days.

---

* This case was prepared by Sam Yerkes Simpson under the tutelage of Dr. Gurpreet Dhillon.

During the 2000s, amid ongoing war and battles on the frontlines in the Middle East and other regions, increasing numbers of injured veterans were coming home to receive the care they are entitled to from the VA. In addition to younger veterans, 43% of U.S. veterans are 65 years or older and require specialized ongoing care. Despite the increase in both younger and older patients and no major increase in funding, in 2011 the VA shortened their goal to fulfill all primary care appointments in just 14 days. This 14-day appointment goal set by former Veterans Affairs Secretary Eric Shinseki and VA leadership was largely unattainable and as a result set off a disruption that trickled down to the appointment makers, pressuring them to do anything they could to make that goal, even if it involved deceptive actions.

In the spring of 2014, multiple news articles surfaced depicting long patient wait lists and mismanagement of scheduling systems. In some cases appointment schedulers were asked by their managers to keep alternative secret patient schedules to exhibit good scores—an alternative to the official appointment scheduling software, a custom legacy application developed 25 years ago by the VA's Office of Information and Technology unit (OI&T). This mismanagement resulted in worsening patient health outcomes and death for some veterans while waiting for care. Further investigation revealed not only that VA facilities were not utilizing the VA's central scheduling software but that they were also not following other agency standard scheduling procedures. These non-standard operating procedures were an effort to disguise long patient wait times from leadership and the American public to meet their measurement target.

In June of 2014 an official internal audit revealed that 120,000 veterans were left without care. One of those was 71-year-old Navy veteran Thomas Breen. In September of 2013 Breen was examined at the Phoenix, Arizona, VA hospital and diagnosed with an *urgent* stage-four bladder cancer. After unsuccessful attempts at scheduling follow-up care at the VA for his immediate issues and being told he was on "a list," Breen died in November 2013. The VA followed up with Breen's family in December 2013 to schedule his next appointment; tragically, Breen's family had to explain that he had already died.

The scheduling scandal was quickly attributed to the VA's rigid and complex home-grown scheduling VistA software, increased patient loads, and insufficient funding for care, due to the increase in the number of veterans from the Iraq and Afghanistan wars. The VA's Office of Information and Technology unit (OI&T) was portrayed as incompetent and wasteful of taxpayer dollars after it was exposed there had been numerous failed attempts at replacing the software prior to the scandal.

Upon reading the previous reports and investigation from the VA Office of Inspector General (OIG) and national scrutiny, the Federal Bureau of Investigation (FBI) opened a criminal investigation against the VA to investigate the negligence. One of the issues the reports noted was that in 2014 all 470 senior VA executives were awarded at least "fully successful" progress reports. Not one manager provided any negative feedback, which some questioned as odd given the large pool of senior managers and the many failing incidents.

The 14-day scheduling goal appeared as an issue due to the fact that this goal was made without any analysis or comparison across different VA hospitals or private hospitals. The lack of centralized reporting from the regional VA hospitals to the central command was reviewed as a misstep. The new goal was arbitrarily put in place by management, but left up to lower office workers to meet. At the regional level, the 14-day goal was often unachievable and unrealistic given their current patient load. The independent culture of each VA hospital

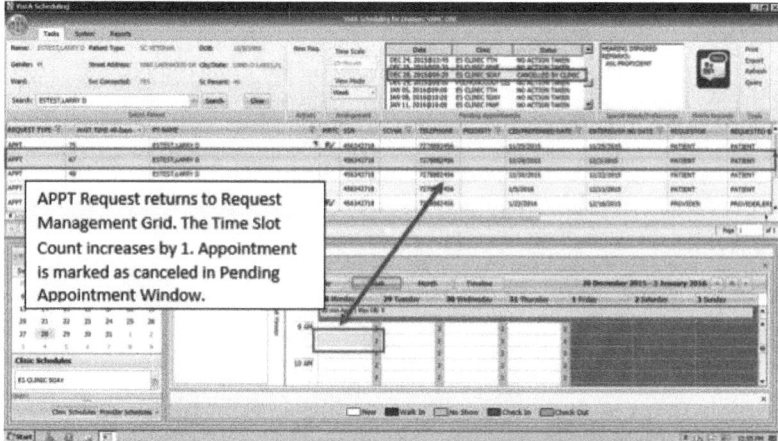

**Figure C.2.** Screenshot of the VistA scheduling software user interface

managed expectations differently and in some cases even encouraged manipulation of the data. Without regular checks, comparison reports with other VA hospitals, or mandatory use of centralized technologies, regional VA hospitals were often encouraged to bluff the system to make their hospital appear to be meeting their goal.

The VA's OI&T unit is made up of 16,000 employees who are 50% permanent employees and 50% contract employees. Following the allegations, the OI&T unit admitted that they had been trying to replace the old system since 2000, but due to IT skills gaps within the department they needed to look to using a commercial product instead of developing a proprietary application. The failed implementation of replacing the rigid legacy system cost $127 million dollars, money that comes from taxpayers. Additionally, it has been widely speculated that the ability to fire and rehire employees within the federal government is too inflexible, allowing unmotivated and unsatisfactory employees steady employment despite ongoing performance issues. Another contributing factor is perhaps the ratio of contract employees to VA employees, which should be shifted due to the fact that in-house employees are typically more loyal and understand the organizational culture (see Table C.2).

**Table C.2a.** Perceived loyalty

|  |  | IT Specialization | |
|---|---|---|---|
|  |  | Low | High |
| Perceived loyalty to a division | High | Permanent employee | Opportunity to hire innovative permanent employees |
|  | Low | Opportunity for the VA to rehire motivated employees | Outside contractor |

## Reform

As a recovery period emerged, Congress and the president of the United States at that time, Barack Obama, emerged as obvious opinion leaders to support the VA to change. The VA brought in new senior leadership (many of their predecessors were forcefully pressured to resign following the scandal), including the Veterans Health Administration's top health official, Dr. Robert Petzel. New leadership was also brought into the OI&T unit when LaVerne Council was hired to strengthen and improve several areas of the IT infrastructure of the VA, including overall portfolio management.

On May 21st, 2014, Congress passed the "Department of Veterans Affairs Management Accountability Act of 2014" which allows for easier removal or demotion from leadership roles for senior management officials based on performance assessment. On June 10, 2014, Congress also passed a bill that would allow veterans to receive care from private non-VA facilities under certain conditions. The costs would be covered by the VA and would be used in situations of over-capacity to combat long patient wait times.

As the software was written 25 years ago, it is likely many of the employees who originally wrote the code and who were most familiar with the system have since moved to a different job or retired. This leaves to new employees the daunting task of not only modernizing the code but enhancing it with new requirements. Improving legacy systems is incredibly complex and after many failed attempts to incrementally improve the code the VA has admitted they do not have the necessary experience to rewrite the code base to the scheduling application. As a results the OI&T has started the process of procuring bids from outside agencies for a commercial off-the-shelf product. Large corporations such as Booz Allen Hamilton, CACI International, IBM, and Microsoft showed interest in winning the contract. This approach would also limit ongoing maintenance costs for many years to come.

In addition to the implementation of just the scheduling software development, the VA began a pioneering initiative, which incorporates the scheduling software upgrade, called Medical Appointment Scheduling System (MASS), to provide state-of-the-art electronic health records, scheduling, workflow management and analytics capabilities to frontline caregivers. MASS aims to web-enable all aspects of the VA's health care in one centralized portal to quickly and efficiently share patient records and data with stakeholders—a difficult undertaking considering how non-decomposable the current legacy systems are.

On June 26, 2015, a five-year, $624 million dollar contract was awarded to Epic Systems, a subsidiary of Lockheed Martin, by the VA and the Department of Defense. The goal of the contract would be to replace the current VistA Scheduling Software with a new implementation of MASS software by 2017.

By April 15, 2016, the contract was put on hold and the MASS project was suspended as the VA tested in-house fixes to their current scheduling software in a trial rollout of 10 locations. This despite the previous admission that they were not able to make the necessary adjustments to the scheduling system themselves. If the fixes are successful they are planned to be rolled out to all VA hospitals and will save nearly $618 million. Successful implementation of the in-house fixes would obviously save a lot of money and provide the VA with the opportunity to align their business practices with their IT solutions. Completing the smaller fixes would also allow the VA to not have to address any supplemental systems that integrate or communicate with the scheduling software. Clearly this plan is an interesting opportunity, but the back and forth in directions from the new VA CIO, LaVerne Council, does not help manage change or improve confidence in the VA's ability to manage information systems and policies.

# Implications

As the trouble that plagues the VA has not only been related to information technology but also focused on meeting business objectives, the agency would benefit from strategic information systems planning recommendations. Such planning would help the agency to ensure that process and data integrity issues are adequately addressed. Flaws in the scheduling system have the potential to be a major cybersecurity challenge. This is because broken processes are a means by which perpetrators exploit systems. Some management considerations could be:

1. Technical implementation strategies coupled with the agency's return on investment
2. Individual employee skills and training programs alongside organizational process as defined by software workflows
3. New project management and agile development methods
4. New leadership that creates an opportunity to improve cultural and organizational forms, and established perceptions
5. Social responsibilities to stakeholders

The VA's Office of Information Technology was faced with a difficult task of replacing a widely used key operational system. Due to the mission-critical nature of the hospital system running 24 hours, 7 days a week, they would not be able to afford any downtime or lag during any type of transition to a new system. As the system was built many years ago, new technologies, policies, and business objectives have emerged. It would be difficult to make even incremental improvements to the system or completely replace the system while

**Table C.2b.** Timeline of major events

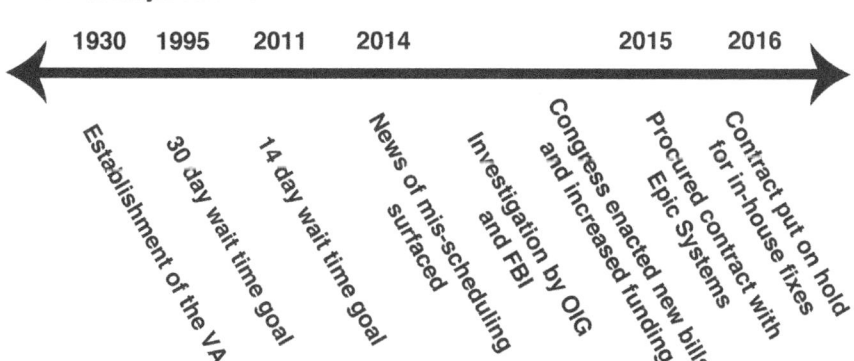

ensuring that all the needs are met for multiple hospitals, hospital units, and stakeholders. Not having the ability to decouple the system and incrementally improve smaller portions at a time creates a tremendous task.

It is also not clear what improvements or replacement to the current VistA software could be made that would equate to a quantifiable return on investment. Changing or replacing the current system with a similar scheduling system lacks impact and exhibits more risk than the perceived reward. Therefore, it is possible the VA would be best off not making any changes to the software, but instead making adjustments in two areas: training users with the current software and managing organizational expectations.

Oftentimes, training is done right after a launch of the software and employees who inherit the system after starting a new position are left to simply "figure it out for themselves" or refer to the documentation. As the VistA software was originally launched in 1997, it is likely many users started using the system after the initial launch and training. Ongoing new user training and self-service documentation modules should be implemented to maintain the integrity of the system and scheduling process. An interactive module or method such as open forums or regular online webinars should be added so employees can be encouraged to ask questions and collaborate around issues with the software. Continuous training and subsequent feedback will ensure that employee morale and customer service remain high, while computing mistakes and computer deception remain low.

Due to different patient issues and calculating critical patient necessity, scheduling patient appointments is not always a well-defined process. The VA needs to hire employees skilled at scheduling, and adept at understanding medical terminology and patient ailments, managing goals, overseeing multiple appointment calendars, and assessing availability of resources, such as equipment, rooms, and personnel. While an integrated computer process in the scheduling software could do an adequate baseline job at this, optimally the software should allow for schedulers to overwrite the default schedule and use their best judgment when making appointments. Baseline knowledge, skills and abilities documentation, and supplemental interview questions would be helpful to establish uniformity in employee skills when hiring for this role across the entire VA agency.

As replacing the legacy scheduling software has been an ongoing project spanning multiple attempts and contracts, it would benefit the OI&T to practice new project management 2.0 techniques and agile development methods. Utilization of these techniques would increase collaboration and communication in an office that is made up of up to 50% of contract workers. It is possible that increased communication would improve morale, build employee relationships, and encourage innovation through employee knowledge transfer. The utilization of agile development methods would allow the office to be flexible when new requirements or policies are set by VA leadership or additions to federal guidelines.

During the reform period following the scandal, LaVerne Council was hired as a new VA Chief Information Officer. Bringing in new leadership such as Council can revive unmotivated employees and create a more innovative culture and workspace. As with any new position of leadership, one must question her own leadership competencies with regard to where the OI&T unit currently is positioned and where they will be in the future. Council should also assess the competencies of her individual staff and also the organizational processes to judge the value that the OI&T can bring to American veterans.

Another important aspect of the VA is its established policies, structure, and culture. Council should be conscious that, while policies for governmental agencies such as the VA are often highly formal and established many years ago, she should recognize her opportunity to create new informal norms and a new culture within the OI&T from the beginning of her tenure.

Generally, the VA follows a very traditional hierarchical organizational model, but with the ongoing changes this could potentially be a positive time to restructure the OI&T to utilize smaller modularized reporting structures. While government agencies are commonly categorized as behind the times in technological innovations and very slow to make decisions, a more agile organizational structure could result in units improving internal decision-making skills, create faster turnaround times with regard to decision-making, become more forward-thinking, and advance their ability to be creative problem solvers. The OI&T should work to change their slow and generally negative perception and become leaders among other federal agencies' information and technology units. Employee learning opportunities such as technology industry conferences and engagements should be encouraged in an effort to inspire employees to experiment with new technology, as experimental technologies have the potential to be used in the future to improve efficiencies through utilization.

As a federal agency that receives federal funds from taxpayers, the VA has a social responsibility to fulfill its mission as efficiently and economically as possible. While many checks and balances are put in place to assure there is no deception, that was not the case during this scandal. During the reform period, the VA made considerable efforts to communicate its malpractice through many subsequent reports and a strong public relations effort, but continuation of this practice is needed. The VA leadership should inspire and instill the values of honesty and transparency within their agency so as to transmit them through the entire organization as a whole.

## Summary

The increase in patient load and management expectations combined with minimal funding and a legacy software infrastructure all fuelled the scheduling scandal in 2014. However, since then, the VA has made considerable efforts through significant reform actions to ameliorate and resolve these issues. Although coverage of the case has already reached a climax in the media, currently the VA is still discussing how to implement a new solution to their scheduling and medical records data system challenges. With new leadership and support from our nation's leaders in Congress and the president, it will be interesting to see if the VA will be able to decide on an implementation method. Although the scheduling and software project implementation has been ongoing for years, there is no better time to implement strategic information systems planning competencies and skills.

## Questions

1. VA had several issues. Some were related to business problems, while others were concerned with data and the systems in use. Suggest how each of the problems should have been addressed.

2. Draw out implications for possible security threats resulting from "broken processes" at the agency.

3. If you were to be brought in as a cybersecurity consultant, how would you go about addressing the concerns?

## Sources Used

Boyd, A. 2014. VA issues RFI for commercial scheduling system. *Federal Times*, November 26.

Bronstein, S., and G. Drew. 2014. A fatal wait: Veterans languish and die on a VA hospital's secret list." *CNN*, April 23.

Conn, J. 2016. Epic stands to lose part of $642 million VA patient-scheduling system contract. *Modern Healthcare*, April 14.

McElhatton, J. 2014. White House warned about "antiquated" VA scheduling system 5 years ago. *Washington Times*, July 20.

Reynolds, T., and L. Allison. 2014. Performance mismanagement: How an unrealistic goal fueled VA scandal. *NBC News*, June 25.

Slabodkin, G. 2016. VA delays contract for appointment scheduling system. *Health Data Management*, April 18.

Sullivan, T. 2015. Epic grabs VA software contract. *Healthcare IT News*. August 27.

US Department of Veterans Affairs. Department of Veterans Affairs Office of Information and Technology (OI&T). 2016. *Vista scheduling graphical user interface user guide*, May.

US Department of Veterans Affairs. National Center for Veterans Analysis and Statistics. 2016. *Department of Veterans Affairs statistics at a glance*, May.

# CASE STUDY 3

# Case of a Computer Hack*

This case study is based on a series of events which occurred over a period of two years at the Stellar University (SU), which is an urban university. SU caters primarily to commuter students and offers a variety of available majors, including engineering, theater, arts, business, and education.

SU is a public educational institution which contains a diverse range of technologies. In general, if it exists in the information systems realm, at least one example of the technology can be located somewhere on campus. Mainframe, AS400, Linux, VAX, Unix, AIX, Windows (versions 3.1 to 2003 inclusive), Apple, RISC boxes, SANs (storage area networks), NASs (network attached storage), and whatever else has been recently developed is functioning in some capacity. The networking infrastructure ranges from a few remaining token ring locations to 10/100/1000 Mbps Ethernet networks, wireless, and even some locations with dial-up lines. A VPN (virtual private network) is in place for some of the systems shared with the medical portion of the university, primarily due to HIPAA (Health Insurance Portability and Accountability Act of 1996) requirements.

In this open and diverse environment, security is maintained at the highest overall level possible. The computer center network connections are protected by a firewall. Cisco routers are configured as "deny all except," thus only opening the required ports for the applications to work. IDS (intrusion detection systems) devices are installed at various locations to monitor network activity and analyze possible incidents. The systems that are located in the computer room are monitored by network and operating system specialists whose only job is the "care and feeding" of the equipment.

Servers may be set up by any department or individual under the guise of "educational freedom" and to provide a variety of available technologies to students. For this purpose, many systems are administered by personnel who have other primary responsibilities, or do not have adequate time, resources, or training. If the system is not reported as a server to the network group, no firewall or port restrictions are put into place. This creates an open, vulnerable internal network, as it enables a weakly secured system to act as a portal from the outside environment to the more secured part of the internal network.

---

* This case was prepared by Sharon Perez under the supervision of Professor Gurpreet Dhillon. The purpose of the case is for class discussion only; it is not intended to demonstrate the effective or ineffective handling of the situation. The case was first published in the *Journal of Information System Security* 1(3). Reproduced with permission.

The corporate culture is as diverse as the computer systems. Some departments work cooperatively, sharing information, workloads, standards, and other important criteria freely with peers. Other areas are "towers of power" that prefer no interaction of any kind outside the group. This creates a lack of standards and an emphasis on finger-pointing and blame assignment instead of an integrated team approach. Some systems have password expirations and tight restrictions (i.e. mainframe) and some have none in place (domain passwords never expire, complex passwords not enforced, no password histories maintained, etc.).

## Computer System

The server in this situation (let's call it server_1) was running Windows NT 4.0 with service pack 5 and Internet Explorer 4. Multiple roles were assigned to this system. It functioned as the domain Primary Domain Controller (PDC) (no backup domain controllers (BDCs) were installed or running), WINS (Windows Internet Naming Service) server, and primary file and print server for several departments. In addition, several mission-critical applications were installed on the server. There were few contingencies in place if this server crashed, though the server was a critical part of the university functionality. For example, if the PDC was lost, the entire domain of 800+ workstations would have to be recreated since there was no backup copy of the defined domain security (i.e. no BDC).

To complicate matters, a lack of communication and standards caused an additional twist to the naming convention. On paper, the difference between a dash and an underscore is minimal; in the reality of static DNS (domain name system) running on a Unix server, it is huge. The system administrator included an underscore in the system name (i.e. server_1) per his interpretation of the network suggestion. The operating system and applications (including SQL 7.0 with no security patches) were then installed and the server was deemed production ready.

As an older version of Unix bind was utilized for the primary static DNS server by the networking group, the underscore was unsupported. There were possible modifications and updates that would allow an underscore to be supported, but these were rejected by the networking group. This technical information was not clearly communicated between the two groups. Once the groups realized the inconsistency, it was too late to easily make major changes to the configuration. Lack of cooperation and communication resulted in each faction coming to its own conclusion: the system administrator could not change the server name without reinstallation of SQL (version 7.0 did not allow for name changes) and a reconfiguration of the 800+ systems that were in the domain. The network group would not make a bind configuration change that allowed for underscores, and instead berated the name server_1, indicating it should have been named server-1 as dashes are supported.

This miscommunication led to further complications of the server and domain structure. Neither group would concede, but the system administrator for the domain had to ensure that the mission-critical applications would continue to function. To this end, the server was further configured to also be a WINS server to facilitate NetBIOS name resolution. As there was no negotiation between the groups, and the server name could not be easily changed, this became a quick fix to allow the production functionality to continue. The actual reason for this fix was not clearly communicated between the groups, thus adding to the misunderstandings.

For various reasons, this server was now running WINS, file and print sharing, PDC (no BDC in the domain), and mission-critical applications. In addition, the personnel conflicts resulted in the server being on an unsecured subnet. In other words, there was no firewall. It was wide open to whoever was interested in hacking it. No one group was at fault for this situation; it was the result of a series of circumstances, technical limitations, and a disparate corporate culture.

The server is an IBM Netfinity, which was built in 1999. At the time of its purchase, it was considered top of the line. As with most hardware, over the years it became inadequate for the needs of the users. Upgrades were made to the server, such as adding an external disk drive enclosure for more storage space and memory.

The manufacturer's hardware warranty expired on the server and was not extended. After this occurred, one of the hard drives in the RAID 5 (Redundant Array of Inexpensive Disks) array went bad (defunct). The time and effort required to research and locate a replacement drive was considerable. A decision was finally made to retroactively extend the warranty, and have the drive replaced as a warranty repair. The delay of several days to accomplish this could have been catastrophic. RAID 5 is redundant, as the name suggests, and can tolerate one lost drive while still functioning at a degraded level. If two drives are defunct, the information on the entire array is lost, and must be restored from backups. Backups are accomplished nightly, but there is still the worst-case scenario of losing up to 23.5 hours of updates if a second drive goes bad just before the backup job is submitted.

## Changes

Several factors changed during this two-year period. A shift in management focus to group roles and responsibilities, as well as a departmental reorganization caused several of the "towers of power" to be restructured. These intentional changes were combined with the financial difficulties of the province and its resulting decrease in contributions to public educational institutions. The university was forced to deal with severe budgetary constraints and cutbacks.

University management had determined that all servers in the department (regardless of operating system) should be located at the computer center. This aligned with the roles and responsibilities of the computer center to provide an appropriate environment for the servers and employ qualified technical personnel to provide operating system support. Other groups (i.e. application development, database administration, client support) were to concentrate on their appropriate roles, which were much different than server administration. The resistance to change was department-wide, as many people felt that part of their job responsibilities was taken from them.

Moving the servers to a different physical location meant that a different subnet would be used, as subnets are assigned to a building or geographical area. The existing subnet, as it was not located at the computer center, did not have a firewall. That fact, combined with personnel resistance and discord between the groups, resulted in quite limited cooperation.

For this reason (more politically driven than "best practices" inspired), the unsecured subnet was relocated to the computer center intact as a temporary situation.

By the same token, the existing system administrators were not very forthcoming about the current state of the systems, and continued to monitor and administer them remotely. This was adverse to the management edict, but allowed to continue. On a very gradual scale, system administration was transferred to the computer center personnel. Due to lack of previous interaction between the groups, trust had to be earned as the original system administrators were still held accountable by their users. They would be the ones to face the users if or when the system went down, not the server personnel at the computer center.

## History of the System

The server (server_1) was relocated on an "as is" basis, and the accountability for the server was transferred. Minimal system documentation and history were included, and since the new system administrators had not built the systems, reverse engineering was necessary to determine what software was installed and how the hardware and software was configured. Minor modifications were made to the servers, with appropriate permission, to bring them in line with current standards. Some of the changes broke applications temporarily, as it was a learning process for the new administrators.

For instance, Windows NT 4.0 service pack 6a was not originally applied. This service pack had several patches to eliminate huge security holes that were inherent in NT 4.0. As there was a tenuous working relationship between the groups, all proposed changes had to be reviewed and approved so trust could be established. Each scheduled maintenance window provided its own challenge, as the system was considered by most of the technicians involved to be "temperamental."

Simple modifications were made with approval, which did not cause a system outage. These changes were designed to decrease the intrinsic vulnerability of the server. The changes were not implemented previously, as the original system administrator had considerably more work to do than one person could handle. His priority was to "fire fight" and keep everything running. Items such as IIS were installed, and services like FTP, WWW, and DHCP were set to "automatic" and "started." These were removed or disabled since they were not being utilized; they only wasted resources and created additional security challenges.

The first off-hours maintenance attempt was quite disastrous. Windows NT 4.0 service pack 6a would not apply (error message of "could not find setup.log file in repair directory"), and had to be aborted. Subsequent operating system critical updates would not apply for the same reason. SQL 7.0 was also behind on maintenance patches, and the installation of SQL 7.0 service pack 4 was flawless until it hit 57%. At that point it would not continue because there was "not enough room to install," and it would not uninstall at that point either. The critical application that used SQL would not launch when signed on locally to the server as an administrator, and had an error. The server was restarted, and was available to the users the next day, though the status of the application was still in question. According to the users, however, the application worked fine the next morning, even though it could not be opened locally on the server.

Research was accomplished to determine how to correct these error messages. Microsoft knowledge base article 175960 had a suggested corrective action for the "could not find setup.log file" error. Another maintenance window was scheduled and service pack 6a was finally applied to the server. But the version of Internet Explorer (IE) then reverted back to IE version 2.0 and the server "forgot" it had more than one processor. Further research and off-hours attempts finally allowed all service packs and security patches to be applied. The single processor problem was corrected via Microsoft knowledge base article 168132. The IE rollback was corrected by reapplying the 6a service pack.

## Other Issues

To complicate matters further, the provincial government had a serious financial crisis. Budgets were severely cut, and for the first time in recent memory, many state employees were laid off. This reduction of manpower caused numerous departments to eliminate their information systems (IS) support personnel and rely on the university technical infrastructure that was already in place. This strained the departments that had the "roles and responsibilities" of the support areas further, as they had decreased their manpower also. This resulted in frustration, heavy workloads, and a change in procedures for many departments.

One of the suggestions for an improved operating environment was to replace the current temperamental system (server_1) with new hardware that had an active hardware warranty and ran a current server operating system. This avenue initially met with a considerable number of obstacles, including the fact that the original system administrators were unfamiliar with the new version of operating system, questions as to whether legacy programs were compatible with the new operating system, and the complication of replacing a temperamental system that was functioning in numerous critical roles.

A joint decision was made between the groups to replace the legacy hardware and restructure the environment in a more stable fashion. Several replacement servers were agreed upon, and a "best practices" approach was determined. The hardware was then purchased, received, and installed in a rack in the computer center. At that point, lack of manpower, new priorities, resistance to change, and reluctance to modify what currently functioned caused a delay of several months. The project's scope also grew, as the system replacements became linked to a migration to the university active directory (AD) forest.

## Hack Discovered

On a Monday morning in February, the system administrator was logged onto the server with a domain administrator account via a remote control product. He noticed a new folder on the desktop, and called the operating system administrator at the computer center. Upon signing on locally with a unique domain administrator level user ID (i.e. ABJones) and password, there were several suspicious activities that occurred. Multiple DOS windows popped up in succession, the suspect folder was recreated and the processor usage spiked higher than normal. The new folder was named identically to the one that had just been deleted by the other system administrator during his remote session.

As server_1 was previously set up to audit and log specific events (per the article "Level One Benchmark; Windows 2000 Operating System v1.1.7" located at www.cisecurity.org), the Windows event log was helpful in determining the cause of the activities. Several entries for "privileged use" of the user ID that was currently logged on as a domain administrator (ABJones) were listed in the security logs. During the few minutes that the server was being examined, no security settings were knowingly modified. These circumstances raised further questions, as the more in-depth the system was examined, the more unusual events were encountered.

A user ID of "Ken" was created sometime during the prior weekend, and granted administrative rights. No server maintenance (hardware or software) was scheduled, and none of the system administrators had remotely accessed the server during that time. Event logs indicated that Ken had accessed the server via the TAPI2 service, which was not a commonly used service at the university. The user ID was not formatted in the standard fashion (i.e. first initial, middle initial, first six characters of the last name), and was therefore even more suspect.

Antivirus definitions and the antivirus scan engine on the system were current; however, the process to examine open files was disabled (Symantec refers to this service as: file system realtime protection). The assumption was that this may have been the first action a hacker took so that the antivirus product did not interfere with the malware application installation. All of these circumstances added up to one immediate conclusion: that the system had most likely been compromised.

## Immediate Response

Both system administrators had previously read extensively on hacks, security, reactions to compromises, best practices, and much of the other volumes of technical information available. This, however, did not change the initial reaction of panic, anger, and dread. Email is too slow to resolve critical issues such as these. The system administrators discussed the situation via phone and came to the following conclusions: disconnect the system from the network to prevent the spread of a possible compromise, notify the security team at the university, and further review the system to determine the scope and severity of the incident.

Each administrator researched the situation and examined the chain of events. It was finally determined that a Trojan was installed on server_1 which exploited the buffer overrun vulnerability that was fixed by Windows critical update MS04-007. This vulnerability was created by Microsoft patch MS03-0041-823182-RPC-Activex which corrected the Blaster vulnerability. Once the compromise was confirmed, a broader range of personnel were notified, including networking technicians, managers, and technical individuals subscribed to a university security list-serve. A maintenance window had been previously approved to apply the new Microsoft patches to this and several other servers on Thursday, in three days.

## Further Research and Additional Symptoms

Continued examination of the system event logs indicated that a password crack program was executed on the previous Saturday evening using TAPI2 and a valid user ID on the system. Since this server was a domain controller, all other Windows servers were examined. Two additional servers were found to be compromised: one was a member server in a workgroup, and one was a domain controller for a Windows NT 4.0 domain that had a trust relationship with the hacked domain.

Upon closer examination, several additional changes were noted on the server:

- A scheduled task (At1.job) was created. This task seemed to be set to delete itself once it ran (it was set to run "one time only" and to "Delete the task if it is not scheduled to run again") to remove the hack traces. The job content was one line of code: "run cmd /c nc.exe –l –p 20000 –e cmd.exe."
- A new directory was created on the system (c:\winnt\system32\asy).
- When any administrator logged onto the system locally, DOS windows flashed momentarily while the Trojan executed the commands regedit.exe and hiderun.exe.
- Services called Gopher and Web were started; normally these were configured as disabled or manual.

Extensive examination of client computer systems within the domain indicated that the attack could have been relayed through another compromised machine at the university. A client system that was located in another area of the campus had the TAPI2 service compromised. The user of this particular system had set the user account password to be the same as the user account ID (i.e. user ID of jksmith has a password of jksmith). This was most likely the weak link that was exploited to gain access to the server.

The DameWare Trojan program DNTUS26 was eventually located on server_1. DameWare provides useful tools for network and system administrators. However, they admit:

> With the increased popularity of internet access, more and more computer systems are being connected to the internet with little or no system security. Most commonly the computer's owner fails to create a password for the Administrator's account. This makes it very easy for novice hackers ("script kiddies") to gain unauthorized access to a machine. DameWare Development products have become attractive tools to these so called "script kiddies" because the software simplifies remote access to machines where the Username & Password are already known.... Please understand that the DNTU and/or DMRC Client Agent Services cannot be installed on a computer unless the person installing the software has already gained Administrative access privileges to the machine (http://www.dameware.com/support/kb/article.asp?ID=DW100005).

There are several websites that discuss this Trojan, and offer methods of removing it, two examples are www.net-integration.net/zeroscripts/dntus26.html and www.st-andrews.ac.uk/lis/help/virus/dec20.html.

The overall symptoms of the hack were consistent with the BAT/mumu.worm.c virus (http://vil.nai.com/vil/content/print100530.htm). Netcat (nc.exe) was an active process, which may have been used to open a backdoor and gain access to the system. An FTP server was installed and configured to listen for connections on random ports over 1024. A directory was created on server_1 (c:\winnt\system32\inetsrv\data) and several files were created and placed there. The files in this directory contained information such as user names, passwords, group names, and computer browse lists from other network machines that could be seen from that server. The assumption was that this information was collected for eventual transmission to the hacker(s) to gain additional knowledge of the network environment. Additionally, a key was added to the registry that would reinstall the malware if it was located and removed by a system administrator.

### Additional Vulnerable Systems

The compromise of server_1 was a major security breach at the university. There are approximately 20 member servers and 800 client workstations in that particular domain. Since the primary domain controller was hacked and all of the domain security information was amassed in a hacker-created directory, it was assumed that the entire domain had been compromised. Once the domain administrator account was known, the hacker had full control of all systems within that domain. By default, the domain administrators group was placed into the local administrator group on all client workstations. This is Microsoft's default action and is accomplished for valid security reasons, such as a user leaving the company and not disclosing his/her password to the system. A domain administrator can, for example, access all of the information on that system and retrieve it for business continuity purposes.

In addition, since there was an explicit two-way trust relationship between this domain and another, the PDC in the second domain was also compromised. Security files were found on this second system in similar locations and containing similar types of information. Again, with the domain controller compromised, the two member servers and 100+ workstations that were a part of that domain were also suspect.

### Immediate Counter-Attack Actions Taken

Before server_1 was reconnected to the network, several actions had to be taken immediately to ensure that the system would not cause any additional security-related problems. The initial task was to clean the servers so that they could be brought back up. System administrators removed all of the malware that had been identified. A list of required ports was compiled to facilitate the firewall configuration by the networking group.

As indicated earlier, there were no password restrictions applied at the domain level, nor any password expiration time period established (GPOs, group policy objects, were not used to set this either, as it was a Windows NT 4.0 domain and GPOs cannot be used on an NT domain). Many of the users had the same password that was given to them when their account was created! A password policy was enabled (minimum password length of six, maintain history for five passwords, and a one-hour account lockout after five invalid sign-on attempts) and all user IDs were set to "user must change password at next logon." This had to be done manually (open the properties of each user ID and click the appropriate selections, and then click okay) as there were no login scripts, policies, or other means to globally apply the change.

These processes had to be accomplished on each infected system and on each of the compromised domains. The process, however, still left the system administrators uncomfortable as there was insufficient experience in forensics to ensure that all the remnants of the attack were removed. For this reason, an external vendor with the appropriate experience was contracted and requested to certify that the systems were completely cleaned. A computer forensic expert was brought in to accomplish this task and to ensure the return of the systems to full functionality. The vendor developed a series of steps to disable the Trojan and remove the infected files. The procedure was accomplished on the infected servers, as well as about 12 client workstations in the associated domains.

### Long-Term Counter-Attack Actions Taken

Once the immediate issues were corrected and the systems were brought back on line, there still remained the post-mortem examination to determine what went wrong and why. In this instance, the post-mortem was handled informally, and consisted of a summary write-up (for management) and an analysis of how to more effectively block against this type of attack in the future (for system administrators).

Several steps were taken to modify the standard server configurations in an attempt to avoid the same type of compromise in the future. First, the configuration for the open-source monitoring tool (Big Brother, http://bb4.com) that is used to report the system status was modified. Most incidents that were reviewed during the research phase of the hack began with a hacker disabling the antivirus product once he/she had gained access to the server. For this reason, the Symantec process that is responsible for real-time file protection was added to the list of services that were monitored. This change would cause system administrators to be paged or emailed if the service was stopped, regardless of the reason. It would not prevent intrusion, but would be an early notification tool that something may be amiss.

The temporary password policy changes were made permanent. A university policy change of this magnitude requires approval from several areas within the university. With the recent glaring example of what had happened when passwords were not restricted, the policy was approved rather quickly. In addition, the domain accounts are being further reviewed by security personnel to eliminate invalid accounts. Some users have been found with two or three IDs, thus increasing the number of "valid" IDs that can be used as means of attack. This would especially be true if the ID still had its original password.

One of the suggestions from http://vil.nai.com/vil/content/print100530.htm was to delete the administrative shares that are automatically created on each server. The shares are recreated after each system restart, but a batch file can be scripted to disable them upon boot each time. The site suggests:

Such worms often rely on the presence of default, administrative shares. It is a good idea to remove the administrative shares (C$, IPC$, ADMIN$) on all systems to prevent such spreading. A simple batch file containing the following commands may be of help, especially when run from a logon script, or placed in the startup folder.

- net share c$ /delete
- net share d$ /delete
- net share ipc$ /delete
- net share admin$ /delete"

Each server is currently configured with a batch file that runs on startup. The batch file gathers system information and places it in a text file on the hard drive, which is backed up nightly. The deletions for the net shares could be tailored to each server and placed in that batch file with minimal effort. This suggestion is still being reviewed by the system administrators.

## Summary

This particular incident was an "eye opener" for all involved. It was a shock to see how quickly, easily, and stealthily the systems were taken over. The tools that were utilized were all readily available on the Internet. The fact that the password policy was inadequate was already known, though the ramifications of such a decision were not fully explored. It was originally deemed easier to set no password policy than to educate the users, though that opinion drastically changed over the course of a few days.

The financial cost of this compromise has not been calculated, but it would be quite interesting to try to do so, factoring in lost time due to the servers being down, vendor contract fees, overtime for technicians (actually, it is compensation time, but it does affect how much additional time the technicians will be out of the office), delays in previously scheduled activities, meetings to determine notification and discussion of actions to be taken.

Computer forensics, in this case, was used to explore and document what the hacker had done, but not to track down who had gotten into the system. There was not enough knowledge on the system administrators' or contractor's part to begin to track down the culprit. This case was more one of "get it back up and running quickly and securely" than it was to prosecute the hacker. More surprisingly, there is a general knowledge of what type of information the servers held, but no concrete idea of what (if anything) was compromised. The details of what was actually compromised may not be apparent until sometime in the future.

# CASE STUDY 4

# Critical Infrastructure Protection: The Big Ten Power Company*

A common understanding of critical infrastructure industries is that a vast majority of them are owned and operated by private companies. This creates a vital need for federal and state cybersecurity and critical infrastructure authorities to engage in coordination with private sector partners. In planning and implementing cybersecurity initiatives, it is necessary for governments to engage with private stakeholders to identify their needs and concerns in terms of security, as any formal action by the government will likely impact the business operations of the private sector organization. In this uncertain climate with pending legislation, government security initiatives, and increasingly sophisticated attacks, companies must be vigilant in their business practices and more importantly their efforts to protect their critical systems from attack.

## The Big Ten Power Company

The Big Ten Power Company (BTP) is an intrastate energy company that provides electricity services to Indiana, Michigan, and Ohio. While centrally located in Indianapolis, Indiana, BTP is the primary provider of energy to the three states, including five major metropolitan areas, which are spread between them. In order to provide an adequate amount of power to support its region of operations, BTP has strategically placed power stations in each state. Due to significant coverage area and equally significant demand for electricity, BTP generates its power from a variety of sources, with 50% from coal, 40% from nuclear power, 8% from natural gas, and the remaining 2% from wind and oil. Since BTP covers a three-state region, it has regional offices located in each state to provide customer service and support, as well as manage power and restoration services in each respective area. Each regional office

---

* This case was prepared by Isaac Janak under the tutelage of Dr. Gurpreet Dhillon.

is in charge of affairs in its respective state but is responsible for any directives and reporting requirements by the main office in Indianapolis.

Big Ten Power began in Indiana in late 1988 but has since spread to two other states, beginning with Ohio in 2002. In the late 1990s, BTP transitioned from a privately held company to a publicly shared one, and is now publicly traded on the New York Stock Exchange (NYSE). In 1995, Ray Finkle took the helm of BTP as its chief executive officer (CEO) and was seen as the major contributing force behind the success and subsequent expansion of the company in the early 2000s. Due to BTP's performance, board members and stock holders continue to hold faith in Mr. Finkle and his efforts to expand the size of the corporation. Today Finkle remains at the helm of BTP, which employs more than 16,000 employees, with a reported $15 million in annual revenue.

## IT at Big Ten Power

In order to maintain the growth of BTP and pursue more robust forms of digital infrastructure, Mr. Finkle, with the support of the board, sought to designate a chief information officer (CIO) to be in charge of all information systems (IS) within BTP. Traditionally IS responsibilities were maintained by the chief financial officer (CFO); however, with BTP's rapid growth in size and geographical service area, Mr. Finkle was forced to designate a CIO to handle all IS/information technology (IT) responsibilities. In 2002, John Anderson, a former senior consultant with an East Coast–based IT consultancy firm, was appointed to the position of CIO. By designation, Mr. Anderson's duties were to revitalize existing IT infrastructure within BTP and to maintain and institute any new IT and industrial control system (ICS) infrastructure involved with business operations, customer service, and power generation and distribution. While not explicitly stated, a large portion of Mr. Anderson's responsibilities dealt with the security of systems and information that resided on systems.

Given the timing of his appointment, Mr. Anderson was forced to maintain and "reinvigorate" antiquated systems within all operations, generation, and distribution facilities in Indiana, while simultaneously implementing information systems into the new facilities in Ohio. BTP's expansion into Ohio was the result of an acquisition of a smaller energy company in an effort to not only expand geographically but to enter into the nuclear energy market. As a result of the acquisition, BTP incorporated a single nuclear facility into their arsenal of energy production and generation, and Mr. Anderson only had to establish IT infrastructure in the new facilities and integrate any of the existing infrastructure at the nuclear and distribution facilities into BTP's control.

While the integration of the out-of-scope infrastructure was feasible, costs for providing infrastructure and services for the new sites in Ohio proved to be too costly for BTP's budget. This forced Mr. Andersen to seek a third-party vendor to provide data center capabilities to maintain IT functionality off-site. This alleviated enough of the budget to continue to provide IT support services in-house. Although not ideal, this methodology seemed to be most effective for rapid growth and a constrained IS/IT budget. As BTP continued to expand into Michigan in 2008, its leadership chose to follow a similar model of retaining existing infrastructure from facilities it incorporated and outsourcing all new infrastructure needs to the same third-party vendor. However, the intention of this methodology was never to sustain infrastructure services through a contracted partnership. In 2009, as business began to stabilize for the Ohio branch, BTP shifted from third-party managed infrastructure to their own, with the intention being that, as the Michigan branch began to stabilize, BTP would establish its own IT infrastructure. Due to an unstable economy and less revenue, however, BTP elected to continue to sustain its infrastructure through a third-party vendor to allow for greater flexibility.

## Fears of Attack

Fears of an impending cyberattack began in 2006 when disgruntled employees successfully intruded into traffic control systems for the city of Los Angeles. According to reports, although management knew of the potential threat and took action to bar employees from the system, two engineers were able to hack in and reprogram several traffic lights to cause major congestion, which spanned for several days. While the attack was against a different type of control system, it indicated to the board and some investors that perhaps the resiliency of BTP's IT and ICS should be reviewed. Furthermore, it prompted the CEO to request a review be done of internal threats to IS.

By 2010, while hundreds of thousands of attempts were made daily to break into BTP's network, none had been successful against both IT and ICS infrastructure. This, however, did not prevent BTP's board from fearing the worst of their internal network and ICS security. Throughout 2010 and into 2011, the number of reported attack attempts and successful data breaches seemed to plague the media, creating a level of mistrust in the state of BTP's IS security among the board. Furthermore, several high profile attacks against ICS indicated that not even the systems controlling the generation and distribution of power were secure from malicious actors in a virtual environment.

In response, Mr. Anderson reviewed the risk associated with internal threats to BTP and elected to conduct an internal training and awareness campaign. Mr. Anderson and his IS department conducted a number of training seminars at each regional office and several of the power plants to inform employees of the types of threats, which were of concern, and to educate them to be aware of improper uses of technology at BTP. In addition, an awareness campaign was conducted in which Mr. Anderson used the Department of Homeland Security's "See Something, Say Something" antiterrorism campaign and applied it to internal threat awareness for employees. These actions satisfied executive leadership as well as the board and investors.

Unfortunately, however, by 2010 evidence of attacks were becoming more prominent throughout mainstream media, the most notable of which was Stuxnet. In 2010, the United States and Israel reportedly unleashed a malicious computer worm known as Stuxnet on Iran's Natanz nuclear power plant. The worm was inserted into the nuclear facility's systems through a USB flash drive and propagated itself throughout the ICS of the Natanz plant, attacking its programmable logic controllers (PLCs) in order to destroy the plant's centrifuges, while remaining entirely undetected by facility operators. This attack prompted several fears in the critical infrastructure industries and, more specifically, hit home for BTP because of its reliance on a nuclear facility for 40% of its power generation. Furthermore, the attack reinstated anxiety within leadership of an insider threat and the physical consequences of an attack against ICS.

From 2010 to 2013, concerns mounted over threats to internal network and ICS security within BTP. According to the Symantec 2014 Internet Security Threat Report, between 2012 and 2013, the number of data breaches increased by 62% to a total of 253%. More telling was that in 2013, 256 incidents were reported to DHS's Industrial Control Systems Cyber Emergency Response Team (ICS-CERT), of which 59% were reported by the energy sector and 3% by the nuclear sector.[1] As a result of the increased threat to BPT's digital infrastructure, its leadership issued a series of directives aimed at securing its IT and ICS against cyberattack. Attacks have been on the rise, though *The Symantic* 2017 report noted that an average cost of ransomware attack alone rose to $1,077, from $294 a year earlier.

## Information Systems Security

The first of such directives was to instate a chief information security officer (CISO), who would be in charge of all IS security aspects for both the IT and ICS domains. The CISO would work under the CIO but was given direct authority to communicate critical information to the CEO in the event of a breach. The decision of hiring a CISO was given to Mr. Anderson, pending the approval of the CEO, and in February 2013, Chris Tucker, a former senior security analyst of a major national bank, was hired for the position. While Mr. Tucker had no experience with ICS and ICS security, he had extensive expertise with IT security and understood the growing trends in the IS security field. Under his management, BTP's leadership directed that changes in IS security be made by the end of the year. During the remainder of 2013, Mr. Tucker and Mr. Anderson instituted a number of improvements to security capabilities at BTP while operating under the same operations budget.

Due to a constrained time frame and budget, Mr. Tucker and Mr. Anderson were limited in the means in which they bolstered existing IS security capabilities. In order to identify weaknesses in the existing IT and ICS infrastructures, both the CIO and CISO elected to evaluate the systems based upon the SANS top 20 controls. The SANS list covers a range of controls including technical, formal, and informal, allowing BTP's IS staff to conduct a comprehensive review of the infrastructure's security. While the review was conducted individually between each of the three divisions, the final assessment was made on the posture of the entire organization.

In concluding the assessment, several observations were made: (1) account monitoring and control needed to be increased to mitigate insider threats; (2) higher standards of data encryption and continuity needed to be established; and (3) an air gap would need to be implemented between ICS and IT infrastructure at all BTP generation and distribution centers. In terms of effecting more stringent account monitoring and control procedures, BTP's IS staff established a process for revoking digital credentials as an integrated step with human resources' (HR) procedures for terminating employees. Much like revoking physical credentials and processing termination papers, HR was now required to draft and submit paperwork to BTP's IS staff, indicating that they were no longer with the company and that their digital credentials would need to be revoked.

Likely the most time-consuming and costly effort for the IS department involved enforcing higher standards of data protection and continuity. In order to examine which portion of BTP's data was sensitive information, and thus worthy of encrypting, the IS department

---

[1] http://ics-cert.us-cert.gov/sites/default/files/Monitors/ICS-CERT_Monitor_Oct-Dec2013.pdf

made a concerted effort to evaluate all data that resided on BTP's internal IT infrastructure. After several months of coordinating with customer service, operations, HR, and executive departments, the IS department was able to effectively set encryption standards on all data deemed sensitive. In order to ensure data continuity and integrity, however, the CIO attempted to convey the need for additional operations funding to Mr. Finkle so that he might purchase additional data systems for operations in Indiana and Ohio to act as backup centers for existing data systems. Unfortunately, his appeal was denied due to a constrained budget as BTP looked to construct a new natural gas plant in Ohio. While the board and CEO urged Mr. Anderson and his team to pursue the construction of a continuity data center utilizing existing funds, this significantly constrained both the IS department's budget and personnel resources.

In reviewing past incidents of ICS attacks against similar companies, the CISO determined that all ICS and IT functions would need to be separated in order to mitigate the risk of malware or other malicious threats presiding on IT infrastructure from transmitting to ICS infrastructure. Fortunately, much of the newly built infrastructure in Indiana and Ohio had already been designed in this manner, though much of the infrastructure that was later incorporated through acquisition was not designed in light of security. Efforts to establish a buffer between IT and ICS systems were extended through 2014 due to existing demands on IS resources and personnel.

# Federal and State Efforts to Secure Critical Infrastructure

While many in the federal government speak of the need to pass comprehensive cybersecurity legislation and establish mechanisms to thwart cyberattacks against critical infrastructure, little has been done to this effect, leaving assistance from the government to the private sector few and far between. A failure to produce legislation is not to say that nothing has been done in terms of cybersecurity and critical infrastructure security and resilience (CISR), as several initiatives have been established through executive order and executive branch agencies such as the Department of Homeland Security (DHS). While these initiatives do support efforts in terms of bolstering cybersecurity, they do not comprehensively address the issue. Inaction on the part of the federal government has led state governments to take action unilaterally. In the case of BTP, impacts of state government action were felt in both Ohio and Indiana, as their states were proactive in partnering with federal initiatives and moving forward with state-centric initiatives due to an impasse on any federal legislative action.

February 2012 brought about the first true concerted national effort to bolster cyber security in both public and private sectors. Through executive order, the president issued a directive that a framework be created to be used by owners and operators as a means of improving the cybersecurity posture of critical infrastructure. In line with the directives of the executive order, the National Institute of Standards and Technology (NIST) produced the Framework for Improving Critical Infrastructure Cybersecurity, or "Framework," which "consists of standards, guidelines, and practices to promote the protection of critical infrastructure." While the federal government intended for the framework to be leveraged independently by owners and operators, it did not disallow its use by state governments to incentivize or regulate industries based on its criteria.

## Ohio

In lieu of any comprehensive cybersecurity legislation originating from its federal partners, the governor of Ohio elected to promote the idea of enhancing cybersecurity within the state by leveraging existing initiatives such as the framework and furthering efforts by establishing state-based actions. As cybersecurity was becoming a widely popular issue in government, industry, and mainstream media, the governor, seeking his second term in office, chose to make cybersecurity a key campaign issue, pledging to enact several initiatives if elected back into office. After winning reelection and the release of the NIST Framework, Ohio was one of the first states to adopt the Framework to use as a tool to bolster their systems internally and promote the Framework to businesses within the state.

Soon after taking office, the governor set to work, nominating his director of administrative services and director of public safety to co-chair a panel to address the issue of cybersecurity within the state from both a technology and public safety perspective. Appointments to the panel were chosen by consensus between the cabinet officials, which ultimately formed a group of 15, including the state CIO, CISO, a number of IS security professionals, and CISR experts. Though the panel was formed in early 2014, by the summer they had created an actionable initiative that the governor could provide to the legislature as a means of improving critical infrastructure cybersecurity.

The initiative was aimed at incentivizing industry leaders to incorporate the NIST Framework into their security operations. The intention for this initiative would be to allow IS managers a means of evaluating their current IS security posture and identifying gaps or means of improvement to make for a more robust organization in terms of infrastructure security. News of this initiative was welcomed by many industries who maintained operations in Ohio, as it would provide them with tax credits for their participation and active use of the Framework. The CEO and Board of BTP were enamored with the idea because, theoretically, it would cut operational costs within Ohio and allow for more revenue to be used toward expansion within the state. In turn, the state would benefit, as the risk of cyberattack would be reduced and thus a threat to the critical infrastructure within Ohio would diminish.

## Indiana

While Ohio took an approach of incentivizing private industry into improving their cybersecurity posture through the implementation of the Framework, Indiana chose to leverage regulation. In order to address growing constituent concerns over the security of utility and government services within the state, the Indiana legislature instituted an independent study to determine the best means of moving forward on a cybersecurity initiative. The committee in charge of the study conducted numerous interviews with the governor's cabinet, agency heads, private industry, and concerned citizens. Their conclusion produced two recommended initiatives that were brought before the legislature and eventually passed into law.

The first of these initiatives was a similar adoption of the NIST Framework in that the legislature mandated the executive branch technology agency apply the Framework to its systems. Furthermore, the legislature directed the Indiana government to ensure adoption of the Framework by critical infrastructure industries operating within the state. To do this, one of the recommendations by the legislative committee was to enforce the Framework's adoption through regulations imposed on critical industries. In the recommendation, the committee cited that a failure of the federal government's rollout of the Framework was that it was without teeth and needed to be enforced upon critical industry sectors to enhance cybersecurity of critical infrastructure by any means. Upon the launch of the initiative, the regulation enforced a threshold that critical industries must meet in order to fall in compliance. Otherwise, those industries who failed to comply would be taxed at a higher rate than those who did.

The second initiative was intended to drive the sharing of information by private industry to government authorities relating to imminent or ongoing cyberattacks. Like the former initiative, the information sharing action was driven by regulations in order to ensure that private industries were in compliance and sharing timely and actionable information with authorities, to better provide them with a clear threat picture and enable them to act upon the information to curtail further harm caused by cyberattack.

## Federal Legislation

Despite years of failed legislation, passage of a cybersecurity bill of some form was inevitable. By early 2014 the momentum for cyberlegislation had begun to take shape due to the release of the federal Framework. In late summer of 2014 Congress had a bill that focused on the sharing of critical cyberinformation between public and private sectors. The bill, titled "The CyberIntelligence Sharing Act" (CISA), "incentivizes the sharing of cybersecurity threat information between the private sector and the government and among private sector entities."[2] Though the act addresses widely accepted needs for establishing a means of sharing critical cyberthreat information, it is not without criticism. Critics state that while it intends to facilitate information sharing and dissemination processes for government and private industry, it is overly vague with the types of information it allows the private sector to share with authorities. Privacy advocates contend that if the act passed, it could mean significant harm to US citizens' privacy rights. Currently the bill has been passed by the Senate and the House, with a narrow victory, and is awaiting the approval of the president, which is anticipated.

Although this would be the first cybersecurity-related law that encompasses public and private sector information sharing, the bill has faced a lot of criticism from privacy groups. As a result of past legislative attempts to encompass critical infrastructure regulative aspects, newer legislation such as CISA have been watered down and are more vague than specific with their provisions. In the case of CISA this means a lack of specificity in terms of the types of information that should be shared between private industry and government, leaving the acceptability of sharing personally identifiable information ambiguous. While this draws concerns from many civil liberty groups and citizens, it has not seemed to stop CISA from being the first cybersecurity legislative measure of the 21st century.

---

2 http://www.infoworld.com/t/federal-regulations/cispa-returns-cisa-and-its-just-terrible-privacy-244679

## Impacts and Reaction at BTPf

The initial release of the Framework and its adoption by Ohio and Indiana seemed promising for BTP, especially when Ohio indicated their desire to leverage it as a means of enhancing industry cybersecurity while incentivizing those who adopted it. BTP's executives and board members embraced the initiative upon its launch, as they saw it as a "win-win" to enhance their infrastructure security while accepting tax breaks for doing so. Contrarily, however, when Indiana began to launch its initiatives for the Framework, BTP's attitude toward the Framework adoption changed, as they were now forced to comply with a standard level of security as determined by the state's technology agency. Rather than seeing a net cost benefit toward adoption of the Framework, BTP would now be forced to contend with a cost increase to compensate for security requirements.

The resulting demand of both Ohio and Indiana's initiatives impacted BTP's IS department even more. With an already constrained budget as a result of infrastructure costs for a data center redundancy, the CIO, Mr. Anderson, was now forced to implement the Framework throughout the tristate corporation. Furthermore, despite Ohio's model of incentivizing the Framework's adoption, Indiana required that a specific threshold be met in order to meet regulatory requirements. While this would ideally only affect intrastate infrastructure, the networked architecture meant that the entirety of the company's infrastructure must meet the required regulations. Most notably, BTP still held much of its divisional IT operations on a third-party infrastructure in Michigan and would need to verify that the vendor met the requirements of Indiana's regulations, as sensitive information from Indiana traveled through the vendor's infrastructure on a daily basis.

In terms of Indiana's information dissemination regulations and the upcoming CISA legislation, Mr. Anderson was forced to deal with additional responsibilities of meeting reporting requirements. To date, Mr. Anderson; the CISO, Mr. Tucker; and BTP's IS security team were able to rely upon third-party partnership organizations for external threat information and intraorganizational reporting structures to share threat information throughout the company. However, with the passage of the state and federal reporting requirements, the CIO and CISO were forced to establish new means of collecting and disseminating timely and actionable information to its state and federal partners. This was not only a laborious task, but one that worried Mr. Tucker, as he was aware of the controversy over CISA and felt that sharing information that might contain personal identifying information (PII) would leave the company liable to some extent. Mr. Tucker was also wary of the security standard to which the information would be held, as there had been an increase in breaches and attacks against federal websites. While these attacks were primarily for defacement purposes, it indicated to him that there was a strong interest in targeting federal entities, and any release of sensitive information could be catastrophic for BTP.

# Questions

1. In the current cyberthreat climate, with attacks increasing in quantity and sophistication, there is no question that the government must play a role in assisting critical infrastructure owners and operators with cybersecurity. An unfortunate misstep in planning for cyberinitiatives on a state and federal level was a lack of engagement with industry owners and operators to determine their needs and concerns in terms of the looming threat of cyberattack. How should this have been undertaken in the case of BTP? What generic lessons can be drawn?

2. The initiatives set forth by Ohio, Indiana, and the federal government neglected to engage stakeholders in planning for cyberinitiatives and thus neglected to consider the impact of such actions on business processes. Many cyberinitiatives are now looking to enhance the flow of information between private industry and government. Although this might further the capability of government to respond to and prevent attacks, the oversharing of information and increased transparency could negatively impact business by exposing sensitive information and limiting discretion. Comment.

3. Though the sharing of cyberthreat information can benefit businesses by alerting businesses of new and ongoing threats, providing such information can have consequences to business if not handled correctly. Discuss.

## CASE STUDY 5

# The Case of Sony's PlayStation Network Breach*

Large corporations and institutional investors are highly sensitive to the public perception and impact of a security breach. Publicly traded corporations have a legal responsibility to report security breaches. In May 2011, Sony Corporation experienced a massive intrusion that impacted 25 million customers. The attack came through the company's Playstation Network. Cyber attackers from outside the United States were able to penetrate Sony's Playstation Network and "liberate" 23,400 financial records from 2007. In addition, detailed banking records of 10,700 customers from Germany, Spain, Austria, and the Netherlands were stolen. Sony advised that names, addresses, emails, birth dates, phone numbers, and other unidentified information were stolen from an outdated database server. Credit card information is also suspected of being stolen. The physical location of the attack was Sony's San Diego data center. This was the second attack within a week, and Sony was reeling to determine how it could have experienced a second attack in such a short period. During the first attack, it was alleged that cyber attackers stole more than 77 million Playstation users' information.

The heart of a company is still its people, but the image of a company is its public perception. After the second breach, the public's perception of Sony was its inability to secure the network. The company's leadership was facing extreme criticism over their inability to secure the network and further critique of Sony's failure to manage a crisis. The Federal Bureau of Investigation (FBI) conducted an investigation of the breach. In parallel, Sony advised that their network services were under a basic review to prevent any recurrence, but a "basic review" sounds like media spin to keep the press at bay and down play the criticality of the breach. If this attack had occurred at a financial institution, there would be very strict compliance rules that govern the security of a financial network. Sony is not beholding to the same rigid parameters, but should protect the security of its customer data.

---

* The case was prepared by Kevin Plummer under the tutelage of Dr. Gurpreet Dhillon.

# Sony Gets Attacked

The Sony attack was performed by a security organization called Lulz Security or LulzSec. Sony Pictures confirmed that LulzSec hacked into its websites. During the second attack, personal information of over one million users was stolen. Amy Pascal and Michael Lynton, who are the CEO and co-chairman of Sony, respectively confirmed the unfortunate breach and that Sony Pictures was working to prevent more intrusions. Lynton and Pascal issued a joint statement and apologized to the company's customers, but the news of this second attack came just hours after Sony brought their PlayStation store online after a massive attack just weeks earlier. In addition, Sony executives had a press conference on Capitol Hill to reassure Federal Communication Commission (FCC) officials that there was no current evidence that the hackers had accessed credit card information during the PlayStation Network (PSN) attack.

LulzSec accepted responsible for the attack via a series of Twitter postings. The motivation for the attacks against Sony was said to be the abuse by the United States Federal Court System against George "GeoHot" Hotz. Hotz appeared to the mysterious world of hacking at the age of 17 when he cracked open the iPhone, which caught the attention of the world, and immediately became the poster child for hackers and attracted a rock star fan following. When Hotz unlocked the iPhone bootloader in August of 2008, Apple's famed device was able to be used on any network, not just exclusively with the service provider. Strangely enough, Hotz's accomplishment was basically a hardware takeover. His real modus operandi just involved some soldering and hardware modification. More experienced hackers actually discovered ways to unlock the iPhone by hacking just the software. Hacker groups around the world felt that the courts had invaded Hotz's privacy. In addition, Sony's decision to block people from using Linux on their PlayStation 2 was also a point of contention. In a parallel statement, LulzSec also bragged they had hacked into a Federal Bureau of Investigations (FBI) database. Sony Pictures contracted a team of consultants to conduct a thorough analysis of the attack. Sony also contacted the FBI and worked closely with them to assist in the identification and arrest of the perpetrators for this attack.

LulzSec started out with a Twitter publicized hack on May 7, 2011. With a previously established Twitter presence, LulzSec posted the details of its first hack online. They claimed to have obtained a database of contestants from the Fox TV show *X Factor*. Days later, LulzSec followed up with additional internal data extracted from Fox online. On May 30, 2011, LulzSec successfully completed the first hack of Sony. Sony responded with a lockdown and an analysis of their security practices, only to fall victim to another hack attack on June 2, 2011. LulzSec posted personal data, including passwords, email addresses, and thousands of music codes and coupons, for more than a million users from a sampling of Sony websites, which were concentrated at Sonypictures.com.

LulzSec has claimed responsibility for multiple high profile attacks, in addition to the compromise of user accounts from Sony. LulzSec also claimed responsibility for taking down the CIA website. The Arizona Department of Public Safety has described the group as a cyber terrorism organization, after their systems were compromised and information mysteriously leaked. Some security professionals have applauded LulzSecs's brazen expertise, for drawing attention to marginally secured network systems and the dangers of reusing passwords. They have gained attention due to their high-profile targets and the sarcastic tagging posted in the aftermath of the attacks.

## Technical Details of the Attack

Large corporate networks share some common networking components. External-facing firewalls, switches, routers, web servers, remote access, web applications, and back-end SQL databases are some of the common components. In the case of the attack on Sony, LulzSec admitted to using SQL injection to penetrate Sony's network. SQL injection is a code subset of unverified or unsanitized user input vulnerability. It is rogue SQL code intentionally slipped into the SQL application. The idea is to trick the application into running the SQL code unintentionally. If the application unknowingly creates SQL strings on the fly and then executes the strings, some interesting results occur. Presuming that the LulzSec hacker had some prior knowledge of the application and access to the source code, he would have an idea of how ubiquitous a SQL server would be for a large corporation like Sony (Unixwiz). Microsoft's SQL Server has become the database management system (DBMS) for enterprises requiring a flexible feature set at a lower price point than other DBMS. Since its release, SQL Server DBMS has been steadily increasing in popularity every year.

By 2001 SQL was ranked the number one DBMS for the Windows platform. As corporations have downsized their staff and network purchases, SQL server is an easy choice for budget-conscious purchasing managers. In any large corporation, it is a safe to assume that SQL Server DBMS is resident in the back-end database. If a potential hacker is interested in making a SQL database discovery, all she has to do is poke around in the system. While browsing a corporation's web page, check out the login page for a traditional username-and-password form and also an email-me-my-password link. This can prove to be detrimental and lead to the downfall of the whole system. When entering an email address, the system presumes to look in the user database for the email address, and respond by mailing something to that address. Since the email address is not found, it cannot send anything. One way of determining if SQL DB is running is to continue to poke around in the system. If the server is running Microsoft's web server called Internet Information System 6 along with ASP.NET, this is a clue that the database was Microsoft's SQL server. The common thinking is that these types of techniques are not unique and can apply to nearly any web application backed by any SQL server. In order to determine if a server is a potential victim, the first test in any SQL code form is to enter a single quote as part of the data string. The intention is to determine if the system constructs an SQL string literally without sanitizing the data. After the form is submitted with a quote in the email address, the response is a 500 error or server failure. This indicates that the "broken" input is actually being literally parsed and the SQL injection can begin. This is a typical example of SQL injection, and other hacker types may perform an augmented approach.

The problem with Sony's network is they did not have SQL injection protection set up to prevent the second attack. SQL injection attacks use malicious code inserted into strings that are later passed to an instance of SQL Server for parsing and execution. Any procedures that are allowed to construct SQL statements should be reviewed for injection vulnerabilities, because SQL Server is not that smart and will execute any syntactically valid queries that it receives. A skilled and determined attacker can manipulate parameterized data. In its primary form, a SQL injection consists of the direct insertion of code into user-input variables. The system executes the concatenated variables with SQL commands.

A less direct hacker attack injects malicious code into strings that are en route for storage in a table or as metadata. When the stored strings are eventually concatenated into a dynamic SQL command, the malicious code is executed. The injection process functions by prematurely terminating a text string and then appending a new command. Because the inserted command may have subsequent strings added to it before it is executed, the bad code terminates the injected string with a comment mark. As long as injected SQL code is syntactically correct, an invasive attack cannot be detected programmatically. A prevention standard is to validate all user input and carefully scrutinize code that executes constructed SQL commands in the server being used. Always validate user input by testing data length, format, range, and type. When implementing precautions against malicious code, Sony should consider all of their architecture and deployment scenarios of their customer facing web applications. It is good to remember that programs coded to run in a secure environment can also be copied and run in an unsecured environment.

For protection against injection, there are a series of best practices that should be followed. Network administrators and database administrators (DBA) should make no assumptions about the size, type, or content of the data that is received by resident applications. The DBA should determine how the application will behave if errant or malicious users enter a 10-megabyte MPEG file where your application expects a postal code. Information security resources should expect the unexpected. Through penetration and stress testing, application professionals should know how an application behaves if a drop table statement is embedded in a text field. Appropriate limits should be enforced when testing the size and data type of input. This process can help prevent deliberate buffer overruns by hackers. Content string variables should be tested to accept only expected values. Any entries that contain binary data, escape sequences, and comment characters should be rejected. This can help prevent malicious script injection and protect against attacks using some buffer overrun exploits.

When working with XML documents, data should be validated against the schema as it is entered. Transact-SQL statements should not be built directly from user input. In order to validate user input, stored procedures should be used. In multitier environments, all user data should be validated before allowing access to the trusted zone. If data does not pass the validation process, it should be rejected and an error should be returned to the previous tier. The implementation of multiple layers of validation is a good practice. By taking low-level precautions against casually malicious users, the network may be ineffective against determined attackers. A proven practice is to validate input in the user interface and at all subsequent access points where it traverses a trust boundary. Data validation in a client-side application can prevent a simple SQL script injection. However, if the next tier of boundary assumes that its input has already been validated, any malicious user who can bypass a client can have unrestricted access to a system.

Network boundaries and tiers should also perform some type of authentication. Invalidated user input should never be concatenated. SQL script injection uses string concatenation as the primary point of entry. When working with XML documents, validate all data against its schema as it is entered. For multitiered environments, all data should be authenticated before entry is granted to the trusted zone. Data that does not pass the authentication process should be kicked back, and an error message should be returned to the previous boundary. Multiple layers of validation are a supported practice. Precautions taken against casually malicious users may be ineffective against determined attackers. A better practice is to validate input in the user interface and at all subsequent points where it crosses a trust boundary. Network segmentation for data security is another design option that can limit access, destruction, and devastation brought about by a hack attack. A portion of a computer network that is separated from other parts of the network by an appliance such as a repeater or bridge is a network segment. Each network segment may contain one or multiple computers or servers. The type of segmentation required differs based on the type of device used. Each network segment can have a dedicated hub or switch. In most cases, a contiguous range of IP addresses will be assigned to each segment or subnet.

One of the greatest advantages of having multiple segments rather than having all hosts on a single mammoth segment is that it can increase the amount of traffic that a network can carry. A major consideration in designing segmentation to maximize network capacity is to assign hosts that do not normally communicate with each other on different segments. Another important feature for network segmentation is information security. If a hacker successfully compromises a single host in a network segment, every computer in that subnet is at risk, but the compromise is contained to that segment. Implementation of network switches instead of hubs can reduce the effect of packet sniffing or eavesdropping on packets traversing the network; they also allow potential hackers to circumvent switch security.

Segmentation will, however, allow all the advantages to a smaller business or other organization of having all of its computers on a single network while protecting each part of the network from unauthorized entry. Each individual segment can be protected from the other segments by employing firewall technology, each using its own set of rules. Data moving between segments must pass through each firewall.

## The Emergent Challenges

The new security challenges of today's businesses make traditional security schemes incapable of measuring up. Many security architectures are designed using perimeter-centric rules and lack comprehensive internal control parameters. Most organizations dependent on firewall security technology might be pushing the limits and asking the security mechanics to do more than it was designed to do. Old firewall rule sets contain tables that stay intact for years. This might be a known liability when the firewall rule sets do not represent contemporary business requirements and fall short of protecting critical assets appropriately. One of the first steps in defending a network is an in depth redesign of the corporate network segmentation policy. The policy is implemented to describe which departments, applications, service model, and assets should reside on separate networks. Network segmentation will aid in assuring that potential threats are contained with minimal impact to the corporate network. Standards institutions like the National Institute of Standards and Technology emphasize the importance of network segmentation but fall short of mandating the requirement. Standards committees emphasize new guidelines that compliance capacity can be significantly reduced by placing all the related assets in the same segment. Network segmentation is common sense in today's market and also one of the most effective and economical restrictions to implement. The implementation return on investment (ROI) is maximized with a minimal outlay.

Whether a network has 5 or 500 nodes, there are some basic information security protocols that should always be followed. Protocols are designed to assist the administrator to effectively protect the network infrastructure from the countless attacks. Ensuring *and securing* that a network is free of compromise is crucially important for network customers because it allows for the sustained, continuous operation of the very network they rely upon to perform their job functions. The foundation of all information security programs is an in-depth defense. A comprehensive strategy for protecting a network is protecting each layer. The layers are generally networked areas such as the network perimeter router, DMZ, physical security, authentication mechanisms, auditing, and logging. By introducing multiple layers of defense throughout the network, administrators will increase the intricacy and skills required to hack through the defenses while simultaneously hardening the network defenses. This is by no means all that is required to protect a network.

Corporate networks *must* have experienced staff associates, who clearly comprehend information security and defense in depth for the program to succeed. In-depth defense is a network framework. The majority of recommendations are related to layers within the network. Simply implementing them individually may very well increase your network security position; however, it is worthwhile to implement all measures to protect your network at the maximum degree possible. Perimeter network security is another practice. The proper use of perimeter security is to deny by default and allow by exception.

Strong perimeter defenses are mandatory for a good network. Every network connection interface must have a premise router. This is the router that is connected to the upstream internet service provider. The network premise router should make use of access control lists (ACL) to only allow the minimum required number of TCP/IP connections both in and out of the network. This is referred to as a "deny by default and allow by exception" policy. If your network does not have a web server in place and accessible by the public, there is absolutely no need to allow port 80/tcp inbound from the world. If there is no SSL server installed and accessible by the public, the information security should not allow port 443/tcp inbound. With all probability, port 1024-65535/tcp and port 1024-65535/udp are not required inbound at all. Allowing the remote possibility for these connections to be active is a huge and unnecessary vulnerability. Essentially, you deny all connections by default and build an ACL, which only allows the required users connectivity access to the network.

Along with a well-configured premise router ACL, all networks should employ at least a stateful firewall situated right behind the premise router. The firewall should be configured identically to the premise router, using the same access and following the "deny by default, allow by exception" policy. The reason a stateful firewall is imperative is because the information security administrator needs to be able to inspect the packets, and keep track of the state of the network connections traversing the firewall. This allows the firewall to adequately distinguish between valid and potentially detrimental connections or connection attempts.

Corporate networks have many moving parts. In October 2011, Sony admitted in an official PlayStation blog post that it had detected more unauthorized access attempts on Sony Entertainment Network, PlayStation Network, and Sony Online Entertainment. The recent attempts manifested as a massive set of sign-in IDs and passwords against the Sony network database. Sony advised that the IDs and passwords may have come from one or more compromised lists from other companies, sites, or other sources. Sony's response to the recent incident seems to have missing details. Sony's interpretation and their communication to the general public did not refer to the unauthorized access as a hack. Only unauthorized access to data in a system is considered a hack. The October 2011 incident involved individuals or an individual gaining momentary access to accounts that contained the purchasing information of a relatively small number of Sony online accounts. Sony advised that since "the overwhelming majority of the pairs resulted in failed matching attempts; it is likely the data came from another source and not from our Networks."[1] The company has taken additional steps to mitigate the activity. Specific mitigation would not be revealed, but may include IP blocking and securing a copy of the list to protect potentially affected user accounts. The only sure way to firewall a system from someone trying a stolen ID or password would be to disable all logins. Sony has admitted that an additional firewall exists between online accounts and backend services. Sony has conceded that less than 0.1% of online users may have been impacted. The potentially impacted users totaled 93,000 accounts globally in which the login attempts succeeded, but the company has already taken action and locked down the affected accounts. Of the 93,000 accounts, there was only a small fraction that displayed additional activity prior to the network lockdown.

Sony continues to mitigate the recent incident by noting that anyone with credit card information on file is safe. They will work with any individual who discovers their account was used to make unauthorized purchases or compromised. Customers will be identified and notified by Sony via email to reset their passwords. At this stage of the game, whether Sony's network was compromised repeatedly or not is small print. The bigger story is the detail of how it happened three times in one year. If Sony was a large US investment company or bank and terrorists were identified as hacking into the systems, this would equate to a national threat.

As Sony and other organizations experience rapid technology contractions brought on by volatile economic uncertainly, excess and vulnerable networks are the reality for many entities. Sony projects the hacker attack on its PlayStation network to cost the company about $170 million. The company said it expects a "significant" decline in operating profits for its networked products and services unit, which includes gaming, over the coming fiscal year (April 1, 2011 to April 1, 2012). The report was presented to shareholders as the company told analysts that it anticipates posting a $975 million profit for fiscal 2012. That would be Sony's first annual net profit in four years. For fiscal year 2011, Sony posted a net loss of $3.2 billion, which is its biggest drop since 1995. Sony's expected forecast, which some analysts declared too optimistic, for 15 million units of PlayStation 3 to be sold over the coming fiscal year. This is the same number as in fiscal 2011 and 27 million TV sets, compared with 22.4 million in fiscal 2011. Sony also posted $274 million in charges related to the March earthquake in Japan and estimated it lost $63 million in sales during that same period. The company expects to recover some of those costs from insurance. When Hackers hit Sony's PlayStation Network in late April, some 77 million users' private information was compromised. Since the second attack, the network has been up and down several times. Sony's network service's status is so unpredictable that an enterprising developer has created an iPhone app that informs users whether PSN is on or off.

---

1 http://latimesblogs.latimes.com/technology/2011/10/sony-93000-playstation-soe-accounts-possibly-hacked.html

A corporate network is very expensive to maintain, and organizations are shifting to distributed and virtual environments. While distributed and virtual servers are more efficient environments, they still face the challenges of secure computing and how to manage growth and contractions. A secure infrastructure will require strong policies from all of the trusted partners. The partners include telecommunication companies, vendors, and financial institutions to cooperate and allocate funding for firewalls and test environments, in addition to an experienced support staff. As part of a documented contingency plan, there should be a contingency alternate site. The alternate site needs to be independent of any current network environment in place today. For a plan to be effective, government, private sector, and banking entities should agree on a cohesive secret security strategic agenda. The agenda needs to be progressive, dynamic, and scalable. Unfortunately, politicians become interested and realign their agendas to match the most popular trend. Establishing strong network security would help withstand a hit by the most technically prepared hacker or terrorist organization in the world. Reacting to a problem, as opposed to planning a proactive strategy, does not provide the best solution. Primarily, corporations need to strategically rewrite network security polices to withstand multiple types of threats and remediate the short- and long-term impact of a hack attack with stronger alternate and contingency environments.

## Questions

1. How should companies such as Sony protect their reputation? Sony, in particular, has had bad press in recent years. Following the PlayStation breach, Sony also experienced a breach after the movie *The Interview* was released in 2015. Sketch out a reputation protection strategy following a breach.

2. Express your opinion on how well the cyber security incident was handled by Sony. Given that the Sony PlayStation breach did occur, what incident-handling steps should have been followed?

3. Compare and contrast the Sony breach with one other breach of your choice. Comment on how the two breaches were handled. Identify the similarities and differences.

# Index

Note: Page numbers in *italics* refer to a figure or a table.

access control, 187
access control lists (ACL), 397
account harvesting, 86
accountability
  assignment of, 11
  culture of, 350
  ensuring, 252
  of executive management, 303
  individual, 27, 112
  lack of, 96, 125, 140
  as organization value 355, 359
  in security systems, 175, 178, 194, 195, 257, 259, 300, 347
  structures of, 7, 114, 115, 209, 305
  of users and programs, 35
  *See also* Health Insurance Portability and Accountability Act (HIPAA)
acknowledgement (ACK), 74–75, 83
ACL (access control lists), 397
ACM. *See* Association for Computing Machinery
adhocracy culture, 263–264
Advanced Encryption Standard (AES), 55
adware infection, 84
AES (Advanced Encryption Standard), 55
Analytical Hierarchical Modeling, 114
Anthem Data Breach, 354–355
  Anthem's history and background of the industry, 355–356
  lawsuits and policy issues, 357
  significance of the problem and what Anthem is doing about ti, 357–359
  strategic alignment--possible cause of the breach, 359–361
  strategic choices, 361–364
  who is at stake? 356–357
Anti-Phishing Act of 2005, 88
API (application programming interface), 93
application controls, 159
application programming interface (API), 93
asset management, 186

Association for Computing Machinery (ACM), Code of Ethics and Professional Practice, 279–280
asymmetric encryption, 55–57
  authentication of the sender, 57–58
  RSA, 58–59
auditing, 216–217
  internal, 302–303
  of security standards, 158–159
authentication
  of communication, 23
  of data, 20, 22–23
  IP address-based, 82
  mechanisms for, 6, 176, 187, 193, 304, 336, 397
  password-based, 85
  protocols, 132
  secure, 72
  and security, 20, 178, 180, 183, 193, 261, 297, 395
  of the sender, 20, 56–58
  servers, 93
  testing, 160
authenticity
  of communication, 23
  of the individual, 58
  of information, 187
  in information systems security, 24, 38, 39
  and integrity, 21
  of the key, 57
  of message transmissions, 44, 63
  of the sender, 228
  tests for, 315
authority structures, 101–102
availability of data, 19, 22, 347–348. *See also* CIA (confidentiality, integrity, and availability)
awareness, 3, 6, 7, 13, 38, 60, 103, 104, 114, 116, 129, 174, 186, 189, 194, 207, 208, 210, 217, 225, 252, 258, 260, 267, 295, 300, 301, 304, 350, 352, 385
  executive, 247
  public, 247, 292, 339, 340

Bank of Credit and Commerce International (BCCI), 277
baseline assessment, 297

401

402 • Index

behavioral analytics, 224–225
   cybercrime, 230–239
   employee threats, 225–227
   social engineering attacks, 227–229
Bell-La Padula model, 27–29, 31–32, 33, 36, 37, 264
Biba Model for Integrity, 32, 33, 35, 36, 264
Big Ten Power Company (BTP), 383–384
   fears of attack, 385–386
   federal efforts to secure infrastructure, 387–388, 389
   impacts and reaction, 390
   information systems security, 386–387
   internal threats, 385
   IT at, 384–385
   state efforts to secure infrastructure, 387–389
bisexuality/gender, 256–257
Bitcoin, 60–61, 61–62
black markets, 232–233
Blaster worm, 259
block ciphers, 52–53
blockchains, 60–61, 187, *188*
   alternative, 62
   application of technology, 61–62
   creation and validation process, *61*
   and Visa, 65–67
Blowfish, 53
border searches, 324
botnets, 96, 233, 234, 241
Brewer-Nash Chinese Wall Security Policy, 36, 37
buffer overflow, 84

C2W (command and control warfare), 235
Caesar Cipher, 48
Canadian Trusted Computer Product Evaluation Criteria, 26
CAP (Certified Authorization Professional), 283
Capability Maturity Model (CMM), 157, 169–170. *See also* System Security Engineering Capability Maturity Model (SSE-CCM)
CAR (Computers at Risk), 194
Carbanak, 204
Case Studies
   Anthem data breach, 205, 206, 214, 354–364
   computer forensics (electronic eavesdropping), 343–344
   computer forensics (hacking), 342–343
   computer hack, 373–382
   critical infrastructure protection, 383–391
   cyberterrorism, 240–242
   cyberwarfare weapon: Stuxnet, 242–244
   Department of Homeland Security, 200
   distributed denial of service (DDoS), 95–96
   DoubleClick ethics issues, 287–288
   *Federal Trade Commission v. Wyndham Worldwide Corporation*, 307–309
   hacking at the University of California Berkeley, 125–126
   Home Depot, 205, 206
   Mitsui Bank, 154–155
   PGP Attack, 65
   process and data Integrity concerns in a scheduling system, 365–372
   security breach via phone call, 41
   security culture, 271
   Sony Pictures Entertainment, 203–204, 205, 215
   Sony's PlayStation network breach, 392–399
   Target, 214
   virtual companies and identity theft, 11–13
   Visa goes blockchain, 65–67
CAST (Carlisle Adams Stafford Tavares), 55
CCFP (Certified Cyber-Forensics Professional), 283
CCSP (Certified Cloud Security Professional), 282
CDI (Constrained Data Item), 34, 35
CDM (Continuous Diagnosis and Mitigation) program, 211, *211*
Center for Strategic and International Studies (CSIS), 312
CERT (Computer Emergency Response Teams), 157, 226
Certified Authorization Professional (CAP), 283
Certified Cloud Security Professional (CCSP), 282
Certified Cyber-Forensics Professional (CCFP), 283
Certified in the Governance of Enterprise IT (CGEIT), 284
Certified Information Security Manager (CISM), 284
Certified Information Systems Auditor (CISA), 284
Certified Information Systems Security Professional (CISSP) certification, 282–823
Certified Secure Software Lifecycle Professional (CSSLP), 283
CFAA (Computer Fraud and Abuse Act), 293–294
CGEIT (Certified in the Governance of Enterprise IT), 284
Chain, 66
Chain Core, 66–67
chain of custody, 334–335
"challenge-response box" technology, 6
Chief Information Officer (CIO), 384, 390
Chief Information Security Officer (CISO), *215*, 252, 386, 388, 390
Chinese Wall Security Policy, 36, 37
CIA (confidentiality, integrity, and availability), 19–20, 27, 39, 44, 131, 132, 136, 175, 187, 188, 193, 200, 266, 283, 347, 350, 352
CIO. *See* Chief Information Officer
ciphertext attacks, 46
ciphertext encryption, 44

Index • 403

CISA. *See* Certified Information Systems Auditor (CISA); Cyber Intelligence Sharing Act (CISA)
CISM (Certified Information Security Manager), 284
CISO. *See* Chief Information Security Officer
CISSP (Certified Information Systems Security Professional) certification, 282–823
clan culture, 264–265
Clark-Wilson model, 21–22, 32, 34–36, 37
click fraud, 234
cloud-based network security, 93
CMM. *See* Capability Maturity Model
COBRA, 143–145, 152
codes of conduct
    ACM Code of Ethics and Professional Practice, 279–280
    ISSA Code of Ethics, 281
    SANS IT Code of Ethics, 280–281
columnar transposition, 50–51
command and control warfare (C2W), 235
Common Criteria, 26, 31, 32, 37, 179–181
    identification of the product, TOE, and TSF, 182–183
    problems with, 181–185
communication, 2
    authentication of, 23
    workstation-to-workstation, 363
communication security, 188–189
Computer Emergency Response Teams (CERT), 157, 226
computer forensics, 310–312, 340–341
    defined, 314–315
    emergent issues, 337–340
    formal procedure for gathering data, 317–320
    gathering evidence forensically, 315–317
    lack of uniform law, 313–314
    law dictating formal procedure, 321
    law governing analysis and presentation of evidence, 334–337
    law governing seizure of evidence, 321–333
    types and scope of crimes, 312–313
Computer Fraud and Abuse Act (CFAA), 293–294
Computer Security Act (CSA), 294–295
computer use and misuse policy, 68
Computers at Risk (CAR), 194
confidentiality
    of access control, 187–188, 264
    of data, 3, 5, 19, 20–21, 24, 26, 27, 44, 55, 175, 346, 347
    organizational, 266, *267*
    of patient information, 281, 282, 297, 301
    of personal information, 307
    *See also* CIA (confidentiality, integrity, and availability)

confidentiality principle, 31, 32, 34, 35, 36, 38, 347–348
consent, 323
Constrained Data Item (CDI), 34, 35
contingency plans, 398–399
Continuous Diagnosis and Mitigation (CDM) program, 211, *211*
coordination, 2, 250, 264, 383
    of security, 174
coordination in threes, 3–5, 9–10, 11
CoreRESTORE ASIC microchip, 316
corporate assets, responsibility for, 186
countermeasures, 121, 150–151, 283
Counterparty, 62
credentialing
    (ISC)$^2$, 282–283
    Global Information Assurance Certification (GIAC), 284–285
    ISACA certifications, 284
    Security+, 285
criminal penalties, 303
critical infrastructure protection
    Big Ten Power Company (BTP), 383–384
    fears of attack, 385–386
    federal efforts to secure infrastructure, 387–388, 389
    impacts and reaction at BTP, 390
    information systems security, 386–387
    IT at Big Ten Power, 384–385
    state efforts to secure infrastructure, 387–389
cryptanalysis, 46–51
    using diagrams, 51–52
cryptocurrency, 60–62, 276
cryptography, 43, 44–46, 187
    public-key, 83
CSA (Computer Security Act), 294–295
CSIS (Center for Strategic and International Studies), 312
CSSLP (Certified Secure Software Lifecycle Professional), 283
CSX and CSX-P (Cybersecurity Nexus certification), 284
cyberattacks, 271
    individual motivation to prevent, 229–230
cybercrime, 230, 232–233, 312–314. *See also* computer forensics
cyberespionage, 233–234
CyberIntelligence Sharing Act (CISA), 389
cyber-jihad, 241
cybersecurity, 358
    different postures, *362*
    in the health care industry, 361–364
    and unemployment, 275
Cybersecurity Enhancement Act of 2014, 292

Cybersecurity Information Sharing Act (CISA), 292
cybersecurity legislation, 292
Cybersecurity Nexus certification (CSX and CSX-P), 284
cyberstalking, 236–238, 236–239
  defense against, 238–239
  examples of cases, *239*
cyberterrorism, 214, 234–236, 393
cyberwarfare, 236, 241
  malware as weapon, 242–244

DameWare, 379
Dark Web, 202, 203, 232–233
data authentication, 20, 22–23. *See also* authentication
data confidentiality. *See* confidentiality
data center infrastructure, *89*
Data Encryption Algorithm, 53
  details of a given cycle, *54*
  single cycle, *53*
Data Encryption Standard (DES), 52, 53–54
data gathering
  formal procedure for, 317–318
  policy step, 318–320
  political step, 318
  training step, 320
  voluntary disclosure of data, 330–331
Data Security Standard, 211
data sharing, 31
data stewardship, 102
data vulnerabilities, 18–19
*Daubert v. Merrill Dow Pharmaceuticals*, 334
DDoS (distributed denial of service), 74, 84, 95–96
decryption algorithm, 44, 45
Deep Web, 202, 203
defense methods, 24, 103, 259, 363, 396–397
  against cyberstalking, 238
  encryption, 24–25
  firewall, 76, 78
  multilayer, 79
  packet filtering, 82
  perimeter, 188
  physical and hardware controls, 25–26
  software controls, 25
defensive options, 149–150
delay and drop attack, 275
demilitarized zone (DMZ), 78
democracy, 268
denial-of-service (DoS) attack, 83–84, 234
Denning Information Flow model, 29–30, 36, 37, 264
Department of Homeland Security (DHS), 211, 385, 387

Industrial Control Systems Cyber Emergency Response Team (ICS-CERT), 386
Department of Veterans Affairs (VA), 365
  implications, 369–371
  Management Accountability Act of 2014, 368
  Office of Information and Technology (OI&T), 366–368, 370–371
  scheduling issues, 366–367
  scheduling reform, 368–369
  scheduling software user interface, *367*
  summary, 371
  timeline of major events, *369*
DES (Data Encryption Standard), 52, 53–54
design-based threats, 148
destruction, 18–19, 234, 235, 237, 259
detection mechanisms, 16–17, 21, 70, 78, 79, 125, 214, 216, 226, 229, 241, 261, 283, 285, 310, 340, 342, 354, 373
DHCP (Dynamic Host Configuration Protocol), 50, 80
DHS. *See* Department of Homeland Security
dictionary attacks, 85
disclosure, 6, 18, 20, 94, 291, 295, 329–330
  financial, 303
  unauthorized, 148, 213
  voluntary, 330
  of vulnerabilities, 189
distributed denial of service (DDoS), 74, 84, 95–96
DMZ (demilitarized zone), 78
DNS (domain name service/system), 73–74, *74*, 95–96
documentation controls, 160–161
Domain Name Service/System (DNS), 73–74, *74*, 95–96
double loop learning security framework, 110, *111*
DoubleClick ethics issues, 287–288
DYN attack, 95–96
Dynamic Host Configuration Protocol (DHCP), 50, 80

economic information warfare (EIW), 236
ECPA (Electronic Communications Privacy Act), 321
education, 6, 7, 103, 186, 194, 195, 207, 247, 275, 280, 282, 283, 340–341, 350, 352
egress filtering, 84
EIW (economic information warfare), 236
Electronic Communications Privacy Act (ECPA), 321
electronic surveillance, 331
electronic warfare, 236
email
  encryption software for, 65
  spamming, 86–88
  suspicious, *87*
  use of encrypted logins, 82

EMET (Enhanced Mitigation Experience Toolkit), 363
employee threats, 225
  extortion, 227
  hacking, 226
  human error, 227
  sabotage, 225–226
  theft, 226–227
employees
  and risk mitigation, 217
  screening of, 210
  *See also* education; training
encryption, 24–25, 43, *56*
  asymmetric, 55–59
  cypher text, 44
  digital signature process, *57*
  email, 65
  future of, 59–62
  process, *45*
  public-key, 25, 42
  by substitution, 47–50
  by transposition, 50–52
encryption algorithms, 44, 45
  block ciphers, 52–53
  stream ciphers, 52
endogenous threats, 148
Enhanced Mitigation Experience Toolkit (EMET), 363
environmental security, *188*
environmental threats, 147
espoused theories, 110
ethics, 268, 273–274
  codes of conduct, 280–281
  credentialing, 282–285
  intellectual property, 275–276
  managing ethical behavior, 277–279
  unemployment and underemployment, 274–275
  white collar crime, 276–277
European Information Technology Security Evaluation Criteria, 26
evidence
  admissibility of, 335–336
  and the chain of custody, 334–335
  testimonial, 336–337
executive leadership buy-in, 103
exigent circumstances, 324
exogenous threats, 147–148
exploitation, 259, 270, 275, 276
extortion, 227

fabrication, 18
Federal Exchange Data Breach Notification Act of 2015, 292
Federal Information Processing Standards (FIPS), 305

Federal Information Security Management Act (FISMA), 292, 304–306
federal legislation, 387–388, 389–390
Federal Rules of Evidence (FRE), 334–335
Federal Trade Commission (FTC), 307–309
federally funded research and development centers (FFRDCs), 68n1
FFRDCs (federally funded research and development centers), 68n1
Fidelity National Information Services, 226–227
filtering, 77–78, 92. *See also* egress filtering; ingress filtering; packet filters
financial disclosure, 303
FIPS (Federal Information Processing Standards), 305
firewalls, 16, 76–77, 79
FISMA (Federal Information Security Management Act), 292, 304–306
flooding, 83–84
Forensic Replicator, 316
Fourth Amendment to the US Constitution, 317, 321, 338–339, 343
  exceptions to search limitations, 322–327
fragment, 72
framing, 70, 71
FRE (Federal Rules of Evidence), 334–335
FRED hardware/software devices, 316
"fried egg" analogy, *5*
FTC (Federal Trade Commission), 307–309

gap analysis, 297
gender differences, 256–257
Generally Accepted Information Security Principles (GAISP), 194
GAISP (Generally Accepted Information Security Principles), 194
GIAC (Global Information Assurance Certification), 284–285
Global Information Assurance Certification (GIAC), 284–285
globalization, 274
GMITS (Guidelines for the management of IT Security), 193–194
Gramm-Leach-Bliley Act (1999), 292
Guidelines for the management of IT Security (GMITS), 193–194

hacker warfare (HW), 236
hackers and hacking, 11–13, 16, 125–126, 226, 271, 342–343
  of multi-function printers, 355–356
hacktivism, 240
Hague Convention, 338–339
hardware controls, 25
hardware vulnerabilities, 18–19

406 • Index

Harrah's, 265
hashing, 56, 60–61, 85
Hawley Committee, 186, 187
HCISPP (Health Care Information Security and Privacy Practitioner), 283
header checksum, 72
Health Care Information Security and Privacy Practitioner (HCISPP), 283
Health Insurance Portability and Accountability Act (HIPAA), 292, 296, 357, 373
  compliance and recommended protection, 297–299
  compliance/risk management officer, 299
  negative aspects of, 300–302
  positive aspects of, 299–300
  requirements, 296–297
hierarchy culture, 264
HIPAA. *See* Health Insurance Portability and Accountability Act (HIPAA)
HIPS (Host Intrusion Prevention System), 363
hoaxes, 228–229
Homeland Security Act (2002), 292. *See also* Department of Homeland Security (DHS)
Homeland Security Cultural Bureau (HSCB), 246–247
Host Intrusion Prevention System (HIPS), 363
host scanning, 80–81
Hotz, George "GeoHot," 393
HSCB (Homeland Security Cultural Bureau), 246–247
HTTPS. *See* Secure HTTP (S-HTTP or HTTPS)
human element, 361
human error, 129, 186, 209, 225, 227, 252
human resources issues, 68–69, 186
HW (hacker warfare), 236
hypervisor, 88–90

I2S2 risk management model, 146, 152
  component 1: threat definition, 147–148
  component 2: information acquisition requirements, 148–149
  component 3: scripting of defensive options, 149–150
  component 4: threat recognition and assessment, 150
  component 5: countermeasure selection, 150–151
  component 6: post-implementation activities, 151
  level one, 146–147, *146*
I²SF (International Information Security Foundation), 194
IBW (intelligence-based warfare), 236
ICMP echo, 83

IDE (integrated development environment), 317
IDEA (International Data Encryption Algorithm), 52, 55
identity theft, 11, 13, 41, 129, 232, 236, 237, 257, 354
IDS (intrusion detection systems/server), 16–17, 70, 79, 373
IHL (Internet header length), 72
*Illinois v. Andreas*, 342
implied consent, 323
incident management, 284
Indiana legislation, 388–390
information
  access to, 1–2, 191
  acquisition of, 148–149
  flow of, 4, 29, 40, 100, 111–112, 119, 236
  handling of, 2–3, 8–9, *9*
information systems (IS), 2, 3
  access to, 187, 226, 304
  activity review, 299
  assessment of, 297
  categorizing, 305–306
  CIO for, 384
  data sharing within, 31
  defining, 3
  designing, 148
  implementation of, 384
  integrity check of, 34
  life cycle of, 194
  managing, 10, 27, 100, 247, 255, 369
  protecting 363
  risks to, 284
  secure, 121, 216
  strategic planning for 369, 371
  support personnel for, 377
  updating, 148
information systems (IS) security issues, 3, 5, 7–9, 21, 22, 24, 25, 26, 36, 38, 70, *107,* 189, 195, 236, 248–249, 267, 268
  10 deadly sins of IS security management, *99*
  classes of security decisions, 108–115, *109*
  formal aspects, 348–350
  formal dimensions of, 99–106
  four classes of, 98
  governance of, 284
  ideal and overengineered situations, *349*
  informal aspects, 350–351
  management commitment to, 259–262
  organization of, 185–186
  planning framework, *120*
  planning principles, 119–122
  planning process, 115–119
  policies, 185
  strategy levels, 106–107

Information Systems Audit and Control Association (ISACA) certifications, 284
Information Systems Security Architecture Professional (ISSAP), 283
Information Systems Security Association (ISSA), 194
  Code of Ethics, 282
Information Systems Security Engineering Professional (ISSEP), 283
Information Systems Security Management Professional (ISSMP), 283
information technology (IT), 2, 119, 157, 261, 296, 365, 369, 384
information technology security (ITS), 163
Information Technology Security Evaluation Criteria (ITSEC), 37, 177–179
  classes of evaluation, *178*
infrastructure
  network, 373
  traditional data center, *89*
  virtualized server, 88–90, *89*
  *See also* critical infrastructure protection
ingress filtering, 82
integrated development environment (IDE), 317
integrity of data, 3, 19, 21–22, 37, 347–348
  and the Biba model, 33
  and the Clark-Wilson model, 34–36
  *See also* CIA (confidentiality, integrity, and availability)
Integrity Validation Procedures, 35
intellectual property, 275–276
intelligence-based warfare (IBW), 236
interception, 18, 19, 24, 59, 311, 332
internal threats, 154, 227, 385
International Court of Justice, 291
International Data Encryption Algorithm (IDEA), 52, 55
international gangs, 232
International Information Security Foundation (I²SF), 194
international law, 337
  and common interest, 338
  Hague Convention, 338–339
  and political good will, 338
Internet, 70, 75–78, 79, 81, 84, 87, 90, 91, 96, 104, 126, 163, 217, 227, 235–238, 240, 241, 242, 287, 293, 296, 298, 315, 348, 382
Internet header length (IHL), 72
Internet of Things, 93, 232
Internet Protocol Address (IP address), 71, 73, 76–78, 87
interruption, 18, 19, 70, 190
intrusion detection, 21, 78, 125, 214, 216, 226, 229, 241, 261, 283, 285, 310, 340, 342, 354

intrusion detection systems/server (IDS), 16–17, 70, 79, 373
intrusion protection systems/server (IPS), 70, 79
inventory searches, 324
IP address spoofing, 81
IP protocols, 72
IPS (intrusion prevention/protection systems/server), 70, 79
IPv4 address scheme, 72–73, *73*
IPv6 address scheme, 72–73, *73*
ISACA (Information Systems Audit and Control Association) certifications, 284
(ISC)², 282–283
ISO/IEC 27002:2013, 185–190
  summary of controls and objectives, *190–192*
ISO/IEC TR 13335 guidelines, 193–194
ISSA (Information Systems Security Association), 194
  Code of Ethics, 282
ISSAP (Information Systems Security Architecture Professional), 283
ISSEP (Information Systems Security Engineering Professional), 283
ISSMP (Information Systems Security Management Professional), 283
IT (information technology), 2
  department support, 103
  specialization, *367*
ITSEC (Information Technology Security Evaluation Criteria), 37, 177–179
  classes of evaluation, *178*

JetBlue, 11–13
job market, 274–275

Kasiski method, 50
key loggers, 154–155
Kidder Peabody, 253
KLM Royal Dutch Airlines, 255–256

LAN (local area networks), 70, 71, 75, 291
Lawrence Berkeley Laboratory, 16
La'Zooz, 62
leadership, and security culture, 259–262, *261*
legal issues, 290–292
  international, 337–339
  national, 339–340
  *See also* seizure of evidence
Level of Risk Matrix, 137–138, *138*
local area networks (LAN), 70, 71, 75, 291
local conditions, 347
logic bombs, 19
Lulz Security (LulzSec), 393

MAC (media access control) address, 71
malware, 12, 204, 210, 242–244, 363, 378–380
man-in-the-middle (MITM) attack, 82–83, 275
market culture, 265–266
MASS (Medical Appointment Scheduling System), 368
media access control (MAC) address, 71
Medical Appointment Scheduling System (MASS), 368
message dropping, 275
message spoofing, 275
micro-segmentation, 90–92
middle boxes, 76
middleware devices, 76–78
Mirai (botnet virus), 96
MITM (man-in-the-middle) attack, 82–83, 275
Mitnick, Kevin, 41
modeling controls, 160
modification(s), 18, 19, 20, 24, 33, 52, 126, 130, 148, 189, 295, 304, 342, 374, 376, 393
money laundering, 276, 277
monitoring, 216–217
Morris worm, 259

National Association of Insurance Commissioners, 357
National Cybersecurity Protection Advancement Act of 2015, 292
National Guard, 292
National Institute of Standards and Technology (NIST), 103–104, 295, 305
    Framework for Improving Critical Infrastructure Cybersecurity, 387–390
    security documents, *196*
national law, 339–340
National Security Agency (NSA), 53, 165, 175, 295
NATs (network address translations), 72–73, 77
need-to-know principle, 20
need-to-withhold principle, 21, 32
network address translations (NATs), 72–73, 77
network attacks, 79
    flooding, 83–84
    IP address spoofing, 82–83
    operating system attacks, 84
    packet sniffing, 81–82
    password, 85–86
    reconnaissance, 80
    scanning, 80–81
    stack-based buffer overflow, 84
    web applications, 86–88
network connections
    east-west traffic attack, *91*
    four directions of, 90–92, *90*
    possible secure configuration, *92*

network interface card (NIC), 88
network security, 68–69
    cloud-based, 93
    in health care, *360*
    middleware devices, 76–78
    new trends, 88–93
    TCP/UDP/IP protocol architecture, 70–76
    types of network attacks, 79–88
    VPN and proxy, 78–79
    *See also* security
network segmentation, 216, 395–396
network signaling and the layers, *72*
network structures, 70, *77*
network traffic inspection, 216
NIC (network interface card), 88
NIST. *See* National Institute of Standards and Technology
no write down rule, 31
non-repudiation of data, 20, 23–24, 38, 39, 44
North American Aerospace Defense Command (NORAD), 41
NSA (National Security Agency), 53, 165, 175, 295

OCTAVE (Operationally Critical Threat, Asset, and Vulnerability Evaluation), 183
OECD Guidelines, 194, 195
    for security culture, 266–268
Ohio legislation, 388
operating system attacks, 84
operating system controls, 25
operational security, *188*
optical splitters, 71
organizational buy-in, 103–104
Orion security strategy, 115–119, *117*
OSI model, 70–71
outsourcing, 276
overformalization, 348

packet captures, 76
packet filters, 77, 82
packet sniffing, 81–82
Palo Alto Network, 79
Pass-the-Hash (PtH) mitigations, 363
passwords, 24–25
    attacks, 85–86
    policies, 16, 80, 85–86
Patriot Act, 321
PayPal, 187
    public policy on blockchains, *188*
PCAOB (Public Company Accounting Oversight Board), 302
peer-to-peer currency systems, 276
Pen Register and Trap and Trace Devices statute (Pen/Trap Statute), 321, 331–333

Index • 409

penetration testing, 216
Pentagon, 41
personal devices, 217–218
personal history information (PHI), 296, 298–301
PGP (pretty good privacy) software, 42–43
pharming, 86–87
PHI (personal history information), 296, 298–301
phishing, 41, 86, 228
physical security, *188*
ping broadcasts, 83–84
plain view, 324
plaintext attacks, 46
plaintext encryption, 44
policy development, 104–106, *105*
port filtering, 77
port forwarding, 77
port scanning, 80–81, *81*
PPA (Privacy Protection Act), 321
pragmatics, 248–249
prevention mechanisms, 21
principle of easiest penetration, 17, 39
privacy issues, 23, 40, 42, 88, 180, 211, 230, 239, 255, 257, 281, 288, 291, 294, 295, 389
  expectation of privacy, 322–323, 325, 328, 342
  and HIPAA, 297
  and personal health information (PHI), 296, 297, 299, 300
  *See also* Health Insurance Portability and Accountability Act (HIPAA)
Privacy Protection Act (PPA), 321
process improvement software, 161–162
program controls, 25
proxies, 72, 73, 76, 78, 363
psychological warfare (PSYW), 236
Public Company Accounting Oversight Board (PCAOB), 302
public key encryption, 25, 42
public-key cryptography, 83

Quantum Cryptography, 59–60
Quantum Key Distribution (QKD), 59–60
QUIC protocol, 75, 78

radar maps, 266, *267*
RC4 cipher, 52–53
reassessment, 105, 195, 268
reconnaissance attacks, 80
Reference Monitor concept, 30–31
replay attacks, 275
response
  corporate, 11, 193, 218
  echo, 83–84
  immediate, 213, 378

policy for, 105, 121, 122, 204, 207, 211, 214, 218
  to security incidents, 192, 193, 268, 283
responsibility, 252, 267–268
  for corporate assets, 186
  structures of, 101–102
revenge porn, 237–238
RFC 2196 site security handbook, 192–193
risk assessment/analysis, 130–139, 141–145, 268, 297, 298
  classes of controls, *135*
  control analysis, 135–136
  control recommendations, 138
  likelihood determination and impact analysis, 136–137, *136*
  magnitude of impact, *137*
  results documentation, 138–139
  risk determination, 137–138
  sample interview questions, 132
  sample risk assessment report, 139
  system characterization, 131–132
  threat classes and resultant actions, *133*
  threat identification, 132–133
  vulnerability identification, 133–135
  *See also* risk evaluation
risk evaluation, 130, 141–145. *See also* risk assessment/analysis
risk management, 128–130, 152, 284, 298
  COBRA model, 143–145
  I2S2 model, 145–151
  risk evaluation and assessment, 131–139, 141–145
  risk mitigation, 130, 139–141
risk management process model, 145
  I2S2 model at level one, 146–147, *146*
  six components of I2S2 model, 147–151
  three levels of I2S2 model, 146–147
risk mitigation, 130, 139–141, 208
  control categories, 141, 143
  flow of activities, *141*
  options for, 140
  technical, formal, and informal controls, *142*
RITE (responsibility, integrity, trust, and ethicality) principles, 350–351
robot networks (botnets), 96, 233, 234, 241
routers, 70, 71, 73, 77, 80, 82, 84, 88–89, 93, 205, 273, 394, 397
RSA, 58–59
Rushby Separation Model, 30–31, 36, 264
Russian Business Network (RBN), 232

sabotage, 225–226
SAFER, 53
sanctions for noncompliance, 299

sandboxing, 363
SANS IT Code of Ethics, 280–281
Sarbanes-Oxley Act (SOX), 108, 111, 302–304
scanning, 80–81
SEAL (Software Optimized Encryption Algorithm), 52
search and seizure of computers, 325–327
search incident to a lawful arrest, 324
search warrants, 325–327
SEC (Securities and Exchange Commission), 302
secret key encryption, 44, 45
Secure HTTP (S-HTTP or HTTPS), 87–88
Secure Socket Layer (SSL), *87*
Securities and Exchange Commission (SEC), 302
security
    design, 268
    engineering, 162–165
    implementation, 268
    incident management, 189–190
    literacy, 252
    metrics, 252
    planning principles, 119–122
    program development and management, 284
    requirements for, *184*
    spending for, 252
    strategies for, 98
    targets for, 180
    *See also* network security
security breaches
    best practices: be prepared, 214–218
    continuous diagnosis and mitigation framework, *211*
    governance, 208–209
    must do's in managing, *215*
    policy considerations, 207
    reputation and responsiveness, 207–208
    response to, 212–214, *214*
    risk and resilience, 208
    steps to avoid a potential attack, 209–212
    technicalities of, 206
    *See also* Case Studies
security compliance management, 190
security continuity management, 190
security culture, 249–252, 269–270, 360–361, 374
    at Harrah's, 265–266
    leadership, 259–262
    obedient, 260–262
    OEDC principles for, 266–269
    role of leadership, *261*
    silent messages, 254–259
security culture framework
    adhocracy culture, 263–264
    clan culture, 264–265
    hierarchy culture, 264

    market culture, 265–266
    types of culture, 262–263, *263*, 269
security decisions
    administrative, 111–113
    classes of security decisions, *109*
    network of means and objectives, *114*
    operational, 113–114
    prioritizing, 114–115
    strategic, 108, 110
security in threes, 5–8
    institutionalization of, 8–10
security information and event management (SIEMs), 79, 93
Security Onion, 79
security policy/ies, 183
    compliance research, *231*
    development of, 104–106, *105*
    management considerations, 230
security specification models
    away from the military, 31
    Bell-La Padula model, 27–29
    Denning Information Flow model, 29–30
    historical review, 26–27
    Reference Monitor and Rushby's Solution, 30–31
security standards, 156–157, 197
    application controls, 159
    auditing, 158–159
    cluster elements and security implementation, *251*
    documentation controls, 160–161
    evaluation process, *181*
    evolution of evaluation criteria, *179*
    Generally Accepted Information Security Principles (GAISP), 194
    international harmonization, 179–185
    ISO/IEC 27002:2013, 185–192
    ISO/IEC TR 13335 guidelines, 193–194
    ITSEC, 177–179
    key constructs and concepts in SSE-CMM, 165–170
    levels of, 176–177, *177*
    modeling controls, 160
    National Institute for Standards and Technology security documents, *196*
    OECD Guidelines, 195
    process improvement software, 161–162
    from Rainbow Series to Common Criteria, 175–177
    RFC 2196 site security handbook, 192–193
    SSE-CMM, 162–165
    SSE-CMM architecture description, 170–175
Security+, 285
segmentation, 395–396

seizure of evidence
  Electronic Communications Privacy Act (ECPA), 321, 328–333
  Fourth Amendment, 321, 322–327
  laws governing, 321–322
  Pen Register and Trap and Trace Devices statute (Pen/Trap Statute), 321
  Privacy Protection Act, 321, 327–328
self-healing systems, 12–13
separation of users, 31
session hijacking, 81
session tracking mechanisms, 88
S-HTTP. *See* Secure HTTP (S-HTTP or HTTPS)
SIEMs (security information and event management), 79, 93
silent messages
  association, 255
  bisexuality/gender, 256–257
  defense/security, 259
  exploitation, 259
  interaction, 254–255
  learning, 258
  play, 258–259
  subsistence, 256
  temporality/time, 257
  territoriality/location, 257
Silk Road, 232, 233
Skipjack, 53
Slammer worm, 259
smart contracts, 60
Smurf attack, 83–84
SoBig worm, 259
social engineering, 41
social engineering attacks, 227–228
  hoaxes, 228–229
  phishing, 228
  spear phishing, 228
sociotechnical systems, 69
socket, 75
socket arrangement, *76*
Soft System Methodology (SSM), 115
software controls, 25
software development controls, 25
Software Engineering Institute, 162
Software Optimized Encryption Algorithm (SEAL), 52
software piracy, 276
Software Publishers Association (SPA), 276
software vulnerabilities, 18–19
Sony Pictures Entertainment, 203–204
Sony PlayStation Network Breach, 392–399
SOX (Sarbanes-Oxley Act), 108, 111, 302–304
SPA (Software Publishers Association), 276
spam, 86, 88, 212, 218, 228, 234, 290

spear phishing, 228, 354
spoofing, 82–83, 86
Sprint, 41
spyware, 234
SQL injection, 394–395
  protection against, 395–396
SSCP (Systems Security Certified Practitioner), 283
SSE-CCM. *See* System Security Engineering Capability Maturity Model
SSL (Secure Socket Layer), *87*
SSM (Soft System Methodology), 115
stack-based buffer overflow attacks, 84
stream ciphers, 52
Stuxnet, 242–244
  hits by country, *243*
  overview, *244*
subpoenas, 329
substitution encryption, 47–50
supplier relationship, 189
Surveillance Devices Act, 343
switches, 41, 45, 70–71, 84, 88, 89, 93, 205, 394, 396
Sync (SYN), 74–75, 83
  SYN attack, 74
  SYN flooding, 83
Sync/Acknowledge (SYN/ACK), 74–75
system acquisition, development, and management, 189
System Security Engineering Capability Maturity Model (SSE-CCM), 162–165
  architecture description, 170–175
  base practices, 171
  capability levels, 172–175
  CMM levels, *163*
  generic practices, 171–172
  key constructs and concepts, 165–170
  level five, *175*
  level three, *173*
  level two, *173*
  process areas, *168*
  SSE-CMM levels, *170*
system server ports, *75*
Systems Security Certified Practitioner (SSCP), 283

Tails, 202
taps, 76
target of evaluation (TOE), 180, 182–183
TCP (Transport Control Protocol), 72, 74–75, 76, 78
TCP/IP model, 70–71
TCP/UDP ports, 71–72

TCSEC (Trusted Computer System Evaluation Criteria), 26, 27, 30, 32, 35, 36, 37, 175–176, 179, 264
technical controls, 6–7
technical security, 38
  and changing local conditions, 347
  confidentiality, integrity, and availability of information, 347–348
technical systems, 4–5
temporality/time, 257
territoriality/location, 257
terrorism, 234–235. *See also* cyberterrorism
The Onion Router (TOR), 202
theft, 226–227, 234, 291
theories-in-use, 110
threat definition
  design-based threats, 148
  endogenous threats, 148
  environmental threats, 147
  exogenous threats, 147–148
threats
  characterization of, 183
  recognition and assessment of, 150, 216
T-Mobile, 271
TOE (target of evaluation), 180, 182–183
TOR (The Onion Router), 202
TP (Transformation Procedure), 34, 35
training
  continuous/ongoing, 7, 207, 295, 370
  for cybersecurity, 213, 225, 228, 290, 291, 357, 360, 362
  for employees, 174, 186, 212, 300, 369
  for engineering groups, 165
  ethical, 278, 279, 286
  forensic, 339
  as informal control, 6
  internal, 385
  relevant, 252
  for security, 6, 103–104, 112, 119, 186, 195, 210, 230, 252, 258, 280, 295, 299, 301, 320, 350
  for socio-technical systems, 69
  on software use, 370
  specialized, 210
  for system developers, 148
  for user support, 103
  well-defined, 189
Transformation Procedure (TP), 34, 35
Transport Control Protocol (TCP), 74–75, 76, 78
transposition encryption, 50–52
Trojan Horse protection, 16, 19
Trojan program, 81, 378, 379

Trusted Computer System Evaluation Criteria (TCSEC), 26, 27, 30, 32, 35, 36, 37, 175–176, 179, 264
Trusted Network Initiative, 32
TSF, 182–183
tunneling, 75

UDP (User Datagram Protocol), 74–75, 76, 78
Unconstrained Data Item, 35
underemployment, 274–275
unemployment, 274–275
University of California Berkeley, 125–126
urban university computer hack, 373–374
  changes, 375–376
  computer system, 374–375
  further research and added symptoms, 378–382
  hack discovered, 377–378
  history of the system, 376–377
  immediate response, 378
  other issues, 377
  summary, 382
URL spoofing, 82–83, 86
US Department of Defense, 16, 26, 27, 37, 175, 264
US Federal Criteria for Information Technology Security, 26
US National Institute of Standards and Technology, 131
US National Security Agency. *See* National Security Agency (NSA)
*US v. Barth*, 342
USB devices, 154–155, 202, 242, 244, 385
User Datagram Protocol (UDP), 74–75, 76, 78
user support, 103

Vernam cipher, 52–53
Vigenère cipher, 48–49
virtual local area network (VLAN), 71, 88
virtual private network (VPN), 75, 78–79, 373
virtualized network traffic, 90–92
virtualized server infrastructure, 88–90, *89*
virus protection, 16, 90, 103, 363, 378
viruses, 2, 16, 17, 19, 94, 96, 129, 210, 229, 234, 236, 313
Visa B2B Connect, 65–67
VLAN (virtual local area network), 71, 88
VPN (virtual private network), 75, 78–79, 373
VPN tunneling, 75
vulnerabilities, 18–19, 39, 131
  behavioral and attitudinal, 133–134
  classes of, *134*
  coding problems, 134–135
  data, 18–19
  misinterpretation, 134
  physical, 135

WAN (wide area networks), 70, 75
web application attacks, 86–88
web application session hijacking, 86–87
web-of-trust authentication model, 57–58
whaling, 228
white collar crime, 276–277
wide area networks (WAN), 70, 75
wireless networks, 71, 93
Wiretap Statute (Title III), 332–333

workplace searches, 325
Wyndham Worldwide Corporation, 307–309

X.509, 57–58, *58*

Zappos, 249
zero-day vulnerabilities, 204
Zimmerman, Phil, 42–43
zombies, 234
Zynga, 225

# APPENDIX A

# Additional End-of-Chapter Short Questions

## Chapter 6

1. _____ helps in recommending risk-reducing strategies.

2. _____ deals with the continuous evaluation of the risk management process, such that ultimately successful risk management is achieved.

3. The risks identified at the requirements assessment stage feed into _____ and _____ trade-offs in systems development.

4. In the _____ stage, re-accreditation and re-authorizations are considered.

5. Any system migration needs to take place in a secure and _____ manner.

6. _____ reside(s) in the motivations of humans to undertake potentially harmful activities.

7. _____ attacks can occur because of a malicious attempt to gain unauthorized entry to a system.

8. The nature and significance of certain kinds of attacks keeps on changing with _____.

9. Accurate _____ is the first step toward accomplishing proper planning and accurate budgeting for software development projects.

10. There are _____ major types of cost estimation techniques available today.

11. The _____ make use of extensive past project data.

12. The _____ approach has led to overestimation or underestimation, each of which translates into a negative impact on the success of the project.

13. Software reliability, software usability, and software efficiency emerged as the factors that would cause the greatest impact on _____.
14. Besides an understanding of _____ aspects of the system, the related roles and responsibilities need to be understood.
15. _____ information relates to the functional requirements of the system, the stakeholders of the system, and security policies and architectures governing the IT system.

## Chapter 9

1. Sabotage typically results in _____ forms.
2. _____ sabotage can be thought of as intentionally not doing something you should be doing, which through this interaction results in harm to the organization.
3. An employee who is _____ and _____ in the workplace increases the chances of them committing an act of sabotage.
4. Acts of _____ sabotage among employees are rare.
5. _____ is a dangerous threat to any organization with respect to their own employees and can lead to tremendous devastation due to an employee's familiarity with their information system and authorized access.
6. _____, even if it has no real evidence of wrongdoing, can still be harmful to an organization.
7. With the confidential customer data and intellectual property just the slip of a keystroke away from the _____, every organization should be considered at risk.
8. A careless or _____ employee may simply think they are doing their job or speeding up the process by sending secure information through an unsecured email attachment.
9. _____ techniques are particularly effective because they are based on what can be considered a bug or flaw in the decision-making processes of humans.
10. While most organizations have _____ filters intended to prevent phishing, it is impossible to stop all attempts, and it only takes one mistake to expose an entire organization.
11. Criminal organizations are driven by _____, rather than personal ambition or sheer boredom.
12. Parties engaging in the production or distribution of prohibited goods and services are thus considered members of the _____.
13. One of the weapons of choice in twenty-first century espionage is the _____.

14. A large number of _____ acts conducted for social and political reasons have come to light over the past few years.
15. _____ has added a new dimension to persecution and makes victims feel a palpable sense of fear.

## Chapter 10

1. Security of informal systems is, simply put, ensuring that the integrity of the _____ stays intact.
2. In terms of managing information systems security, it is important that we focus our attention on maintaining the behavior, values, and _____ of the people.
3. _____ has been considered the single most important factor leading to the success or failure of a firm.
4. A proper mix of _____ and _____ makes the difference in the success or failure of a firm.
5. Often _____ are considered without an appreciation of the context, and hence solutions are developed often without a complete understanding of the problem.
6. Lack of _____ of control structures also results in security problems.
7. While the role of technical controls cannot be underestimated, true _____ can only be achieved if the technical controls have been adequately institutionalized.
8. Security culture is the totality of patterns of behavior that come together to ensure the protection of _____ of a firm.
9. Once well-formed, security culture acts as the glue that brings together the _____ of different stakeholders.
10. In many cases, _____ and _____ programs do not necessarily relate to the task at hand.
11. Security _____ ensure that assets remain protected and the business flourishes.
12. The mismatch between corporate _____ and professional practices leads to divergent viewpoints.
13. IT-based systems adversely affect the subsistence issues related to different _____ groups.
14. Many organizations use play as a means to prepare for possible _____.
15. Organizations such as the _____ and government agencies are typically hierarchical in nature.

# Chapter 11

1. _____ are the moral principles that guide individual behavior.
2. In order for a sound _____ strategy, the recognition of ethical principles is important.
3. While cybersecurity represents a blooming industry, there is a shortfall of _____ graduates.
4. While outsourcing and offshoring may have played a role, a majority of the job losses can be attributed to _____ and automation.
5. The consequences of _____ and _____ are known to motivate individuals to circumvent controls and find shortcuts for monetary gain.
6. In robotic systems, it is possible to orchestrate an _____ that exploits intermediate network nodes.
7. While many do not acknowledge the ill effects of something like software piracy, illicit sales are generating millions of dollars in sales, which is contributing to underground _____ markets.
8. Because of the distributed nature of the _____, there are concerns that money launderers use them systematically.
9. Deviants have been characterized as greedy individuals who typically lack _____.
10. The lack of _____ poses interesting challenges in terms of white-collar crime, money laundering, and the role of centralized clearing houses.
11. Simply issuing a _____ or a credo, without following it up with training and publicity to reinforce the message, will not be enough to prevent employee transgressions.
12. Ethical training and _____ campaigns can have the desired effect on the bulk of the employees of an organization.
13. Some employees evade _____ controls because of cultural idiosyncrasies.
14. Many organizations depend upon the _____ to ascertain the proficiency level of an individual.
15. Currently, many organizations and institutions have made it mandatory for new _____ hires to acquire some form of certification.

## Chapter 12

1. The fact that many computer and servers can be accessed via the _____ increases the risk of theft and misuse of data by anyone with sufficient skills in accessing and bypassing security safeguards.

2. The _____ passed a ruling that one country's territory cannot be used to carry out acts that harm another country.

3. The key elements of the Computer Fraud and Abuse Act (CFAA) are to provide _____ and _____ for violating the law.

4. The implementation of the CFAA made it easier to prosecute complaints of _____ of sensitive information.

5. The CFAA allowed a double whammy against the _____ and allowed the _____ to attempt to recover more in damages.

6. *Damage* is defined as any impairment to the _____ of computer data or information.

7. Motivation for the Computer Security Act (CSA) was sparked by the escalating use of computer systems by the _____ and the requirement to ensure the security and privacy of unclassified, sensitive information in those systems.

8. The purpose of the CSA was to _____ and tighten security controls on computers in use throughout the federal government and those in use by federal contractors.

9. Whether you are a large health care provider, an insurance company, a small rural physician, or a benefits consulting firm, you will have to consider a _____ for personal history information to be in compliance with the Health Insurance Portability and Accountability Act (HIPAA).

10. HIPAA has important implications for all health care providers, payers, patients, and other _____.

11. The _____ assessment inventories an organization's current security environment with respect to policies, processes, and technology.

12. The goal of the _____ is to complete the current environment with the proposed regulatory one in terms of the level of readiness and to determine whether gaps exist and, if so, how large they are.

13. A _____ should provide an analysis of both likely and unlikely scenarios in terms of probability of occurrences and their impact on the organization.

14. The Sarbanes-Oxley (SOX) legislation increases the requirements that organizations must adhere to for financial _____.

15. The Federal Information Security Management Act (FISMA) security program requires the _____ of each federal agency to define and implement an information security program.

## Chapter 13

1. Computer forensics is a(n) _____ attempt to balance society's need to protect itself and the rights of the individuals who are perceived as threatening society's survival and prosperity.

2. It is almost impossible to enact effective _____ until the scope of the problem is revealed through real-life, everyday experience.

3. Like any other forensic science, computer forensics deals with the application of a science to a question of _____.

4. Computer forensics deals with the preservation, discovery, identification, extraction, and documentation of computer _____.

5. The forensic process applies _____ to the examination of computer-related evidence.

6. Forensics is done through managing the investigation of the crime scene and its evidentiary aspects through a thorough, efficient, secure, and documented _____.

7. The logical side of computer forensics deals with the extraction of _____ from any relevant information resource.

8. Computer forensics does not _____ or _____ the evidence that was seized at the crime scene.

9. Computer forensics makes an exact _____ copy of the evidence, and analysis is performed on the mirror-image copy while the original evidence is held in a secure environment.

10. Forensically, _____ data is a mirror image of the humanly understandable information, such as time and date stamps.

11. The process of making a forensically acceptable copy of data for _____ involves a whole host of steps to ensure that the data copy is forensically identical to the data source.

12. Law enforcement officials have _____ responses when asked about how forensic evidence should be gathered.

13. "Don't touch anything; call us!" This response springs forth from law enforcement's commitment to help those in need, uphold the law, and put their expertise to _____ use.

14. The nuances of the law and the _____ of the treatment do not allow a full discussion of either the law or the procedures dictated by the law.

15. There are exceptions to _____ that allow investigators to perform warrantless searches.

## Chapter 14

1. While it may be prudent to focus on the ease of use and functionality in some cases, in others, maintaining _____ of private data may be the foremost objective.

2. Although _____ is a goal that most organizations aspire to achieve, it is often not possible to have it.

3. Traditionally organizations have been viewed as _____, and security has for the most part not been considered part of the useful system designed for purposeful activities.

4. Exclusive emphasis on _____ and designing it in a top-down manner is counterproductive.

5. Changes in technology and the _____ nature of business makes most standards obsolete even before they get published.

6. The _____ of a model is judged on the basis of the extent to which it represents a given subset of the reality.

7. Security problems arise as a consequence of _____ and managerial inability to balance the rule- and norm-based aspects of work.

8. Problems of overformalization are usually a consequence of the _____ syndrome.

9. In practice, controls have dysfunctional effects because _____ solutions are proposed for specific problems.

10. A mismatch between the _____ and _____ of the organization could potentially be detrimental to the health of an organization and to the information systems in place.